Foundation of Offshore Wind Turbine

Shengdong Zhang
Xiuying Yang Kaicheng Yan Jungkeun Oh
Dae-Won Seo Jackyou Noh Kiweon Kang

도서출판 GS인터비전

Preface

This book mainly introduces the basic principles, structure and application of offshore wind turbines. Chapters 1 and 2 mainly introduce "Current situation and future of offshore wind power" and "Energy Utilization and Global Economic and Social Development". Chapter 3 describes "Overview of Worldwide Wind Power Industry", Chapter 4 describes "Environmental loads on offshore wind turbine infrastructure", and Chapters 5 to 7 respectively describe the composition of the floating system, single pile foundation, and gravity foundation and working principle. Chapter 8 describes the Influence of wind load on offshore wind power foundations, where wind load focuses on the influence of wind load on single pile foundations, bucket foundations, large-diameter steel pipe pile foundations, and high pile cap foundations. Chapter 9 deals with the Influence of wave load on wind power structure foundation, and Chapter 10 deals with the Influence of wind, wave, current, ice cone and other coupling or combined actions on offshore wind power foundation. The last appendix part of this book introduces wind power generation software. This monograph is positioned as a researcher, and can also be used to train personnel engaged in the manufacture and use of wind turbines. It can also be used as teaching materials or teaching reference materials for related majors in colleges and universities. Self-study reading material for technicians and wind power enthusiasts.

February 2023

Shengdong Zhang

Acknowledgments

This book was supported by the following funds: Jining University's "Hundred Outstanding Talents" Support Program Cultivation Project, Jiangxi Provincial Natural Science Foundation Key Research and Development Project (No.20192BBEL50028), Humanities and Social Sciences Project of Universities in Jiangxi Province (No. JY17235), Jiujiang University School-level Teaching Reform Key Project(No. XJJGZD-20-03) and Korea Institute of Energy Technology Evaluation and Planning(KETEP) grand funded by the Korea Government (MOTIE)(20224000000040)

Contents

Current Situation and Future of Offshore Wind Power

1

1.1 The importance of foundation design

1.2 Development Status and Benefits Improvement Analysis of Offshore Wind Power in China

Wind energy can be considered nowadays a main participant in the energy market and power generation, (Mahdy & Bahaj, 2018). In 2019, it overtook coal as the second-largest form of generation capacity in Europe, only being overpassed by gas, and showing the highest development rate of every source considered (*Offshore Wind in Europe Key Trends and Statistics* 2020, 2021). Although the offshore wind energy sector is still far from the stage of maturity of its older brother (onshore wind) (Esteban et al., 2011a), the similarities between both have allowed offshore wind turbines (Hevia-Koch & Klinge Jacobsen, 2019) to become efficient and commercial-ready in a relatively short time (Esteban et al., 2019). In 2018, a total of 4496 MW were installed in six different countries (Global wind Energy Council, 2019). China became the largest investor in this field (1800 MW), overtaking the United Kingdom for the first time (1312 MW), which had led the market during the past years. Germany (969 MW), Belgium (309 MW), Denmark (61 MW), and South Korea (35 MW) were the other participants(Sánchez et al., 2019).

Referring to the total installed capacity, the United Kingdom is still far ahead of any other country, covering 35.0% of the total power. Germany and China stay behind, accounting for 27.3% and 19.6%, respectively. Denmark, Sweden, and the Netherlands, which started designing and testing these installations in the 90s, are still playing an important role as industry manufacturers and technology developers; however, the capacity installed in these countries has not gone up to that scale. [Tab. 1-1] gives some details of the picture at the end of 2018(Sánchez et al., 2019).

According to (Li & Yu, 2018), (Enevoldsen & Valentine, 2016), (Kaldellis et al., 2016), (Wüstemeyer et al., 2015), (Kumar et al., 2016), (Yang et al., 2018), (Gil-García et al., 2019), the main advantages of offshore wind farms over onshore ones can be briefly described as follows:

[Tab. 1-1] Capacity installed by country at the end of 2018 (UK: United Kingdom; GER: Germany; CHI: China; DEN: Denmark; BEL: Belgium; NET: Netherlands; SWE: Sweden; VIE: Vietnam; JAP: Japan; KOR: Korea; FIN: Finland; USA: United States of America; IRE: Ireland; SPA: Spain; NOR: Norway; FRA: France). (Sánchez et al., 2019)

Country	Wind Farms	Power Installed (MW)	%Total Power Installed	Turbines Installed	Average Wind Farm Power	Average Turbine Capacity (MW)
UK	37	8184	35.0%	1920	221.2	4.3
GER	23	6375	27.3%	1314	277.2	4.9
CHI	36	4581	19.6%	1224	127.3	3.7
DEN	14	1327	5.7%	514	94.8	2.6
BEL	8	1187	5.1%	274	148.4	4.3
NET	6	1118	4.8%	365	186.3	3.1
SWE	5	201	0.9%	86	40.2	2.3
VIE	2	99	0.4%	62	49.5	1.6
JAP	11	84	0.4%	35	7.6	2.4
KOR	3	71	0.3%	24	23.7	3.0
FIN	3	68	0.3%	24	22.7	2.8
USA	1	30	0.1%	5	30.0	6.0
IRE	1	25	0.1%	7	25.0	3.6
SPA	2	10	0.0%	2	5.0	5.0
NOR	1	2	0.0%	1	2.0	2.0
FRA	1	2	0.0%	1	2.0	2.0

- The existing wind resource in the sea is higher than on nearby coasts.
- Due to its location offshore, the visual and acoustic impact is lower than wind farms on land, which allows better use of the existing wind resource, with larger turbines and the use of more efficient blade geometries. Likewise, the lower surface roughness in the sea favors the use of lower tower heights.
- It provides a bigger creation of employment in the phases of construction, assembly, and maintenance, due to the greater complexity during installation and exploitation.
- Possibility of integration in mixed marine complexes.
- The spaciousness of the environment.
- More constant and stable energy is generated.

However, these marine facilities also have significant disadvantages concerning terrestrial ones, which are limiting their development: non-existence of electrical infrastructures close to the location; more severe environmental conditions; evaluation of wind resources as more complex and expensive; and above all, its higher investment ratios and operating expenses, needing specific technologies for construction and foundations, transport and assembly at sea, laying of electrical networks underwater and operation and maintenance tasks (Esteban et al., 2009), (Esteban et al., 2011b). Furthermore, there is some research emphasizing the advantages of harnessing wind and waves at the same facility (Pérez-Collazo et al., 2015), (Veigas & Iglesias, 2015), (Astariz et al., 2018), (Sánchez et al., 2019).

To face the investment costs and become more productive and, therefore, more attractive to the market investors, offshore wind farms must generate as much power as possible with the lowest construction and installation costs (Lacal-Arántegui et al., 2018), (Schwanitz & Wierling, 2016), (Dismukes & Upton, 2015). Developing more and more efficient turbines has become the greatest challenge of the sector from the early years, while the costs of foundations and electric installations remain stable (Fathabadi, 2016), (Novaes Menezes et al., 2018). During the last 25 years, the average power capacity of the turbines has increased by 1500% (Fig. 1-1), evolving from the first Vestas and Nordtrank used in 1995, with a capacity of 500 kW (Fig. 1-2), to the 7~8 MW turbines that are already being used on some of the largest and newest projects these days (MHI Vestas, 2019, p. 3). The European Offshore Wind Deployment Centre in Scotland, and Horns Rev 3 in Denmark, are currently using Vestas V164-8.4 MW turbines, the

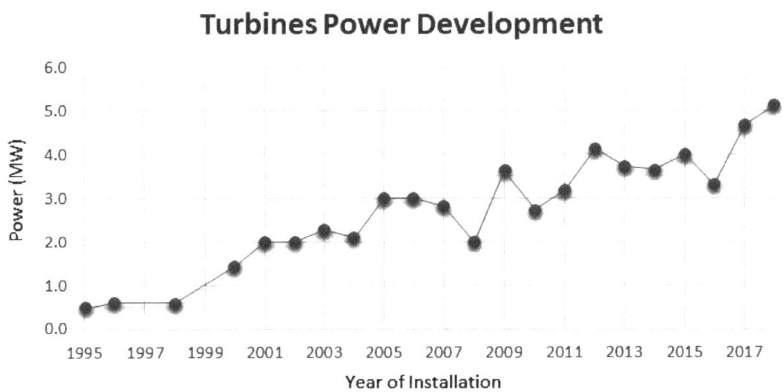

[Fig. 1-1] Average offshore wind turbine power installed every year, source:(Sánchez et al., 2019)

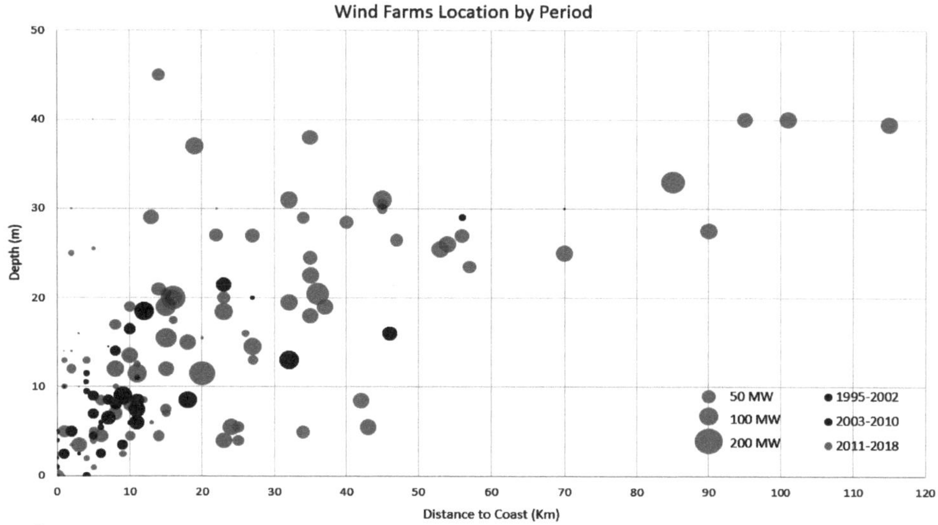

[Fig. 1-2] Offshore wind farms in operation classified by depth and distance from the coast at the end of 2018 (the size of the bubbles represents the capacity installed), source: (Sánchez et al., 2019)

most powerful turbines nowadays for commercial power generation. Currently, the industry is talking about exceeding 10 MW unit power (Velarde & Bachynski, 2017),(Bottasso et al., 2016), (Sánchez et al., 2019).

1.1 The importance of foundation design

Foundation selection plays an important role in the overall concept design for offshore wind farms as there are large financial implications attached to the choices made. Typically, foundations cost 25 to 34% of the overall costs. For the

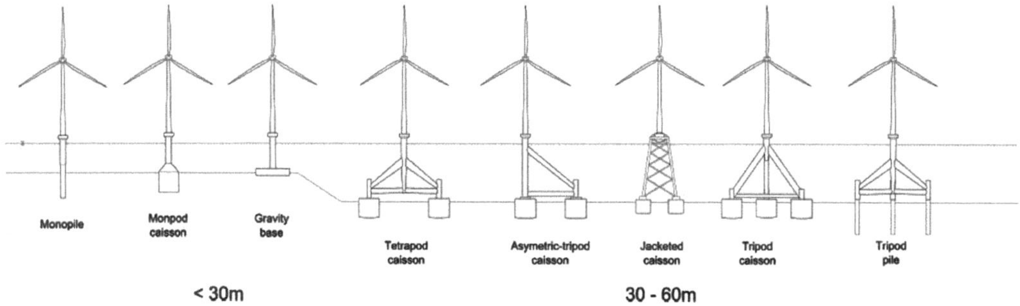

[Fig. 1-3] Figure showing different types of foundations, source: (Bhattacharya, 2014)

North Hoyle project, the cost of the foundation was 34%, Carter (2007) and it has been reported that the development of the Atlantic Array wind farm did not go ahead and one of the main reasons is the expensive foundation. Foundations for wind turbines can be classified into two main types: fixed (or grounded to the seabed) and floating. While most of the currently installed or operating turbines are supported on fixed/grounded foundation systems, research and development of floating foundations are underway. [Fig. 1-3] shows the different types of grounded system (fixed to the seabed) foundation system either in use or proposed and will constitute the main part of this article. (Bhattacharya, 2014)

1.2 Development Status and Benefits Improvement Analysis of Offshore Wind Power in China

China has a long coastline and rich offshore wind energy resources. The offshore wind energy resource reserves are 750 million kW, three times the onshore wind energy resource reserves. The exploitable amount of wind power technology in the offshore area of China is about 190 million kW at 100 m height and 5~25 m water depth, and 320 million kW at 100 m height and 25~50 m water depth (Fang Tao, 2014). However, offshore wind power technology is more complex than onshore wind power, and the construction cost is higher than onshore wind power. China's offshore wind power started late. In 2010, the grid-connected power generation of the Shanghai Donghai Bridge Wind Farm Phase I Project (LIU Baiqiong et al., 2015) marked the official pace of China's offshore wind power development.

With the advantages of rich and stable offshore wind resources and close to the load center, China has preliminarily explored and found that by the end of 2013, the installed capacity of offshore wind power in China had reached 290000 kW (LIU Baiqiong et al., 2015), and by the end of 2017, the installed capacity of offshore wind power had reached 2.79 million kW, and the cost of offshore wind power has also been continuously reduced. The cost of the first phase of the Shanghai Donghai Bridge Wind Power Plant, which was put into operation in 2010, was 23186 yuan/kW. The cost of the second phase of the project, which was completed in 2014, was 19207 yuan/kW (TAO Ye & SHI Jingli, 2013). At present, the average cost of offshore wind power has been reduced to 16000 yuan/kW; However, compared with the current situation where the cost of onshore wind power is 7000 yuan/kW, the price of the offshore wind turbine is about twice that of an onshore wind turbine, the price of electrical equipment such as submarine cables and offshore booster stations is much higher than that of onshore wind farms, and the cost of wind turbine foundation, wind turbine installation, etc. is also far higher than that of onshore wind farms.

This section first reviews the development status of offshore wind power in China, and summarizes the characteristics of offshore wind power in China from the perspective of the development area, water depth, geological conditions, and foundation form; Then, the future cost of offshore wind power and the government's electricity price subsidy policy has prospected; Finally, some new trends related to offshore wind power are listed(WEN Feng, 2016a).

Wind energy is a widely distributed clean energy, which was used for water lifting, flour grinding, and other production activities as early as BC. However, the use of wind energy for power generation and making it an electric power resource originated from the first world oil crisis in 1973. At that time, to get rid of dependence on oil, European developed countries turned to look for new energy and increased the development of wind power, which also became the starting point for the large-scale development of wind power. After more than 40 years of development, wind power has completed the transformation from decentralized small-scale wind power to specialized large-scale wind farms, small-power wind turbines to high-power wind turbines, and from onshore to offshore.

Wind farms can be divided into onshore wind farms and offshore wind farms by location. The installation and maintenance of onshore wind farms are convenient, but there are problems such as land occupation, more quiet periods, high turbulence intensity of wind farms, and affecting the life of wind turbines; In contrast, offshore wind farms have the advantages of high wind speed, few quiet periods, low turbulence intensity and no land occupation. In addition, offshore wind power is generally close to economically developed areas, and power transmission and consumption are guaranteed. There is no need to worry about wind abandonment. At the same time, waterway transportation is convenient for the transportation of high-power and large-diameter fans, and wind turbine transportation is convenient. Offshore wind power with many advantages is considered the development direction of wind power in the future and attracts much attention.

In the 1990s, Europe began to build offshore wind farms and has been at the forefront of global offshore wind power development. By the end of 2014, the total installed capacity of offshore wind power in the world was about 8759 MW, and the vast majority (about 91%) of offshore wind power was distributed in Europe. China installed the first offshore test wind turbine platform in 2007. At present, several offshore wind farms have been put into operation, but on the whole, China's offshore wind power started late, the development of related industries is immature, and challenges and opportunities coexist on the development path.

(1) Development status

The offshore wind power construction areas are intertidal zone, subtidal zone tidal flats, and offshore areas from near to far. Intertidal zone refers to the sea area between the highest tide level during spring tide and the lowest tide level during spring tide; The tidal flat in the subtidal zone generally refers to the sea area from the lowest tide level to 5 m water depth; The offshore generally refers to the sea area 5 m to 50 m below the lowest tide level, and the open sea refers to the sea area 50 m and above below the lowest tide level. At present, all the offshore wind farms that have been built are offshore wind farms.

In April 1986, Shandong Rongcheng onshore wind farm was connected to the grid for power generation, the first wind farm connected to the grid for power generation in China; In 2007, CNOOC built China's first offshore wind turbine with an installed capacity of 1.5 MW by using the discarded jacket foundation of Suizhong offshore oilfield in Bohai Bay. As a pilot offshore wind power project, it opened the prelude to China's offshore wind power development.

China has a long coastline and rich offshore wind energy resources. According to the assessment results of the National Climate Center in 2009, the available wind energy resources within 50 km offshore are 758 million kW. The abundant offshore wind energy provides the possibility for China's offshore wind power development. After several years of development, China's offshore wind power has begun to take shape. See [Tab. 1-1 for the total installed capacity of offshore wind power and the new installed capacity of that year from 2007 to the end of 2015. It can be seen from the table that the annual newly added installed capacity of offshore wind power in China fluctuates greatly, but since 2010, the newly added installed capacity has exceeded 100 MW in all years except 2013, and the newly added installed capacity in 2014 exceeded 200 MW for the first time.

[Tab. 1-2] Accumulated installed capacity of the offshore wind farm in China, source: (WEN Feng, 2016a)

Year	accumulate(MW)	New in the current year(MW)
2007	1.5	1.5
2009	17.5	16
2010	153	135.5
2011	262	109
2012	389	127
2013	428	39
2014	658	230
2015	1004	346

Energy Utilization and Global Economic and Social Development

2

2.1 Challenges in Global Economic and Social Development

2.1.1 Climate Change

As early as the early nineteenth century, French physicist, Joseph Fourier, had recognized the importance of the earth's atmosphere and compared the atmosphere to a greenhouse capable of intercepting part of the solar radiation heat. In 1859, Irish physicist, John Tyndall, discovered that solar short-wave radiation can penetrate the atmosphere and carry energy to the surfaces of the earth, but the earth's long-wave counter-radiation would be trapped by carbon dioxide. Therefore, the earth can maintain temperature and sustain life. Based on this idea, he thought it might explain the climate change in the history of the earth. In 1896, Swedish professor, Svante August Arrhenius, who won the Nobel Prize in Chemistry, estimated that if carbon dioxide in the atmosphere tripled, it would raise the earth's temperature by 5-6℃.

In 1956, Gilbert Plass first elaborated the carbon dioxide theory of climate change using climate models. Then, Charles David Keeling from the University of California, San Diego, published the annual change measurement data of carbon dioxide in the atmosphere and found that atmospheric carbon dioxide concentration continued to rise. At the beginning of the Keeling Series, the concentration of carbon dioxide was 315 ppm and rose to 325 ppm in 1970 and 335 ppm in 1980. Then, the figure continued to rise, from 360 ppm in 1995 to 380 ppm in 2005. This further confirms Professor Arrhenius's view that the increasing carbon dioxide in the atmosphere can lead to global warming. However, Keeling found that carbon dioxide content was increasing much faster than Arrhenius had estimated. This also drove scientists to begin a more detailed examination of the effect of carbon dioxide and other greenhouse gases on the climate, opening the way for scientific analysis of global warming. In 1977, scientists began to reach a new consensus that global warming was indeed a serious threat to human beings. In 1979, the US National Academy of Sciences declared that it was certain that doubling the level of carbon dioxide in the atmosphere would result in a global temperature increase of 1.5~4.5°C

Scientific observations indicate that concentrations of greenhouse gases in the earth's atmosphere are increasing, mainly as a result of the effect of human activities since the industrial revolution, in particular, the continuous increase of greenhouse gases, such as carbon dioxide, due to continuous increase of the consumption of fossil fuels (coal, petroleum, etc.) and massive deforestation of forest vegetation. For example, the annual growth rate of carbon dioxide in the atmosphere is about 1.8 ppm (approximately 0.4%). Although the results of the forecasting models vary slightly, the trends reflected are the same: With the doubling of the concentrations of greenhouse gases such as carbon dioxide, the global atmospheric and soil temperatures will rise by 1.5~4.5°C. This is a gradual change, which is an approximate rise of 0.3~1°C every 10 years under the effect of ocean waters.

At present, there is a consensus among the scientific community on the consequence of global climate change that its effects and harms are mainly reflected as follows: Sea level rise. About a third of the world's population lives in economically developed and city-intensive areas 60 km away from the coastline. The expansion of ocean water and the melting of polar ice and snow caused by global warming may raise the sea level by 50 cm by 2100, threatening the global coastal areas, especially the densely populated and economically developed estuaries and coastal lowlands. These areas may suffer inundation or invasion of seawater, beach erosion in the coastal zone, land deterioration, seawater encroachment, intensified flooding, affect coastal aquaculture, and destroy the water supply and drainage system.

Agriculture and the ecosystem are affected. With the increase in carbon dioxide (CO_2) concentration and climate warming, it is possible to increase plant photosynthesis, prolong the growing season and make some parts of the world more suitable for farming. However, the change in global temperature and rainfall may also enable agriculture in other parts of the world not to adapt to the natural ecosystem, resulting in extensive forest vegetation destruction and agricultural disasters. Natural disasters intensified. Natural disasters caused by climate warming may be a more prominent problem. If the global average temperature rises slightly, it may bring frequent hydrometeorological disasters, such as heavy rain, extensive drought, and persistent high temperature, which will inevitably cause serious losses. Human health is endangered. High temperatures may increase the burden on the human circulatory system, and heat waves may increase mortality. At the same time, as temperatures rise, malaria, lymphatic filariasis, schistosomiasis, leucorrhea, dengue fever, encephalitis, and other infectious diseases may spread in many countries.

As a developing country, China's economy will grow at medium and high speed for a long time in the future. According to the research results of many institutions at home and abroad, in the high economic growth rate scenario, China's energy demand is likely to reach nearly 5 billion tons of standard coal, 6 billion tons of standard coal, and 7 billion tons of standard coal by 2020, 2030 and 2050, respectively. The China White Paper on Energy Policy released in 2012 pointed out that China's energy resources per capita were still at a relatively low level in the world, with the per capita possession of coal, oil, and natural gas amounting to only 67, 5.4, and 7.5% of the average level in the world. Compared with other fossil energy resources, coal resource is relatively rich in China, but their development is restricted by many aspects, such as the number of resources, water resources, ecological environment, safety factors, transportation conditions, and environmental capacity. The quantity of coal resources that can be effectively exploited and utilized is insufficient. China's dependence on foreign oil has soared from 32% at the beginning of the twenty-first century to 60% at present, and its dependence on oil imports is expected to rise further to more than 70% after 2030, while natural gas demand has increased from more than 50 billion cubic meters in 2006 to 300 billion cubic meters by 2030.

As a result of economic and social development and the improvement of people's living standards, China's future energy consumption will also increase substantially. While the pressure on the supply of fossil energy and resource constraints will continue to increase, CO_2 emissions will also increase rapidly. This is bound to have two effects: increased pressure on international climate negotiations. Under the United Nations Framework Convention on Climate Change, most countries are taking active actions to tackle climate change together. The international community has reached a consensus to limit global warming to 2°C by 2050. To that end, global greenhouse gas emission needs to peak around 2020 and reduce by 50% by 2050 than 1990.

Some developed countries and organizations such as the EU, Japan, and Canada, have set their own greenhouse gas emission reduction targets. At the same time, developing countries are called on to bear the corresponding responsibility for emission reduction. China's greenhouse gas emission is large and growing rapidly, and CO_2 emission from fossil energy has ranked first in the world, with per capita emissions exceeding the average level in the world. According to estimates from U.S. Energy Information Administration (EIA), China's CO_2 emission reached 8.55 billion tons in 2012, compared with 5.07 billion tons of the USA, the world's second-largest. China's per capita CO_2 emission is close to 6.3 tons, far above the average level in the world (about 4.51 tons), leading to increasing pressure on China to negotiate internationally. China will suffer the adverse effects of global climate change for a long time. In recent years, extreme weather events have occurred frequently in China, and the losses caused by hydrometeorological disasters have been serious. For example, Southwest China has frequently suffered severe autumn, winter, and spring droughts rare in history; Northeast China and North China have experienced severe winter and spring high temperatures rare in the latest 40 years; the northern part of Xinjiang has had a snow disaster that has been the most severe in history since meteorological records were made; Hainan Province has witnessed a sustained heavy precipitation process rare in history; Gansu, Sichuan, Guizhou, Yunnan, and other places, due to locally heavy rainfall, have suffered serious mountain torrents, debris flows, landslides, and other geological disasters. Based on these circumstances, the Chinese government has promised the international community that, by 2020, the CO_2 emission per unit of GDP will be reduced by 40~45% from that in 2005. Non-fossil energy will account for about 15% of primary energy consumption. By 2030, CO_2 emission will reach its peak and the proportion of non-fossil energy in primary energy consumption will increase to 20% or so. The increasingly stringent and even harsh global climate change targets have put forward more urgent requirements for China to further save energy and reduce emissions. Under the condition that carbon sink potential is limited and carbon capture technology has not yet made a breakthrough, essentially, CO_2 emission reduction is to limit the consumption of high-carbon fossil energy. This poses a severe challenge to China's high-carbon energy structure mainly been based on coal for a long time.

2.1.2 Energy Security

With the growth of the global population and economy, the world energy security situation has attracted the attention of all countries. Especially, on how to realize a stable and reliable supply of clean and efficient energy, both in the long term and in the short term, the countries in the world are facing some difficulties and challenges.

Adequate energy supply, the safety of transportation routes, and affordability of prices are the usual considerations of traditional energy security. Under the traditional energy security framework, a country should try to protect its energy supply from disruption or the risk of prices out of control. As the main energy consumption market in the world, the Organisation for Economic Co-operation and Development(OECD) established the International Energy Agency(IEA) after the oil embargo in the 1970s to protect the energy supply security of the developed countries. By building the capability of oil reserve, fuel replacement and production increase in emergency, IEA member states have jointly set up an emergency reserve capacity to deal with the possible disruption of the oil supply and release the oil tension through the release of strategic oil reserves to the market.

But in the long term, traditional fossil energy has been difficult to support the continuous growth of global energy demand. Global energy consumption is expected to continue to grow over the next 25 years, despite slowing economic growth, rising environmental requirements, volatile energy prices, and other factors. It is predicted that, between 2012 and 2040, total global energy consumption will increase by 48%, with most of it from countries outside the OECD, especially those with relatively strong economic growth, such as China,

India, and other Asian countries. As a result, the contradiction between the limited fossil energy reserves and the continuous growth of energy demand is becoming increasingly prominent. In addition, although the fossil energy reserves-production ratio has increased slightly in recent years, as of 2011, the global oil reserves-production ratio was still less than 50, and the natural gas reserves-production ratio was less than 60. Getting rid of traditional fossil energy resources is a global problem that must be solved thoroughly in the next 20~50 years([Fig. 2-1] and [Fig. 2-2]).

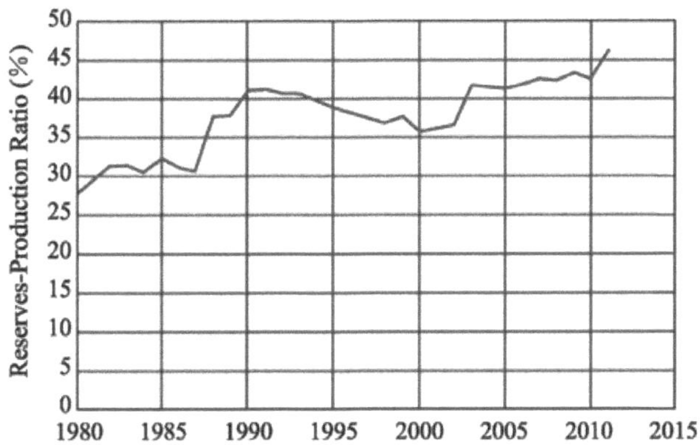

[Fig. 2-1] Global oil reserves-production ratios in 1980~2011. Source US Energy Information Administration (EIA)

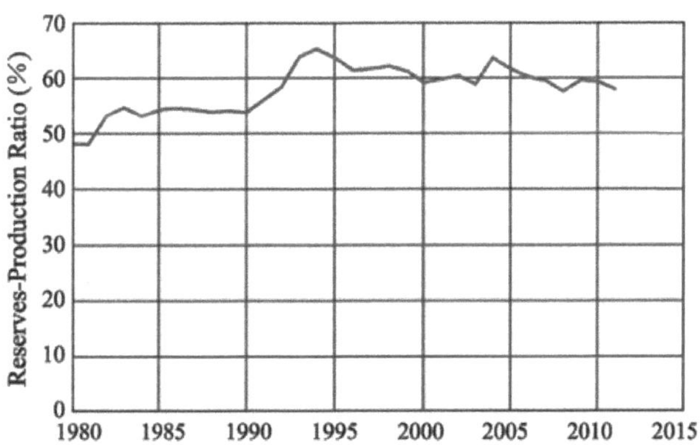

[Fig. 2-2] Global natural gas reserves-production ratios in 1980~2011. Source US Energy Information Administration (EIA)

In addition to resource constraints, fossil energy consumption will increasingly be constrained by carbon emissions in the future, and climate change will challenge the existing energy system in terms of energy supply, infrastructure, energy use method, etc. Therefore, improving energy security requires a strategic vision for the future. Subject to the constraints of energy security, environmental protection, sustainable development, and energy prices, fossil energy growth will gradually slow in the next approximately 30 years. While oil-based liquid fossil fuels remain the world's largest source of fuel, the share of liquid fossil fuels in the global energy consumption market is expected to drop down to 30% by 2040 from 33% in 2012. Under the influence of price and environmental factors, more energy users will prefer more energy-efficient and cleaner energy-use technologies and take this as the goal to promote their respective energy transformation strategies. For example, Japan and other countries have made improving energy efficiency a national policy; Germany, Denmark, and other countries have taken a high proportion of renewable energy as the direction of energy transformation. The Chinese government has also begun to implement an energy policy to reduce coal and carbon dioxide emissions. In a word, to cope with global climate change and accomplish the transformation toward efficient and clean energy, the development and utilization of renewable energy have become important means to ensure global energy security.

Among all kinds of primary energy resources in China, coal resources are rela-tively rich, and coal always occupies a dominant position in the energy consump-tion structure of China, and the absolute consumption of coal resources have been increasing. However, with the development of China's oil and natural gas industry, hydropower, nuclear power, and renewable energy, the proportion of coal in total energy consumption has been decreasing in recent years since 2008. China's coal consumption in 2015 was about 3.97 billion tons, down 1.5 billion tons from the previous year, i.e., with a YOY growth rate of 3.7%, and coal accounted for 64% of total energy consumption, the lowest level since the founding of the People's Republic of China(Fig. 2-3). Compared with coal, the consumption of other energy sources is relatively low; for example, in 2015, the crude oil consumption was equal to 770 million tons of standard coal, accounting for 18% of the total energy consumption; the natural gas consumption was equal to 250 million tons of standard coal, accounting for 5.9%; and non-fossil energy consumption was equal to 590 million tons of standard coal, accounting for 12%. At present, China's coal consumption accounts for about 50% of the total coal consumption of the world, ranking first in the world, and the per capita coal consumption is about 2 tons of standard coal per person, about 2.5 times the average level in the world. On the whole, the increasing proportion of coal consumption occurred during the period of rapid growth of energy consumption in China. This is mainly because coal can adapt to the demand of energy consumption quickly, thus further strengthening its leading position.

However, in the increasingly severe resource and environmental constraints, China's current coal production and consumption mode are difficult to sustain. China's coal reserves are about 114.5 billion tons, and according to the national coal production of about 620 million tons in 1980, the reserves-production ratio would be 185. However, by 2015, the annual coal production had risen to 3.73 billion tons and the coal reserves-production ratio had fallen to 31(Fig. 2-4). In other words, the energy supply system, which is highly dependent on coal, has become more and more difficult to continue for a long time, and the transformation of energy structure is imperative.

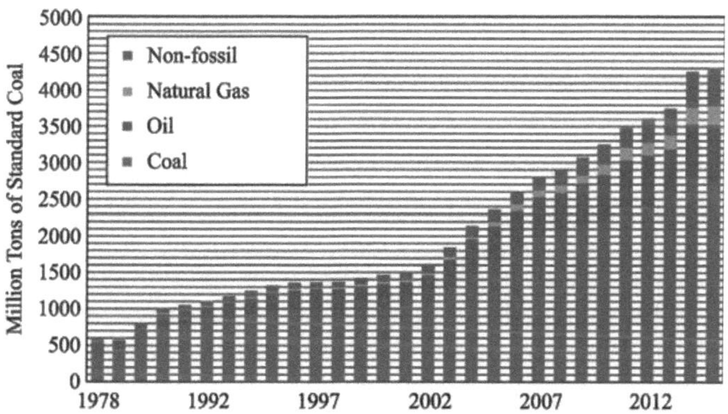

[Fig. 2-3] Primary energy consumption and structure of China in 1978~2015 (attached with color graph). Source China Statistical Yearbook 2016, National Bureau of Statistics of the People's Republic of China

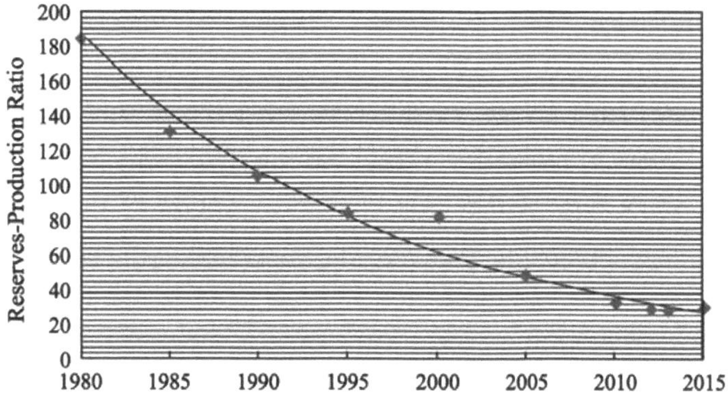

[Fig. 2-4] China coal reserves-production ratios in 1980~2015. Source BP Statistical Review of World Energy 2016, China Statistical Yearbook 2016

With the development of the tertiary industry, especially the transportation indus-try, China's oil consumption is on the rise generally. From 1953 to 1978, the average share of China's oil consumption was 114%. After the reform and opening-up, motor vehi-cles were rapidly popularized, and the oil demand increased rapidly. China's oil consumption averaged 19% from 1980 to 2012. By 2015, oil consumption had risen to 540 million tons. At present, Chinese oil consumption accounts for about 12.9% of the world, ranking second in the world, with per capita oil consumption of about 0.39 tons per person, about 65% of the average level in the world, 14.7% of that of the USA, and 26% of that of Japan. Compared with coal, China's oil resources are more limited and the national reserves are only 2.5 billion tons. According to the oil production of 215 million tons in 2015, the reserves-production ratio is only 116. In the context of the rapid growth of China's oil demand, the degree of dependence on foreign oil is also increasing. From a net oil exporter in the early 1990s to a net importer(Fig. 2-5), China had the proportion of oil imports rise further to 60% by 2015. However, the growing dependence on foreign oil also forces China to speed up the substitution of oil fuels.

Historically, the proportion of natural gas consumption in China was relatively low and had maintained at about 2% of the total energy consumption before 2000. In recent years, with the increase in development and import, the

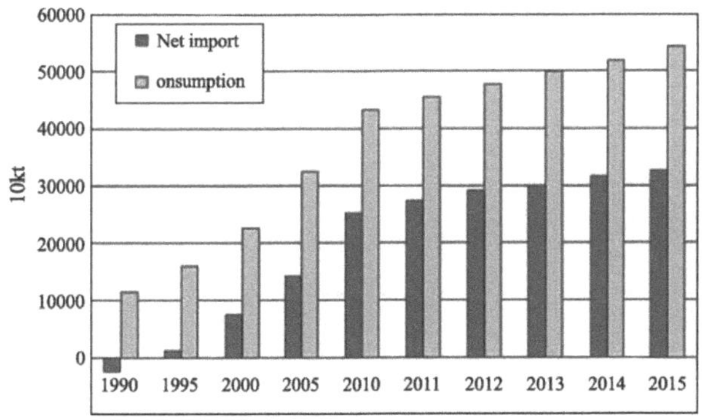

[Fig. 2-5] China oil net import and consumption in 1990~2015. Source National Bureau of Statistics of the People's Republic of China, China Statistical Yearbook 2015

consumption of natural gas has increased rapidly. In 2008, the consumption of natural gas exceeded 100 million tons of standard coal, accounting for 3.7% of primary energy consump-tion. National natural gas consumption reached 188 million tons of standard coal in 2012, and the proportion of natural gas to primary energy further increased to 5.2%. In recent years, China's natural gas consumption has steadily increased(Fig. 2-6). In 2015, China's natural gas consumption was about 190.6 billion cubic Sustainable Development of Wind Energy in China. meters, equivalent to 254 million tons of standard coal. At present, China's natural gas consumption is about 5.7% of the world, ranking third in the world. The per capita natural gas consumption of China is about 139 cubic meters, about 28.9% of the average level in the world, about 5.7% of that of the USA, and 5.1% of that of Russia.

With the advancement of non-fossil energy technologies and the strengthening of non-fossil energy development, the proportion of non-fossil energy consumption in primary energy consumption has risen steadily, from 3.4% in 1978 to 5.1% in 1990 and 6.4% in 2000. Especially since the promulgation of the *Renewable Energy Law* in 2006, the proportion of non-fossil energy consumption has risen rapidly, reaching 8.8% in 2010, and the non-fossil energy consumption reached 559 million tons of standard coal in 2015, accounting for more than 12% of primary energy consumption. However, the proportion of non-fossil energy consumption is still low compared to fossil fuels such as coal(Fig. 2-7).

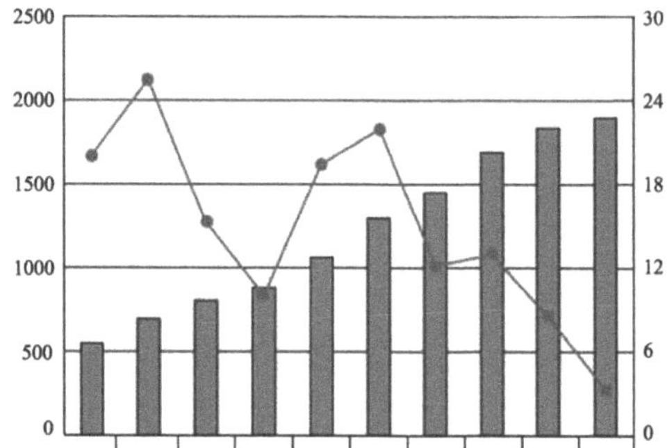

	2007	2008	2009	2010	2011	2012	2013	2014	2015
Natural Gas Consumption (100m m3)	705	813	895	1069	1305	1463	1699	1845	1906
Growth Rate	25.6	15.3	10.1	19.5	22.1	12.1	13.0	8.6	3.3

[Fig. 2-6] China natural gas consumption (100 mm^3) and growth rates in 2006~2015. Source China Statistical Yearbook 2016

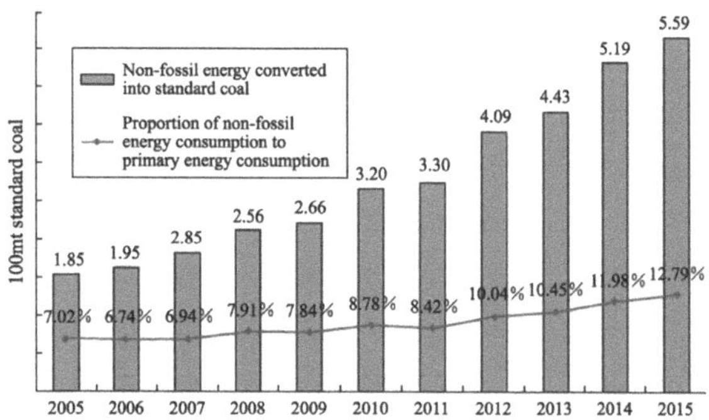

[Fig. 2-7] China non-fossil energy increments and growth rates in 2005~2015

The development of renewable energy is an important approach to achieving the goal of low-carbon development in China. The rapid growth of fossil energy demand such as coal has stimulated the increase in carbon dioxide emission in China(Fig. 2-8).

China overtook the USA to become the world's largest carbon dioxide emitter in 2007. In 2013, China's per capita carbon dioxide emission was 7.2 tons, over that of the EU, and 1.44 times the average level in the world.

Reducing carbon dioxide emissions through energy transformation has become an urgent task for China to promote the target of global greenhouse gas emission

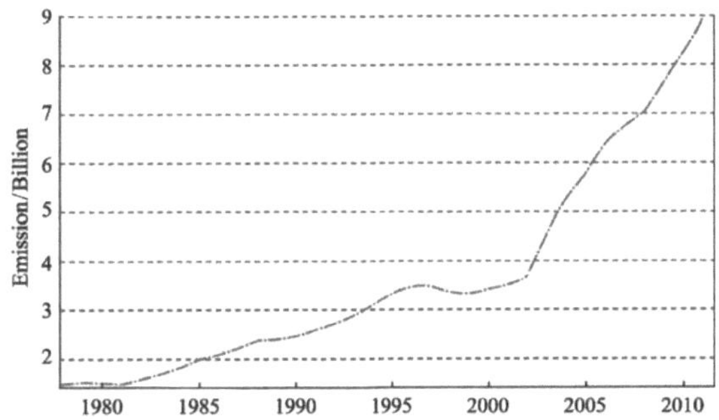

[Fig. 2-8] China CO_2 emissions in 1970~2011 (BT)

reduction. According to the United Nations Intergovernmental Panel on Climate Change (IPCC) report, global emissions should be cut by 15 billion tons by 2030 to reach the target of keeping temperature rise within 2°C. To this end, the *Paris Agreement*, adopted at the Paris Climate Conference at the end of 2015, set the specific target of "keeping the global average temperature within 2°C above the pre-industrial level and striving to keep the temperature rise within 1.5°C." During the 20th summit in September 2016, China signed and submitted the *Paris Agreement* to the UN. It can be predicted that with the *Paris Agreement* coming into effect, the Chinese government's energy structure adjustment will be further strengthened under the pressure of emission reduction.

In short, after decades of development, China has become the world's largest energy producer and consumer, and its energy supply capacity can effectively support economic growth and social development. However, the current energy supply system based on fossil energy resources, such as coal and oil, and natural gas, has led to a series of problems, such as climate change, energy security, environmental deterioration, and it can no longer meet the requirements of the sustainable development of China's social economy in the future. In particular, with the transformation of industrial structure in recent years, the growth rate of energy consumption has slowed down. The YOY growth rate of energy consumption in 2015 was only 0.9%, the lowest since 1998. With the deepening of economic transformation, China's energy consumption growth rate will continue to slow down in the future, and the energy structure will also enter a strategic adjustment period from mainly relying on fossil energy to meeting demand and supply to relying on non-fossil energy to meet demand increment, and it becomes increasingly remarkable to replace coal with oil and fossil energy with non-fossil energy.

2.1.3 Environment Pollution

From the 1930s to the 1960s, environmental pollution incidents that shocked the world occurred frequently. The most serious "world's eight environmental pollution incidents" are as follows: the Belgium Maas Valley Smog Event on December 1~5, 1930; and the Smog Incident in Los Angeles in the USA from May to October 1943, causing a large number of residents to suffer from eye redness, laryngitis, deterioration of respiratory diseases and other diseases and more than 400 people over the age of 65 to die; the Donora Smog Disaster in the USA on October 26~30, 1948; and the Great Smog in London, UK, on December 5~8, 1952, due to smog caused by coal burning in winter, resulting in more than 4000 people's death in 5 days; the Japan Minamata Disease Event occurred in 1953~1968; the Yokkaichi Asthma Episode in Japan in 1955~1961; the Rice Bran Oil Event in Aichi Prefecture, Japan, in March 1963; and Toyama Pain Disease Event in Japan in 1955~1968.

From the 1970s to the 1990s, major pollution incidents occurred frequently around the world. Among them, the famous "ten incidents" are as follows: the Dead Lake Incident in North America beginning in the 1970s; the Cadiz Oil Tanker Incident occurring on March 16, 1978; the Blowout in the Gulf of Mexico occurring on June 3, 1979; and the Kuban Tang "Death Valley" Event occurring in the 1980s; the West Germany Forest Disease Event; the India Bhopal Disaster occurring on December 3, 1984; the Chornobyl Nuclear Leak occurring on April 27, 1986; the Rhine River Pollution Event occurring on November 1, 1986; the Athens "Emergency Event" occurring on November 2, 1989, when the concentration of carbon dioxide in central Athens reached 631 mg/m^3 at noon, surpassing the highest record in history, and the concentration of carbon monoxide also broke the line of danger. Many citizens had toxic symptoms such as headache, fatigue, vomiting, dyspnea, and the Gulf War oil pollution incident.

With the rapid expansion of industrial globalization and the use of a great amount of fossil energy, environmental problems are becoming more and more serious and becoming a major crisis affecting human survival and development: Air pollution. Air pollution is mainly manifested as coal smoke pollution or photochemical smog event and results in the concentration of suspended particles in the urban atmosphere generally exceeding the standard. Sulfur dioxide pollution is at a high level, and motor vehicles' emission of hydrocarbons and nitrogen oxides also exacerbates air pollution. Acid rain is one of the forms of atmospheric pollution and refers to rain, snow, or other forms of precipitation with a pH value below 5.6. The term "acid rain" is now generally used to refer to the transfer of acidic substances from the atmosphere to the surface in the form of wet deposition(rain, snow) or dry deposition(acidic particulate matter). The vast majority of the components of acid rain are sulfuric acid and nitric acid, mainly from fossil fuels. The effects of acid rain on the human environment are various, for example, acid rain changes the ecological environment of rivers and lakes, resulting in the reduction or extinction of fishes and shrimps; acid rain can acidify the soil, destroy nutrients, make the soil barren, and endanger the growth of plants and forests. In addition, acid rain also corrodes building materials and destroys cultural relics. Because of the weak capability of soil in Europe to buffer acidic substances, acid rain has historically affected and degenerated 30% of Europe's forest areas, and the acidification of some lakes has even led to fish extinction. Data from the US National Surface Water Survey shows that acid rain has acidified 75% of the lakes and about half of the rivers in the USA. The Canadian government has also estimated that about 43% of the land of Canada(mainly in the east) has been affected by acid rain and 14,000 lakes are acidic. Ozone hole. 90% of the earth's ozone gas is concentrated in the stratosphere 10~50 km above the earth's surface and forms an ozone layer with a thickness of about 3 mm at 25 km above the earth's surface. However, the ozone layer is very fragile, and if ozone-depleting substances(ODS) such as HCFCs are exposed, the ozone layer is destroyed by chemical action with ozone. Since the ozone hole appeared in Antarctica in 1985, the situation has become more and more serious. By 1994, the ozone depletion over Antarctica had reached 24 million square kilometers. Now, the ozone layer is thinning over the USA, Canada, Western Europe, Russia, China, Japan, and other countries and regions. In fact, before the control of ODS in 1996, the world had released 20 million tons of ODS into the atmosphere, destroying the ozone layer. So, even if the world completely stops releasing ODS, it will take another 20 years for humans to see the sign of ozone layer recovery. The destruction of the ozone layer will increase the surface's exposure to more ultraviolet(UV) radiation, endangering biological proteins and genetic substances, deoxyribonucleic acid (DNA), causing cell death; increasing the incidence of skin cancers in humans; increasing the incidence of cataracts; inhibiting the growth of plants such as soybeans, melons, vegetables. In addition, UV may penetrate the water layer, killing plankton and microorganisms, thus endangering the food chain of organisms in the water and the source of free oxygen, and affecting the ecological balance and self-purification capability of waters. Water pollution. Water is one of the substances that we need most in our daily life. However, due to the random discharge of industrial sewage, agricultural sewage, domestic sewage, and a large number of chemicals, seawater, lakes, and groundwater are seriously polluted. Water pollution has exceeded the environment's purification capability and the situation of water pollution is austere. Many parts of the world are facing a serious water crisis. The World Bank estimates that more than 1 billion people worldwide cannot have safe drinking water due to water pollution and lack of water supply. The health hazards of toxic chemicals and hazardous wastes. Toxic chemicals such as lead, mercury, and some industrial reagents and pesticides produced by human activities are often exposed to the natural environment in the form of waste and have serious effects on human health. Currently, heavy metals and persistent organic pollutants(POPs) are becoming the focus of concern and affect the entire ecosystem through the aggregation of food chains. Long-time exposure to heavy metals will lead to slow movement, a dysfunctional immune system, and even carcinogenesis. POPs will affect human reproduction, development, and intelligence.

The huge population of China, the unique coal-based energy structure, and the process of rapidly realizing industrialization have all posed extraordinary pressure on the environment and resources, making China face a more serious environmental crisis: Serious air pollution. Air pollution in China is concentrated in urban agglomeration. The main pollutants are smoke, carbon dioxide, sulfur dioxide, and so on. Coal combustion and automobile exhaust emission are the main sources of pollution. At present, both sulfur dioxide and carbon dioxide emissions in China rank second in the world, and atmospheric pollution has caused great pressure on China's environment. Since October 2011, smog weather has continued to occur in many places of China, including Beijing and Shanghai, seriously affecting the daily life of residents. One of the most important factors for such smog weather is that the PM2.5 in the suspended matter in the atmosphere exceeds the standard seriously. The pollution sources mainly come from secondary particles such as sulfate aerosol generated by soot, motor vehicle tail gas, and the like. The south of China is the area with the most serious acid rain and has become a major acid rain area after Europe, the USA, and Canada.

Water resource crisis. The distribution of water resources in China is extremely uneven, and 81% of the water resources are concentrated in the Yangtze River basin and the area in its south, which accounts for 36% of the agricultural acreage of the whole country. The Huaihe River basin and the area on its north, which accounts for 64% of the agricultural acreage of the whole country, have only 19% of the water resources of China. The shortage of water supply has resulted in a reduction in agricultural production, and an annual loss of nearly 100 billion yuan in industrial output value is caused by water shortage in China. However, the pollution of water resources brought up by power generation and industry is also very prominent. The area of urban groundwater pollution continues to rise, and a third of the wells in China has not reached the drinking water standard. According to the statistics of the Ministry of Environmental Protection of China, 44.7% of the main water systems, such as the Yangtze River, the Yellow River, the Songhuajiang River, the Liaohe River, the Haihe River, the Huaihe River, and the Taihu Lake, the Chaohu Lake, and the Dianchi

Lake, have lost the drinking function. Disastrous waste pollution. The emission of solid industrial waste and municipal solid waste in China is increasing day by day, but the treatment and utilization rate thereof is still low. In 2009, the production of industrial solid waste in China was 2.04 billion tons, an increase of 7.3% over the previous year. In particular, the toxic solid waste, which has been buried for a long time, has been leaching through rain and snow, and the soluble components have seeped down from the surface of the earth with water to migrate and transform into the soil, making harmful substances concentrated, acidifying, alkalizing, and hardening the soil near the backfilling site, and even forming heavy-metal pollution. For example, in the soil near a general non-ferrous smelter, lead is 10~40 times higher than that in normal soil; copper, 5~200 times; and zinc, 5~50 times. On the one hand, these toxic substances enter waters through the soil; on the other hand, they accumulate in soil and are absorbed by plants, poisoning crops. (Li & He, 2020)

2.2 Worldwide Actions

2.2.1 Climate Change Mitigation

In the 1980s, the impacts and hazards of climate change began to receive global attention. Since the global impact of climate change goes far beyond natural science, global cooperation is needed to tackle climate change. Therefore, under the leadership of the UN, the IPCC was established in Geneva in 1988 to coordinate climate research and climate protection around the world. In March 21, 1994, the *United Nations Framework Convention on Climate Change* came into force. There-after, a conference of the contracting states is held every year. Three of these conferences have had a significant impact on key actions to promote global climate change: First, the Third Conference of the Parties (COP 3), held in Kyoto, Japan, in December 1997, whose largest outcome is that it was proposed that greenhouse gas emission should be bound by the international laws. Countries that signed the *Kyoto*

Protocol promised to reduce their emission of carbon dioxide and other green-house gases by 5.2% from 1990 and achieve this goal by 2008~2012. For example, the EU promised to cut emissions by 8%; the USA, by 7%; and Japan and Canada, by 6%. Because of the global nature of the greenhouse effect, the industrialized countries that exceeded the quota could buy their emission rights into the hands of the developing countries that had never run out of the quota. Second, the Thirteenth Conference of the Parties (COP 13), was held in Bali, Indonesia, in December 2007. This conference adopted a resolution called the "Bali Road Map." The resolution recognized that greenhouse gas emissions must be "drastically reduced" to prevent human activity from exacerbating climate change. Citing scientific research, the document suggested that greenhouse gas emissions should be reduced by 25~40% by 2020 from that in 1990, but the document itself did not have a quantified emission reduction target. It was proposed to hold climate change negotiations for 2 years and a new agreement should be reached by 2009 to reserve sufficient time for the new agreement to take effect by the end of 2012. Four major conferences on climate change were planned to be held in 2008. The negotiations should consider setting greenhouse gas emission reduction targets for the industrialized countries, and the developing countries should take steps to curb the growth of greenhouse gas emissions. The more developed countries should transfer environmental protection technologies to the more backward countries. The negotiators should consider pro-viding urgent support to poorer countries to deal with the inevitable consequences of climate change, such as building breakwaters. Negotiations should consider "Pos-itive Incentives" to encourage developing countries to protect the environment and reduce deforestation. Third, the Twenty-first Conference of the Parties (COP 21), was held in Paris, France, from November 30 to December 11, 2015. During this con-ference, 195 state parties around the world adopted the historically new agreement on global climate change, *Paris Agreement,* which is the first global agreement on climate change in history. Under the Agreement, the parties agreed to strengthen the global response to the threat of climate change in the context of the requirements of sustainable development and efforts to eradicate poverty, limit the rise in global average temperature to less than 2°C from that in the pre-industrial period, and continuously make efforts to limit temperature rise to 1.5°C to significantly reduce the risks and effects of climate change. In addition, the Agreement required that developed countries should continue to take the lead in their efforts to achieve the emission reduction target while developing countries should continue to strengthen emission reduction efforts according to their different national conditions and gradu-ally achieve targets of emission reduction or emission limitation. In respect of funds, the Agreement required that developed countries should assist developing countries by providing financial resources for both mitigation and adaptation. At the same time, with "$100 billion per year after 2020 will be provided to help developing countries cope with climate change" as the bottom line, the Agreement required that the parties should bring forward the new funding targets by 2025 at the latest. On the much-watched issue of the Intended Nationally Determined Contributions (INDCs), under the Agreement, all parties would participate in the global response to climate change in an "INDCs" way. All parties should step up their current INDCs according to different national conditions, do their utmost, and bear common but differentiated responsibilities. The developed countries will continue to take the lead in reducing emissions and strengthening financial, technological, and capacity-building supports for the developing countries to help them mitigate and adapt to climate change. The Agreement also established a checkup mechanism, that is, from 2023, the overall progress of global action would be checked up once every five years to help the countries enhance their efforts and strengthen international cooperation to achieve the long-term goal of the global response to climate change.

2.2.2 Energy Supply Security

In response to energy security, many developed and developing countries have pro-posed their energy transformation plans, all of which focus on renewable energy such as wind and solar energy. Even Canada, Australia, and the countries in the Mid-dle East and North Africa rich in traditional fossil energy have also proposed renewed-able energy development goals to reduce dependence on fossil fuels (Tab. 2-1).

(1) EU

Europe is the world's leader in building a low-carbon, clean, and sustainable energy system. EU has set up a target for energy development to make the proportion of renewable energy in energy consumption reach 20%

[Tab. 2-1] Renewable energy development targets and main fields of main countries (regions)

Country/ region	Renewable energy development targets	Main fields and measures
EU	Renewable energy will account for 20% of total energy consumption by 2020 and 50% by 2050	To promote wind energy, solar energy, biomass energy, smart grid, and implement the Emission Trading Scheme (ETS)
UK	By 2020, renewable energy will account for 15% of total energy consumption, 40% of electric power will come from the green energy sector	To actively develop onshore wind power, offshore wind power, biomass power generation, etc.; to promote intelligent ammeter and demand-side transmission technology; to implement renewable energy generation price contracts for difference (CFD)
Germany	By 2020, 2030, 2040, and 2050, renewable energy will account for 18, 30, 45, and 60% of terminal energy consumption, respectively, renewable energy electric power will account for 35, 50, 65, and 80% of total electricity consumption, respectively	To support wind power, solar power generation, energy storage, expansion of transmission pipeline and grid facilities, expansion of energy storage capacity and to implement renewable energy feed-in tariff (FIT) and fit-in premium (FIP)
Denmark	By 2020, wind power will account for 50% of the total electricity consumption; by 2050, it will completely get rid of fossil energy consumption	To support the development of wind power and green heating system, promote the application of renewable energy in the sectors of construction, industry and transportation, and to promote the development of smart grid
USA	By 2030, carbon dioxide emission from the electric power sector will be cut by 30% from that in 2005	To promote the construction of wind power, solar power generation, biofuels, and smart grid; to implement production tax credit (PTC) and investment tax credit (ITC); and to implement the renewable portfolio standard (RPS) in 29 states, Washington, D.C., and 2 dependent regions
China	By 2020 and 2030, non-fossil energy will account for 15 and 20% of primary energy, respectively	To support hydropower, wind power, solar power generation, renewable energy thermal utilization and fuel; to support FIT for new energy generation and FIP for distributed solar power generation

by 2020 and 27% by 2030, and set a quantitative index to restrict the member states to accelerate the development of renewable energy. Denmark has proposed to get rid of fossil energy completely by 2050; Germany's strategic goal is that renewable energy will account for 60% of energy consumption and 80% of electricity consumption by 2050, and Germany has set up specific goals for every stage.

European countries have set up their own development goals under the guidance of the EU, and many countries have made mandatory purchase requirements for renewed-able energy generation. And they ensure revenues from renewable energy generation through FIT, FIP, or CID and increase the cost of fossil energy through EU ETS, carbon tax, or carbon standard. Guided by strategic goals and driven by policies, despite the stagnation of individual countries due to policy adjustments, Europe as a whole continues the trend of renewable energy development. In 2015, 77% of the 2890 kW new electric power installed capacity in the EU came from renewable energy sources, and the newly installed capacities of wind power and solar power reached 44 and 30%, respectively, both, respectively, exceeding the sum of the newly installed capacities of coal and natural gas. Germany has led the development of renewable energy in Europe for a long time, with the installed capacities of wind energy and solar PV power generation reaching 4500 kW and 4000 kW, respectively, in 2015, bringing the proportion of the power generated with renewable energy to 30% of the total power generating capacity of the country.

(2) USA

The federal government has not directly set up targets for renewable energy. The Clean Power Program (CPP), approved in 2015, requires the US electric power sec-tor to reduce its carbon emission by 32% by 2030 from that in 2005 and requires states to come up with their implementation plans by 2018 at the latest. By that time, it is expected that the proportion of the installed capacity of power generated with renewable energy will increase to 28%. The relevant research by the US Department of Energy suggests that wind power will account for 20% of all the power generating capacity by 2030, and the power generated with renewable energy will account for 80% of all the power generating capacity by 2050. The most important renewable energy support policies in the USA are federal PTC, ITC, and the cash premium in the Economic Stimulation Act 2009. In December 2015, the two parties at the Congress agreed to extend the PTC/ITC policies until 2022. These have laid the policy basis for the USA to continue to promote wind power and solar power generation toward sus-tainable growth. In addition, about 29 states and Washington, D.C., have established the Renewable Energy Certificates (REC) policy, under which renewable energy can be sold and Renewable Energy Certificates can get benefits.

Under the policy impetus, the application scale of US wind energy increases unceasingly. In 2014, in the USA, renewable energy accounted for 9.8% of the total primary energy consumption of the country, and renewable energy generation accounted for 13.2% of the total power generating capacity of the country. In 2015, 65% of the newly installed capacity of the USA came from renewable energy. That year, in the USA, the new installed capacity of wind power reaches up to 8.6 GW; the cumulative installed capacity reached 74.40 GW. In the same year, the power-generating capacity of the USA reached 191 billion kilowatt-hours, ranking the first in the world, and wind power accounted for 4.7% of the total power-generating capacity in the USA. The new installed capacity of solar power reached 7.3 GW, accounting for 28% of all the new installed capacity in the USA that year, higher than that of natural gas.

(3) China

At the meeting of the Central Leading Group on Financial and Economic Affairs held in 2014, Xi Jinping, General Secretary of the CPC, proposed that China should actively promote the revolution in energy production and consumption, control the consumption of fossil energy, and vigorously develop renewable energy sources such as wind energy and solar energy. In recent years, a consensus has been gradually reached among all stakeholders, who have realized that the essence of the "Energy Revolution" is the replacement and substitution of the main energy and the revolution is a revolution of replacing high-carbon fossil energy such as coal with clean low-carbon energy such as renewable energy. In 2015, China committed itself to the peak of greenhouse gas emissions in the international community around 2030, i.e., the proportion of non-fossil energy in primary energy consumption increased to about 20%.

In recent years, China has continuously improved the policy of supporting the electricity price of renewable energy. In the new round of electric power reform, China has brought forward and established the system feeding the power generated with renewable energy into the grid firstly. Driven by a series of policies, China has maintained the largest market size of renewable energy in the world. By the end of 2015, the cumulative installed capacity of hydropower, wind power, and solar power in China had reached 320, 130, and 4300 GW, respectively, continuing to lead the world. The proportion of non-fossil energy consumption reached 12%, realizing the "12th Five-year Plan" target in renewable energy development.

(4) Other Emerging Economies and Developing Countries

India is the second most populous country in the world. Since the twenty-first century, energy demand has kept greatly increasing. The energy supply and environmental protection pressure on India are increasing day by day. In 2010, the Indian govern-ment announced that it would increase the proportion of the power generated with renewable energy from 5% in 2010 to 15% by 2020; in 2014, the government pro-posed that the installed capacity of wind turbines would reach 150 GW by 2027. At present, India implements such policies as renewable energy fixed electricity prices, and a renewable quota system. In 2015, renewable energy accounted for 6.3% of total energy consumption in India, and wind power and solar power reached 25 and 5 GW, respectively.

Energy for Brazil has been dependent on hydropower, oil, and gas for a long time. Hydropower accounts for 28% of all the primary energy consumption and 64% of all the electric power consumption of Brazil. However, in recent years, Brazil has vigorously supported the construction of new energy sources, implemented the renewable energy bidding mechanism and the power purchase agreement (PPA) auction plan, and promoted the development of wind energy, solar energy, and other new sources of energy. By the end of 2015, the wind power installed capacity of Brazil had reached 8.7 GW.

The United Arab Emirates, Saudi Arabia, and other traditional oil-producing coun-tries have also increasingly attached importance to the development of renewable energy. In 2012, Saudi Arabia set up a long-term renewable energy development tar-get and planned to build a 55 GW installed capacity of power generated with renewable energy by 2032. Although the progress is slow at present, the latest release of *Vision 2030* in April 2016 emphasizes improving the investment context by changing laws and liberalizing the fuel market, to ensure the competitiveness of renewable energy. The UAE plans to complete Master City dependent on renewable energy by 2025. Renewable energy is also an important way to solve the shortage of modern energy in Africa. The Africa Renewable Energy Action Initiative (AREI) announced during the Conference of the Parties (COP 21) 2015 in Paris calls for at least 10 GW of new installed capacity of power generated with renewable energy by 2020 and 300 GW newly added by 2030. South Africa has offered to invest $90 billion in renewable energy over the next 20 years and plans to increase its total renewable energy use by 40% and double the country's total power generating capacity.

2.2.3 Climate Change Mitigation

In the face of increasing environmental problems, people begin to solve all kinds of environmental problems around the world and have set off environmental protection campaigns against pollution and public hazard. In the 1960s, the first wave of environmental protection beyond national boundaries opened in human history. In 1967, the Japan Federation of Bar Associations issued a public notice, clearly pointing out that environmental problems affected the production and life of mankind. At the same time, the theory circle intensified the research in this field. As a result, some influential books, such as *Population, Resources, and Environment* by P. R. Ehrlich and *Limits to Growth* by the Club of Rome, were published one after another. On April 22, 1970, a memorable day in the history of environmental protection in the world, a mass environmental campaign swept across the USA, with people from all walks of life having parades, rallies, and speeches, and there were more than 20 million people taking part in this unprecedented campaign. This campaign had a wide influence on the international community and received positive responses from many countries in the world and became a worldwide environmental movement. To commemorate the event, April 22 was designated as "World Earth Day."

In response to the global environmental crisis, countries around the world held an international environmental conference to actively consult and formulate development plans. Since 1972, the UN has convened the following international environmental conferences:

(1) United Nations Conference on Human Environment, Stockholm, 1972

On June 5~16, 1972, the UN convened the first major international conference on the human environment in the history of mankind, i.e., the UN Conference on the Human Environment, in Stockholm, Sweden. The conference was attended by approximately 1200 delegates from 114 states and observers from a large number of intergovernmental and non-governmental organizations.

The purpose of the conference was to discuss jointly the environmental problems in front of human beings, to develop strategies and measures "to achieve a common vision and develop common principles to inspire and guide the peoples of the world to preserve and improve the human environment," and to end and turn round the effects of environmental degradation. The conference lasted for 12 days. At the conference, the theoretical and practical issues concerning the protection of the human environment were fully discussed and summarized; a series of countermeasures and measures were developed; a series of important documents and conventions were formulated, and the slogan of "only one earth" was put forward. The conference also called upon all governments and people to work together to protect and improve the human environment and benefit future generations. This was the first world conference devoted to human environmental issues in the history of the UN. It marked the UN's full involvement in world environmental and development affairs and built up a new milestone in the history of human environmental protection. The conference caused all the countries to reach a consensus and formulated common principles and it played a great role in arousing human environmental consciousness and promoting international cooperation in environmental protection. To commemorate the convening of the conference and arouse public attention and vigilance on environmental issues, the UN General Assembly of the year resolved on setting and the delegates agreed to set June 5 as "World Environment Day." With the unremitting efforts of the participating countries, the conference passed two important documents, i.e., t*he Declaration of the United Nations Conference on the Human Environment and the Action Plan for the Human Environment.*

(2) United Nations Conference on Environment and Development, Rio de Janeiro, 1992

If we say that the greatest achievement of the United Nations Conference on Human Environment, Stockholm, 1972, was to awaken the world to the environment, we have to say that United Nations Conference on Environment and Development, Rio de Janeiro, 1992, after 20 years not only extended the scope and depth of understanding of environmental problems but also combined environmental problems with economic and social development for research to explore their mutual influence and dependence.

On June 3, 1992, the high-profile United Nations Conference on Environment and Development opened in Rio de Janeiro, Brazil, with the participation of delegations from 178 countries, a total of more than 8000 delegates including more than 100 heads of state and governments, more than 9000 journalists, and a large number of observers from international agencies and nongovernmental organizations. To echo with this, the World Civil Environment Assembly, "Global Forum 1992," also opened at the same time.

The purpose of the conference was "to develop strategies and measures to halt and reverse the impact of environmental degradation, in the context of strengthening-in national and international efforts to promote the sustainable and environment-friendly development of all countries." The two-day summit culminated with concrete initiatives by many country leaders of the world. During the general debate of the conference, more than 200 speakers spoke on behalf of governments, intergovernmental agencies, and nongovernmental organizations on the issues of the environment and development they are concerned.

The conference was an international environmental event of unprecedented scale and far-reaching significance. It had set a new record in the history of the UN with its long preparation time, large attendance, and high level and had been recognized as the highest symbol of international cooperation in the field of environmental protection and development. It won high praise from the international community. The success of the conference not only was the result of the sincere cooperation of all the countries in the world but also closely related to the efforts of the UN. From the preparation and holding of the conference to the breaking of the deadlock, the UN carried out a great deal of organization and coordination work to ensure the smooth progress of the conference. Through the joint efforts of many countries, the conference achieved important results and adopted and signed two far-reaching programmatic documents: *The Rio Declaration on Environment and Development and Agenda 21.*

(3) United Nations World Summit on Sustainable Development, Johannesburg, 2002

Although the 1992 Conference on Environment and Development put forward the concept of sustainable development, because of the intricate contradictions in international environmental development, the interests were intertwined. The implementation of important documents such as Agenda 21 aiming at sustainable development was not satisfactory; the global environmental crisis was not reversed; on the contrary, it was worsening, poverty was widespread, and the gap between the North and the South was increasing. In this context, the UN General Assembly Resolution 55/199 decided to convene the United Nations World Summit on Sustainable Development. The World Summit on Sustainable Development held in Johannesburg, South Africa, from August 26 to September 4, 2002, has been the largest international conference on environment and development so far. There are 17,000 delegates from 192 countries, including 104 heads of state and governments. The conference conducted an overall review and evaluation of the implementation of Agenda 21 since the 1992 United Nations Conference on Environment and Development and reaffirmed the global partnership for sustainable development. With concerted efforts of all the parties, the conference adopted two important documents: The *Johannes-burg Declaration on Sustainable Development* and *the Plan of Implementation of the World Summit on Sustainable Development.*

(4) 2015, the "Year of Sustainability of the World"

In 2015, the "Year of Sustainability of the World," the UN held high-level meetings on the topic of "sustainable development." Among these meetings, on August 2, 2015, 193 members of the UN agreed on the future development agenda and formulated a consensus document entitled "Transforming our World: The 2030 Agenda for Sustainable Development." The consensus document reflected both the increasing attention of the international community to environmental issues and the call by the international community represented by the UN for a more effective global environmental governance system. At the UN special summit in September 2015, the leaders of all the member states cosigned and ratified the Post-2015 Development Agenda.

All countries intend to achieve sustainable economic and social development through innovation in concept, policy, technology, industry, or management. To fulfill the environmental protection tasks promised in the international treaties, they also want to realize the adjustment of domestic energy structure and economic structure and bring new economic development models and new economic growth points.

The US government has put forward a clean coal plan based on the Clean Air Act and the Energy Policy Act and the goal of the plan is to make full use of advanced technology to improve efficiency, reduce cost, and reduce emissions. Since 2001, "the U.S. government has invested $2.2 billion in promoting advanced clean coal technologies from R&D to demonstration and marketization. Through the 'Coal Research Program' policy, the National Energy Technology Laboratory of the Department of Energy carries out clean coal technology R&D and develops innovative pollution control technologies, coal gasification technologies, advanced combustion systems, steam turbines, carbon capture, and storage technologies, etc." In addition, the USA adopts a variety of fiscal and tax policies to develop a sustainable economy. For example, it makes tax exemptions to encourage energy conservation, provides tax credits for the purchase of energy-efficient vehicles, and encourages American consumers to buy energy-efficient cars.

Europe, as an established industrial area, is engaged in a series of innovative activities for a sustainable economy in the region, in various fields including concept, policy and system, technology and industry, business operation, and consumer life. These low-carbon activities improve its national competitiveness and build up a good image in environmental protection. In 2006, the EU published the *Stern Report*, which formed the policy basis for the EU's response to the environmental crisis.

As a sponsor and advocate of the *Kyoto Protocol*, Japan has also made great efforts in sustainable economic development. Japan has invested heavily in the development of new and renewable energy, not only in the industrialization of new energy but also in the basic research of solar energy, wind energy, hydrogen energy, fuel cell, tidal energy, water energy, geothermal energy, and so on. In 2007, Japan decided to invest 209 billion yen in clean car technology over the next five years, to reduce not only fuel consumption but also greenhouse gas emission. In June 2008, Japan proposed a "Fukuda Blueprint" to increase its solar power generation by 10 times by 2020 and 40 times by 2030 from the current level.

Brazil, as a developing country, has achieved remarkable results in energy conservation, environmental benefit, and multifaceted resource conservation by adopting the policy supported by the power fund. Brazil, as the world's largest producer of ethanol fuel, will further boost its low-carbon energy growth by working with the USA to promote ethanol and other biofuels to other Latin American countries as an alternative energy source for export.

India has abundant renewable energy resources and regards the development of renewable energy as an important energy strategy of the country. It has set up a special government agency responsible for the management of renewable energy and established the development goals of renewable energy on a large scale. Between 2004 and 2005, India built 100,000 biogas power plants for 16,000 solar lighting systems. In 2015, the Indiana government again raised its renewable energy development target to 175 GW installed capacity by 2022. Rising plans have spurred India to slow its greenhouse gas emission and boost rapid economic growth and sustainability.

The Chinese government also attaches great importance to environmental protection. In 1994, China approved *China's Agenda 21—a White paper on China's Population, Environment and Development in the 21st Century*, which was used to guide the future social and economic development of China. *The Decision of the Central Committee of the Communist Party of China on Some Issues Concerning the Improvement of the Socialist Market Economy*, adopted at the Third Plenary Session of the 16th Central Committee of the

Communist Party of China in October 2003, points out that: "We should uphold the principle of 'people-oriented' and form the concept of overall, coordinated and sustainable development to promote the overall economic, social, and human development." In his speech at the Paris Conference in 2015, President Xi Jinping mentioned that China "wants to achieve a higher level of global sustainable development and build the international relationship of cooperation and common win." The thought of sustainable development has been accepted by more and more people and has gradually become the consensus of the world. It not only affects people's mode of production and thinking but also relates to the future and destiny of mankind.

(5) "Paris Agreement" 2016

On April 22, 2016, more than 100 countries gathered at UN headquarters in New York to finally sign the Paris Agreement, which had become an important chapter in the process of sustainable development of humans. The greatest contribution of the Paris Agreement lies in identifying the global common pursuit of "hard indicators," promoting the active transition to green and sustainable development in a way of "independent contribution," avoiding the continued harm to the natural ecosystem of the growth pattern that had been heavily dependent on petrochemical products over the past decades, and guiding the global shift to green energy and low-carbon economy and environmental governance. (Li & He, 2020)

2.3 Position and Role of Wind Power

2.3.1 Improving Structure of Energy Utilization

At present, the main energy of most countries in the world still comes from fossil energy such as oil and coal. With the greenhouse effect, air pollution, acid rain, and other ecological and environmental problems highlighted, energy transformation has been an increasingly concerned, making countries around the world begin to adjust the direction of energy development, from the traditional old energy system based on natural gas, coal, and oil gradually to the new system based on clean and renewable energy.

As a clean energy, wind energy is an inexhaustible renewable energy and it has also experienced a gradual development process: In the 1980s, wind farms were built in some countries and regions such as California, the USA, and Denmark. In the 1990s, wind turbines began to become large. Around 2002, the mainstream models in the international wind power market reached 1.5 MW. After 2005, the installed capacity of wind power in various countries had an explosive growth; in various countries, with the acceleration of the pace of energy structure adjustment and the breakthrough of wind energy technology, the capacity of a single mainstream wind turbine had increased to 2.0~2.5, and 5 MW wind turbines had been used. In 2015, Denmark, Spain, and Germany had the total generating capacity of wind power account for 42.1, 19.4, and 12.0%, respectively, of their own national total power consumption. By the end of 2015, worldwide, the total installed capacity of wind power had reached 432 GW, accounting for 5.2% of the total power market in the world. Wind power is expected to account for 12% of the world's total power-generating capacity by 2020. After decades of development and utilization, wind energy has become one of the renewable energies with the most mature technology, the lowest development cost, and the highest development value. It is an important choice to achieve the goal of energy conservation and emission reduction.

The development of large-scale wind energy in China started late but grows rapidly. The proportion of wind power consumption in the total power consumption of China increases year by year. The proportion was 3.3% in 2015 and is expected to reach 6~7% by 2020. The use of wind energy reduces the consumption of conventional fossil energy and plays an important role in protecting the ecological environment and saving energy in China. Domestic and foreign research institutions have also carried out prospects for the development of wind power in China. It is estimated that, in the basic scenario, the installed capacity of wind power will reach 200 GW, 400 GW, and 1000 GW by 2020,

2030, and 2050, respectively; in an active scenario, that will reach 300 GW, 1200 GW, and 2000 GW by 2020, 2030, and 2050, respectively; wind power will become one of the five major power sources in China; by 2050, in the two scenarios, wind power will account for 17% and more than 30% of the power demand(Fig. 2-9). Achieving these goals will bring enormous environmental and social benefits. In both scenarios, the carbon dioxide emission in 2050 will be 1.5 billion tons and 3 billion tons, respectively(Fig. 2-10).

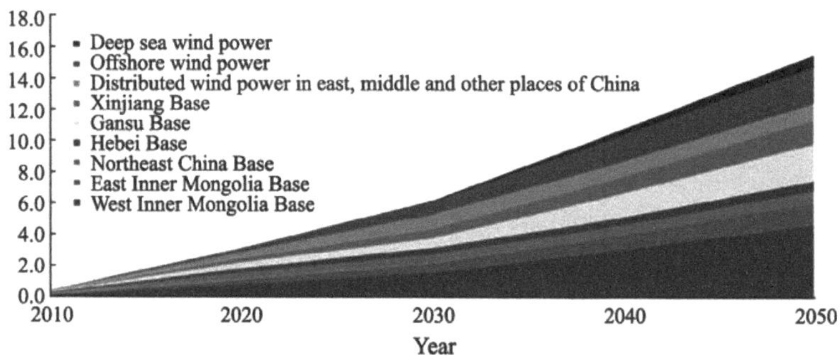

[Fig. 2-9] China's wind power development goal and layout (attached with color graph). Source National Renewable Energy Center

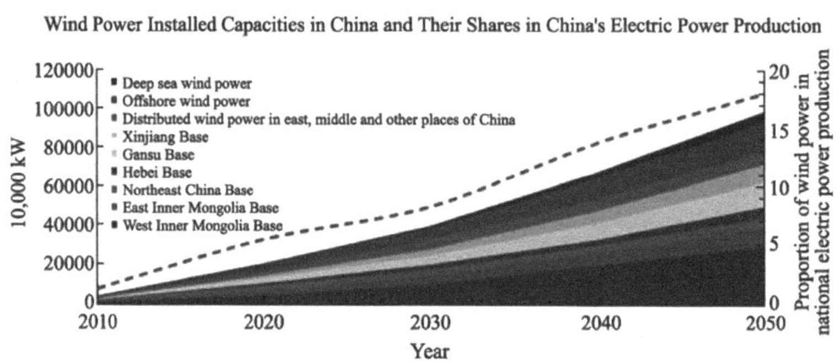

[Fig. 2-10] China's wind power CO_2 emission reduction potential (attached with color graph). Source National Renewable Energy Center

2.3.2 Providing Employment Opportunity

The development of the wind energy industry will have a significant impact on the macroeconomy and employment. The impact of the development of the wind energy industry on the macroeconomy includes several aspects: First, the transfer of related industries, the gradual elimination of backward production capacity, the transfer of employees in the fossil energy sector, the formation into the industry and rapid expansion of new energy and new technology and the remarkable increase of the employees thereof. Second, the impact of energy price changes on the macroeconomy. Replacing fossil energy with wind energy will lead to a change in energy production cost, and the environmental external cost not considered in the past will be reflected in the energy cost in the form of tax, etc. This will lead to the change of production cost of the downstream industries and also bring about an impact on industrial sectors and the macroeconomy. Third, the linkage impact on the industries related to the industry chain. For example, the rapid expansion of the upstream industries, such as wind power equipment production, brings about the expansion of related industries such as raw materials, electronics, and machinery, and correspondingly, the industry and the employees of the relevant industrial chain will change.

The effects of wind power on the social economy can be divided into direct effects and indirect effects. The indirect effect includes not only the positive pull effect of renewable energy investment and operation on other industries but also the negative squeezing effect on the traditional energy supply industry and its upstream and downstream industries. The addition and subtraction of the positive pull effect and the negative squeezing effect above are the comprehensive effects on the overall macro economy. For the wind energy industry chain, the main factors that affect the macroeconomy are power generation, wind farm construction, equipment manufacture and transportation, material production, and so on.

The upstream link of the wind energy industry mainly refers to material production which affects the manufacturing of traditional and emerging materials such as steel, special glass, and cement. The manufacturing of wind power equipment parts includes the manufacturing and R&D of blades, gearboxes, towers, generators, engine compartment covers, bearings, control systems, and other parts. The function of each part is different, and its cost proportion in the unit is also different. The cost of key parts, such as blades, gearboxes, and bearings, accounts for the main part of the total cost of the whole machine. In recent years, China's wind energy industry has been leading in the world. China has a series of vendors engaging in wind power equipment manufacturing and transportation and formed a huge wind power equipment industry.

Wind farm construction involves infrastructure, equipment installation, operation, maintenance, as well as transportation, etc. At present, the models of domestic wind power operation and maintenance market can be roughly divided into three categories: First, the developer sets up their special operation and maintenance company. Second, the whole machine manufacturer sets up an operation and maintenance company to mainly provide after-sales services for owners. Third, a third-party company is engaged to provide operation and maintenance services. The installed capacity of wind power in China has already ranked first in the world, but it still lacks experience in the operation and maintenance management of wind farms, and the wind power service industry is still in its infancy.

Compared with the input-output-table industry category, the main industries corresponding to and affected by the wind power industry chain include electric power production, electric power instrumentation manufacturing, wind power equipment production, infrastructure construction, and manufacturing of related materials(manufacturing, service, and transportation), etc. Researches find that by2030, the values added to wind power in both scenarios will exceed that of hydropower and reach 0.62 and 1.5 trillion yuan, respectively, 11 times those in 2015(Tab. 2-2).

In terms of employment, the industry chain of wind power mainly includes part manufacturing, wind turbine manufacturing, wind farm development, and operation and maintenance. Among them, part manufacturing, wind turbine manufacturing, wind farm development, installation, operation and maintenance, power transmission and distribution, consulting and engineering, research and development, financing,

[Tab. 2-2] Wind power industry values added (at price 2010)/(trillion yuan)

	Wind power value added	2010	2015	2020	2025	2030
The scenario of established policy	Wind power	0.00	0.13	0.32	0.51	0.62
The scenario of a high proportion	Wind power	0.00	0.13	0.36	0.72	1.50

[Tab. 2-3] Employment of the wind power industry

	The scenario of established policy (by 2030)	The scenario of high proportion (by 2030)
Jobs directly created by the wind power industry (10 K persons)	97.99	362.85
Jobs indirectly created by the wind power industry (10 K persons)	136.34	328.22

and many other links can create direct job opportunities. These links can also create indirect job opportunities through their links with other industrial sectors and further produce an induced effect on the entire national economy through consumption behavior. Researches show that wind power and related industries are expected to create 6.9 million jobs by 2030(Tab. 2-3).

2.3.3 Coping with Climate Change

As a clean energy source, wind energy can reduce ecological damage and water consumption caused by coal mining and it cannot only reduce greenhouse gas emissions significantly but also reduce local environmental pollution. It is an important way to deal with global climate change and reduce environmental pollution.

Many research institutions at home and abroad have studied the environmental impact of global wind power development. By 2030, it is estimated that the more than 1 billion kW installed capacity of wind power in the world will produce approximately 2.7 trillion kWh of wind power, equivalent to 9% of global electric power production. Wind power will increase to 5.2 trillion kWh by

2050(accounting for 12%, with an installed capacity of more than 2 billion kilowatts). By 2050, global electric power production will be almost entirely dependent on zero-carbon emission energy technologies including renewable energy such as wind energy and solar energy(46.5%), fossil fuel with carbon dioxide trap and sequestration(26%), and nuclear power(23%).

In 2007, the report of the Chinese Academy of Engineering pointed out that air pollution in China belonged to coal smoke-type pollution, with 85% of sulfur dioxide and carbon dioxide emissions and 70% of fume coming from coal combustion. In China, 63.5% of the air environment is moderately or seriously polluted, 61.8% of the cities in South China have acid rain, and the area of acid rain accounts for one-third of the national territory. The economic losses from air pollution account for 3~7% of GDP. If air pollution is not effectively controlled, the economic cost of diseases from coal combustion pollution only will reach USD 390 billion by 2020. According to the State Environmental Protection Administration, the environmental capacity of China is limited to 16.2 MT sulfur dioxides and 18.8 MT nitrogen oxides. If no effective measures are taken, by 2020, the emission of both will reach 40 and 35 MT, respectively, and greatly exceed the environmental capacity; thereby China will be under increasing international pressure in reducing carbon dioxide emissions. Although China's per capita carbon dioxide emission is low and close to the world average and the accumulative value of emission is not high, its growth rate is relatively fast. China's total carbon dioxide emission is forecast to double between 2004 and 2030 due to strong economic growth and heavy dependence on coal in the power generation sector and industry. China, as the first emitter of carbon dioxide, does not undertake the obligation of reducing greenhouse gas emissions soon, but as a largely responsible power, it has the responsibility and obligation to take measures to reduce greenhouse gas emissions.

This shows that wind energy development has become an important measure in international greenhouse gas emission reduction and environmental protection. *China's National Plan to Deal with Climate Change* issued by China has taken wind power as an important means of reducing greenhouse gas emissions.

Overview of the Worldwide Wind Power Industry

3

Since the beginning of the twenty-first century, the development and utilization of renewable energy have become an important choice for global energy transformation. Wind energy is an important renewable energy source that can be continuously developed and utilized. This chapter introduces the development status of wind power in the world and China from five aspects: wind energy resources, wind energy market, wind energy industry, wind energy technology, and wind energy policy and analyzes the development trend of them.

3.1 Development of Global Wind Power

3.1.1 Global Wind Energy Resources

(1) Total available resource

In 1981, the World Meteorological Organization (WMO) assessed global wind energy resources. According to the average wind energy density and corresponding annual average wind speed, the global wind energy resources were divided into ten grades. Globally, the wind energy density at 10 m height is greater than 150~200 W/m^2, accounting for about 2/3. It is estimated that the total wind energy in the world is 2740 TW, of which about 20 TW can be utilized.

(2) The distribution of global wind energy resources

The World Energy Council estimates that 27% of the earth's 1.07 * 108 km^2 land area is 10 m above the ground, and its annual average wind speed is more than 5 m/s. The high-value areas of wind energy above force 8 are mainly distributed in the middle and high latitudes of the Southern Hemisphere and the North Atlantic, the North Pacific, and the Arctic Ocean of the Northern Hemisphere, while the upper continental wind energy is generally no more than force 7, and among them, the wind speed around western USA, the coastal areas of Northwest Europe, the top of Ural Mountains, and the Black Sea Region is relatively stronger, and the global distribution of land wind energy resources is shown in [Tab. 3-1].

[Tab. 3-1] Global distribution of wind energy resources

Region	Total land area (10,000 km^2)	Land area with wind force 3~7(10,000 km^2)	The proportion of the land area with wind force 3~7 (10,000 km^2) (%)
North America	1933.9	787.6	41
Latin America and the Caribbean	1848.2	331	18
Western Europe	474.2	196.8	42
Eastern Europe and CIS	2304.9	678.3	29
Middle East and North Africa	814.2	256.6	32
Sub-Saharan Africa	725.5	220.9	30
Pacific	2135.4	418.8	20
China	959.7	105.6	11
Central and South Asia	429.9	24.3	6
Total	1066.6	2914.3	27

3.1.2 Present Status of the Global Wind Power Industry

(1) The steady growth of the wind energy market scale

In recent years, the global wind energy market as a whole has maintained steady growth, with a newly installed wind power capacity of about 40 GW. Newly installed capacity in the USA declined in 2013 and resumed growth in 2014. The Global installed capacity of wind power exceeded 60 GW with a year-on-year growth of 22.6%. The cumulative installed capacity of wind power in the world has exceeded 430 GW, with a growth rate of 17%. According to statistics released by the Global Wind Energy Council (GWEC), 52,573 MW of wind power was installed globally in 2017 (Fig. 3-1), with a cumulative installed capacity of 539.581 MW (Fig. 3-2). Among them, China's new installed capacity is 19.66 GW, and the cumulative installed capacity is 190 GW, accounting for 37.4 and 34.8% of the world's total, respectively. With the advantage of technology and policy orientation, Europe and

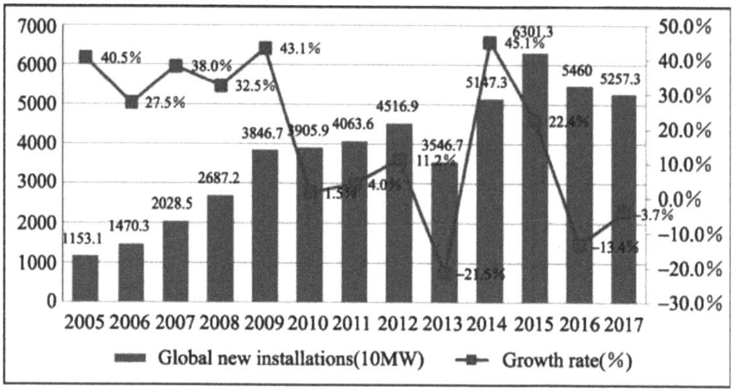

[Fig. 3-1] Global newly installed capacity and growth rate of wind power over the years. Source

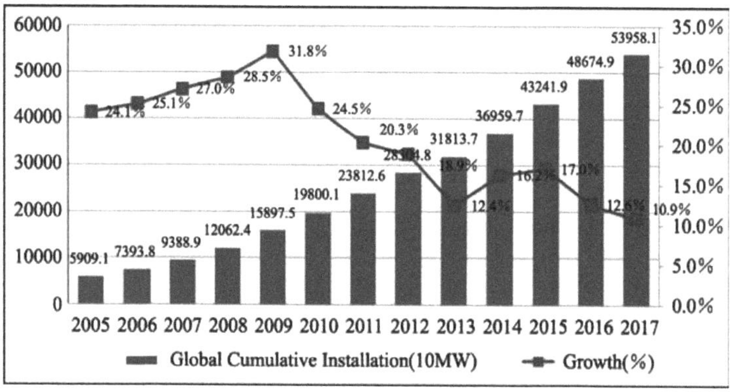

[Fig. 3-2] Global cumulative installed capacity and growth rate of wind power over the years. Source GWEC

America are the earliest areas for global wind power development. In particular, Europe has been strongly developing wind power since the 1990s. At the same time, with the development of wind power in Asian emerging market countries and the increasing emphasis on clean energy, in recent years, the global wind power market has gradually shifted from Europe and the USA to China and India, and other Asian regions. Especially, China has accelerated growth in recent years. In 2010, China's total wind power installed capacity reached 44.7 GW, becoming the world's first. From 2010 to 2015,China's total wind power installed capacity nearly doubled by two times.

In 2017, the top 10 countries leading the way in newly installed wind power capacity were China, the USA, Germany, UK, India, Brazil, France, Turkey, South Africa, and Finland, accounting for 90% of the global market share (Tab. 3-2). By 2017, the top 10 countries in cumulative installed wind power capacity were China, USA, Germany, India, Spain, UK, France, Brazil, Canada, and Italy, taking up 85% of the global market share (Table 3-3).

[Tab. 3-2] Top 10 countries by newly installed wind power capacity in 2017

S/N	Country	Share of newly installed wind power capacity (%)
1	China	37
2	USA	13
3	Germany	12
4	UK	8
5	India	8
6	Brazil	4
7	France	3
8	Turkey	1
9	South Africa	1
10	Finland	1
Others		10

[Tab. 3-3] Mix of the top 10 countries by the global cumulative installed wind power capacity in 2017, *source*:GWEC

S/N	Country	Share of newly installed wind power capacity (%)
1	China	35
2	USA	17
3	Germany	10
4	India	6
5	Spain	4
6	UK	4
7	France	3
8	Brazil	2
9	Canada	2
10	Italy	2
Others		15

In recent years, offshore wind power is an important trend in global wind power development. According to the statistics of the Global Wind Energy Council(GWEC), global offshore wind power projects are mainly concentrated in Britain, Germany, China, Denmark, the Netherlands, and Belgium. The total installed capacity of offshore wind power in 2017 was 4334 MW, with a cumulative installed capacity of 18,813 MW. Among them, the newly installed capacity of offshore wind power in China has increased by 1160 MW, with a cumulative capacity of 2790 MW, up 97% from 2016. The cumulative capacity ranks third in the world, followed by Britain and Germany (Fig. 3-3 and Fig. 3-4).

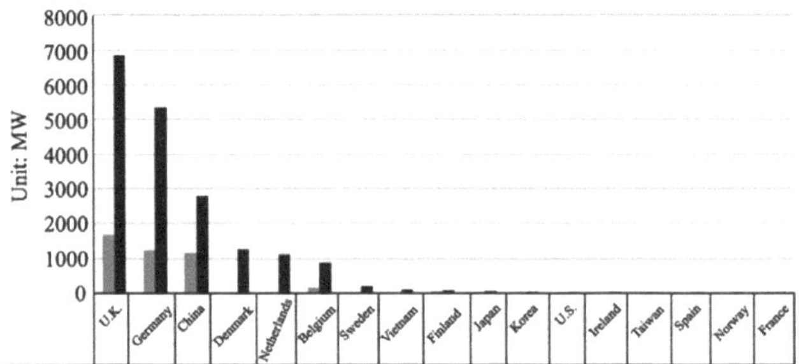

	U.K.	Germany	China	Denmark	Netherlands	Belgium	Sweden	Vietnam	Finland	Japan	Korea	U.S.	Ireland	Taiwan	Spain	Norway	France
▪ New Installations in 2017	1680	1247	1164	0	0	165	0	0	60	5	3	0	0	8	0	0	2
▪ Cumulative Installations by 2017	6836	5355	2788	1271	1118	877	202	99	92	65	38	30	25	8	5	2	2

[Fig. 3-3] Global newly/cumulative installed capacity of offshore wind power in 2017. Source GWEC

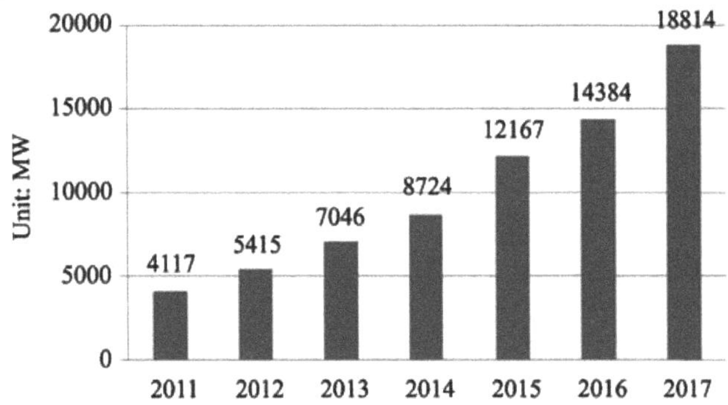

[Fig. 3-4] Global cumulative installed capacity of offshore wind power over the years. Source GWEC

(2) Globalization in wind energy industry

After continuous exploration and development, the global wind energy industry is gradually becoming mature. Either manufacturers, or developers, or operators, there is a clear trend of internationalization and large scale integration.

1) Manufacturer

According to FTI Consulting, the installed capacity of wind power in the world reached 52.2 GW in 2017, among which the top 10 wind power equipment manufacturers accounted for 80.2% of the global market share. In 2017, Vestas ranked first in the world of wind power equipment manufacturers with a market share of 16.7%. In 2017, eight of the top 15 wind turbine manufacturers in the world were Chinese (Tab. 3-4).

In recent years, global large-scale wind power equipment manufacturers have begun to expand to the world. According to FTI Consulting, in 2017, for the world's leading wind turbine manufacturers, the market is located in

[Tab. 3-4] Market share of global wind turbine manufacturers in 2017. source FTI

S/N	Manufacturer	Country	Market share in newly installed wind power capacity (%)
1	Vestas	Denmark	16.7
2	Siemens Gamesa	Germany Spain	16.6
3	Goldwind	China	10.5
4	GE Renewable	USA	7.6
5	ENERCON	Germany	6.6
6	Envision Energy	China	6.0
7	NORDEX Acciona	Germany	5.2
8	Mingyang Wind Power	China	4.7
9	Senvion	Germany	3.7
10	Suzlon	India	2.6
11	United Power	China	2.6
12	CSIC Haizhuang Windpower	China	2.3
13	Shanghai Electric	Germany	2.1
14	XEMC Windpower	China	1.8
15	Dongfang Electric	China	1.6
Others			9.2

[Tab. 3-5] Global market share of the top 5 leading wind turbine manufacturers in 2017. source FTI

	Major market	Total installed capacity in 2017 (10 MW)	Installed capacity in a major market (10 MW)	Share in a major market (%)	Share in other markets (%)
Vestas	USA	8709	2453	28.2	71.8
Siemens Gamesa	UK	8657	1950	22.5	77.5
Goldwind	China	5476	5156	94.2	5.8
GE Renewable	USA	3963	1447	36.5	63.5
Enercon	Germany	3442	2029	58.9	41.1

Asia-Pacific, the Americas, Europe, and other regions. Among the world's top five wind turbine manufacturers, Vestas, GE, Siemens Gamesa, and Enercon account for more than 40% in non-major market countries, while Goldwind occupied the main market in China, only 5.8% in other countries (Tab. 3-5). So far, Vestas has installed wind turbines in nearly 80 countries and regions on six continents worldwide and has become the world's leading "international" wind power equipment manufacturer.

2) Developers

According to FTI, the global top 15 wind power developers accounted for 28.5% of the global market on cumulative installed wind power capacity by the end of 2017. Of the top 15 wind power developers in the world, 9 are Chinese enterprises (Tab. 3-6).

3) Operators

The latest research of GlobalData shows that the global market of wind turbine operation and maintenance is estimated to increase from 13.7 billion US dollars in 2016 to 27.4 US dollars in 2025. The compound annual rate of growth from 2016 to 2025 will be 8%. In 2016, onshore wind power represented the main focus of the market. The contribution of offshore wind power to the market of wind turbine operation and maintenance came slightly over 8% and is expected to hit 18.4% in 2025.

[Tab. 3-6] Global market share of leading wind power developers in 2017. Source FTI

S/N	Developer	Country	Cumulative installed capacity (10 MW) in 2017	Market share (%) in 2017
1	Longyuan	China	1839.5	3.4
2	Iberdrola Renewables	Spain	1598.3	3.0
3	SPIC	China	1367.8	2.5
4	NextEra Energy Resources	USA	1311.1	2.4
5	CGN Wind Power	China	1179	2.2
6	Huaneng Renewables	China	1133.3	2.1
7	EDP Renovaveis	Portugal	1086.2	2.0
8	China Shenhua	China	903.5	1.7
9	CDT Renewable Power	China	882.6	1.6
10	Enel Green Power	Italy	743.1	1.4
11	Acciona Energy	Spain	738.2	1.4
12	Huadian Fuxin Energy	China	716.3	1.3
13	CR Power	China	668.1	1.2
14	EDF Energies Nouvelles	France	648.8	1.2
15	Tianrun (Goldwind)	China	554.7	1.0

At present, operation and maintenance services are provided by wind turbine manufacturers, wind power developers, and independent third-party O&M compa-nies. Vestas, for example, has prioritized wind power operation and maintenance in its future roadmap. As one of the largest service providers, Vestas has offered operation and maintenance services to about 60 GW of wind turbines across the globe.

(3) Technological innovation and progress

With the development of the wind power market, wind power technology, especially offshore wind power, is constantly seeking new technical breakthroughs, and the power and reliability of wind turbines have been improved.

According to statistics from FTI Consulting, 21,691 new wind turbines were installed in the world in 2017. In China, the world's largest wind

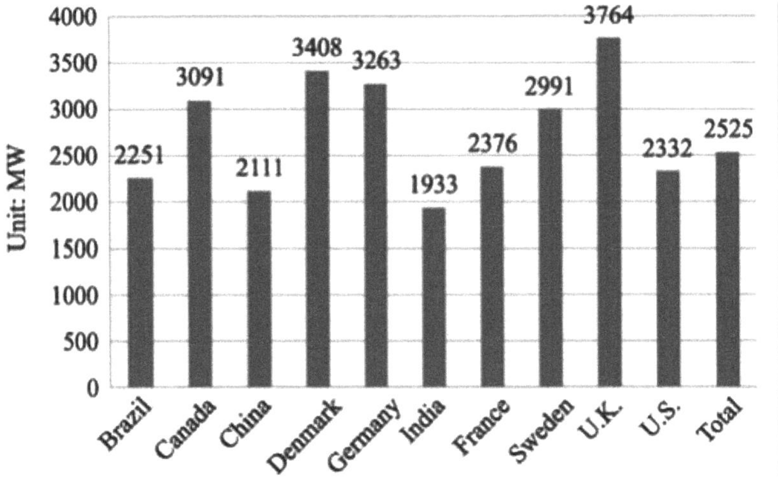

[Fig. 3-5] Average power of new installations in major countries in 2017. Source FTI Consulting, Inc.

power market, the average power of new wind turbines in 2017 reached 2.1 MW. Among them, the average power of the new wind turbines in the UK is the highest, about 3.8 MW (Fig. 3-5).

The analysis of new installations in 2017 reveals that wind turbines below 2 MW took up 4.59% of the market; 2~2.5-MW wind turbines accounted for 60.32%, and wind turbines over 2.5 MW represented a share of 35.09%.

In Europe and America, 3.6~6-MW wind turbines dominate offshore wind farms. Vestas and Siemens have developed 8~10-MW ultra-large offshore wind turbines whose prototypes have passed the test and started commercial operation. As more and more wind farms are built in or relocated from coastal waters to the deep sea, fixed foundations for wind turbines are replaced by floating platforms. Besides, dou-bly fed induction generators, a direct-drive/semi-direct-drive permanent magnet full-power converter, and direct-drive excitation full-power converters are also applied to mainstream 6-MW-plus offshore wind turbines.

(4) Gradual Decline in Wind LCOE

1) Onshore Wind LCOE

These years have witnessed a gradual decline in the global wind Levelized cost of electricity (LCOE) to 6 cents/kWh in 2017. It is estimated that the figure will drop to 5 cents/kWh in 2020, lower than that of fossil energy in some regions.

Wind turbines (including towers and installations), a significant part of wind farm investment, can take up over 60% of the total installation cost of onshore projects. The prices of wind turbines are susceptible to cyclical economic changes and steel and copper price fluctuation. The cost of onshore wind turbines in America was 750 US dollars per kW from 2000 to 2002 and 1800 US dollars per kW in 2008. The hike was explained by the rising cost of such raw materials as steel and cement. As shown in [Fig. 3-6], a sharp decline followed the peak in 2008/2009. It is estimated that the price of wind turbines will continue to fall with further industry integration in the immediate future.

Wind LCOE varies greatly among different projects for a combination of factors: resource conditions, installation cost, and O&M cost. On the whole, wind LCOE came down every year as shown in [Fig. 3-7].

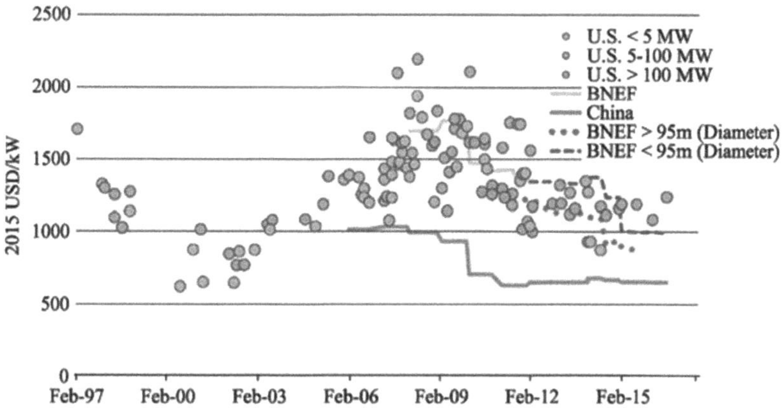

[Fig. 3-6] Price changing trend of wind turbines (1997~2006). Source IRENA

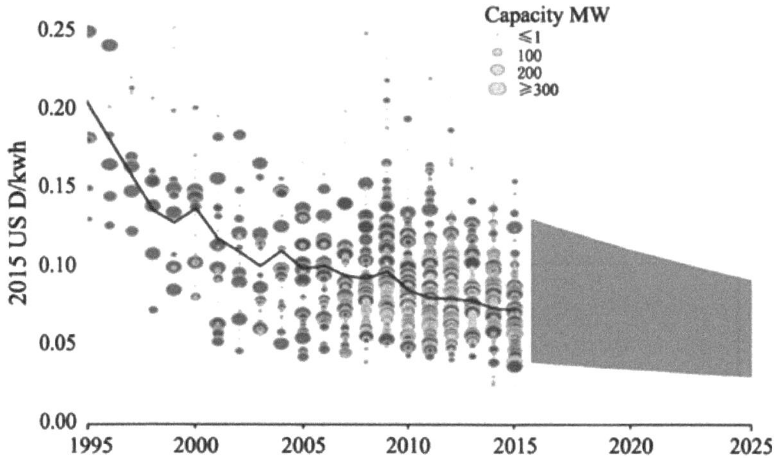

[Fig. 3-7] Changing trend of global onshore wind LCOE. Source IRENA

2) Offshore wind LCOE

Overall, offshore wind power investment cost is about two times that of land wind power. The cost of offshore wind power investment in 2008 was \$3100~4700/kW; in 2010~2013, the cost rose to \$3600~5600/kW; and in 2014, the average cost of investment was \$4950/kW, which is related to offshore distance and water depth of offshore wind farms. [Fig. 3-8] presents the cost components of typical offshore wind power projects.

Bloomberg New Energy Finance (BNEF) issued a “leveling the cost of electricity” report in 2016. It shows that the economy of offshore wind power is rapidly improving; especially in some offshore projects, leveling the cost of electricity has been or is approaching other mature power generation technology. In the second half of 2016, the global offshore wind power technology cost (weighted average) is estimated at \$126 per MW, which dropped about 22% from that in the first half of the year and about 28% from that in the second half of 2015. Among them, the adoption of more powerful offshore wind turbines, offshore wind power project construction technology upgrading, European competitive project bidding, and other factors are the reasons driving the global offshore wind power cost reduction.

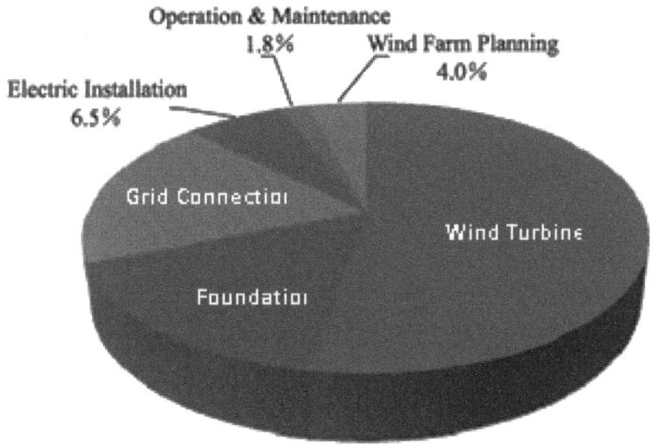

[Fig. 3-8] Cost components of European typical offshore wind power projects. Source CWEA

3.1.3 Perspective

(1) Steady growth of the wind energy market

1) Targets of major countries in wind energy development

According to REN21, by 2015, 64 countries worldwide have invested $286 billion in renewable energy generation, 173 countries have set specific targets for promoting renewable energy and 146 countries have supportive development policies. In addition, 52 countries have adopted financial measures to promote renewable energy projects by providing loans, free aid, and preferential treatment. Vigorously developing renewable energy such as wind power has become a global consensus. The target of wind power development in major countries is shown in [Tab. 3-7].

2) Scale forecast for the wind power market

FTI Consulting Company made a forecast of the global wind power market from 2016 to 2025. The results show that after the newly installed capacity of global wind power reached a new peak in 2015, future growth is expected to fluctuate for some time, showing roller-coaster-like growth. It is estimated that the newly installed capacity of wind power will peak again in 2023. The main reason for the slowdown is the slow pace of

[Tab. 3-7] Planning targets of major countries in wind power development

Country/region	2020 development targets	2050 development targets
USA	Over 10% of electricity is generated from wind energy; the offshore wind power market and supply chain	Over 35% of electricity is generated from wind energy in the American power structure; 86 GW of offshore wind capacity
Germany	35.8 GW of onshore wind and 10 GW of offshore wind capacity	Over 80% of renewables used to meet all energy needs, and the share of renewable energy sources in final energy consumption increased to 60%
Spain	35 GW of onshore and 3 GW of offshore wind capacity, with wind power producing 50% of electricity from renewable energy sources	
Denmark	50% of its electricity is sourced from wind	100% renewable electricity
UK	31 GW of wind power capacity, including 18 GW from offshore and 13 GW from onshore	
Brazil	16 GW of wind capacity; 9% of electricity generated from wind energy	
India	60 GW of wind capacity by 2022	
France	19 GW of onshore and 6 GW of offshore wind capacity	

Source China National Renewable Energy Centre

development of China's wind power. The average annual growth rate is expected to be around 1.4% in 2016~2025 years, with an installed capacity of about 622 GW. The capacity distributions were: Southeast Asia (49.4%), Europe (23.3%), North America (13.0%), Latin America (6.6%), Africa (3.4%), OECD (1.8%), and other countries (2.5%).

The Global Wind Energy Council (GWEC) 2017 annual report predicts that the global wind power market will resume rapid growth in 2019 and 2020, and the cumulative installed capacity of wind power is expected to reach 840 GW by 2022, as shown in [Fig. 3-9].

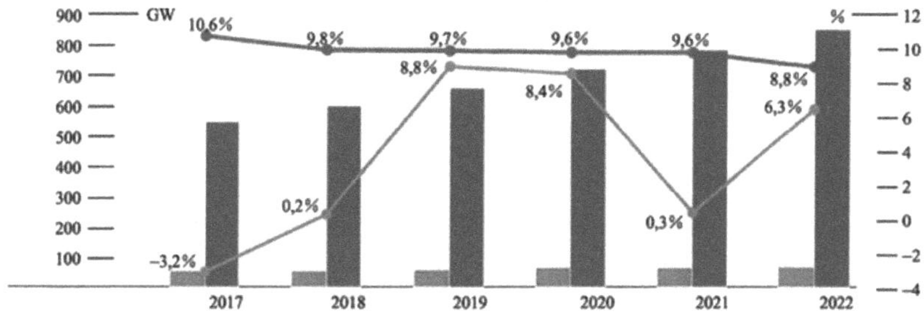

	2017	2018	2019	2020	2021	2022
Cumulative installed capacity	539.1GW	592GW	649.5GW	711.8GW	774.4GW	840.9GW
Cumulative capacity growthrate	10.6%	9.8%	9.7%	9.6%	9.6%	8.8%
Annual installed capacity	52.5GW	52.9GW	57.5GW	62.4GW	62.6GW	66.5GW
Annual installed capacity growth rate	-3.2%	0.2%	8.8%	8.4%	0.3%	6.3%

[Fig. 3-9] 2018~2022 global wind market forecast. Source GWEC

(2) Stronger industry integration

With the increasing number of market competitors, large enterprises are pouring into the wind power industry, and mergers and acquisitions are becoming more and more common. The acquisition of wind turbine manufacturers in the global wind power market can be divided into horizontal acquisition and vertical acquisition. The horizontal acquisition is a wind turbine manufacturer by another wind turbine manufacturer. In June 2016, Siemens from Germany merged with Gamesa from Spain, and Siemens made three big gains: firstly, breakthroughs in emerging markets such as Latin America, India, and China; secondly, breakthroughs in the segmented blade and low-speed wind turbines; and finally, obtaining the ability of O&M services from Gamesa.

A vertical merger is the acquisition of wind turbine manufacturers to the upper and lower reaches of the industrial chain. In October 2016, the US GE acquired LM, the world's largest wind blade manufacturer. The LM acquisition will be of great benefit to extending the US GE wind power industry chain, opening up the markets in Asia, Europe, and America as well as emerging markets, reducing manufacturing costs, and enhancing global wind power competitiveness.

Looking at the foreign market, onshore wind power has been developed for 20 or 30 years, and the market development space is limited. In the next five to ten years, offshore wind power will be the main competition. The combination of the wind power giants generates a great shock to the industry. In 2016, the international wind power giant merged and integrated one after another; such M&A transactions will make the international market competition more intense.

(3) Wind technology innovation

In recent years, with the continuous increase in single-machine capacity, wind power generation technology has become more and more mature, and wind power technology has undergone major changes. In addition to doubly fed asynchronous power generation technology, variable-speed propeller power regulation technology has also developed direct-drive permanent magnet full-power converter technology, direct-drive excitation full-power converter technology, and semi-direct-drive permanent magnet full-power converter technology. From stall power regulation with a fixed propeller to variable-speed power regulation with a variable propeller, the efficiency and reliability of wind turbines are increasing, and the cost of operation and maintenance is also significantly reduced.

Offshore wind power generation remains at the forefront of technological innovation, and offshore wind farms are popping up all over the world. In the future, offshore wind farms will move into deeper waters, over 50 km away from the shore. Floating foundations, integral installation, and self-propelled self-elevating construction platforms will dominate. Wind turbines with 8 MW capacity have been in service, and 10-MW models are in the pipeline. Disruptive superconducting technology is a key path to developing 10-MW or larger wind turbines. A zero-resistance superconductor offers a solution to heat dissipation. Its application will greatly push up both power density and torque density by at least 50% as compared to traditional permanent magnet generators, thus bringing down LCOE by 30%.

In recent years, a new idea of wind power generation appeared: harnessing wind at high altitudes from approximately 480 m to 12,000 m. There are two approaches available: First, a helium-filled shell is used to lift turbines off the ground to higher altitudes, sending electricity back to the ground via its tethering cables; secondly, a high-flying kite is connected to the ground station which collects wind energy through tension reels. These new ideas are still under feasibility study.

(4) Improved wind power policies

Wind power in its infancy is not as competitive as fossil-fired electricity, and fixed feed-in tariff is a major measure to boost renewable energy across the world. Price subsidy has catalyzed rapid wind installation. By the end of 2016, more than 110 countries and regions including China have released national or state/provincial policies on fixed feed-in tariffs.

It is, however, difficult to sustain such subsidy programs given the ever-growing wind capacity. Against economic downturn and mounting fiscal pressure, a growing number of countries have turned to competitive bidding for wind power. Germany, for example, has changed from a fixed feed-in tariff to premium tariff and trial auctions considering the declining cost of renewable energy and more burden on end users from the growing subsidy. It announced to adoption of wind power bidding across the country in 2017. EU also required competitive bidding for subsidies on mature renewable energy projects starting in 2017. Competitive bidding allows market-oriented resource allocation and enables renewable energy to better compete with traditional energy. In Brazil, South Africa, and Egypt, the bidding price for wind power is lower than the feed-in tariff of traditional fossil-fired power. Wind power price locked in through long-term contract in the USA is almost the same as that of fossil energy. Wind power is becoming an economic choice. The competitive bidding for wind power is grabbing eyeballs globally, and it is a key solution for new power subsidy policies.

3.2 Progress in China's Wind Power

3.2.1 China's Wind Energy Resource

(1) Total resources available

According to the *Wind Energy Assessment of China* released by the Wind and Solar Energy Resources Centre at China Meteorological Administration in 2014, the total onshore exploitable wind power at a height of 70 m is 7200, 5000, and 2600 GW for a power density of over 150 W/m^2, at 200 W/m^2, and at least 300 W/m^2 and that at a height of 80 m is 10,200 and 7500 GW for a power density of 150 and 200 W/m^2, respectively. The onshore exploitable wind energy at a height of fewer than 100 m is about 190 and 320 GW for a water depth between 5~25 and 25~50 m, respectively. In 2015, China Meteorological Administration evaluated the low-wind-speed energy potential of 18 provinces and regions in Central and South China, and the results show an increase of 600 GW in the exploitable wind energy.

(2) Wind energy distribution

Inner Mongolia, Xinjiang, and Gansu are the provinces with the most abundant wind energy resources in China, with a technological development capacity of 1460, 440, and 240 GW, respectively. Wind energy resources exceed 10 GW in 14 provinces (regions), including inland Shanxi, Yunnan, Qinghai, and Ningxia (Tab. 3-8).

At present, China's offshore wind energy resources are mainly distributed along the southeastern coast of China and its adjacent islands, with an effective wind energy density of more than 300 W/m^2; the coastline extends outward, the wind speed increases gradually, and the wind power density level increases gradually as well, of which the Taiwan Strait has the most abundant wind resources. The average annual wind speed in the coastal areas of the Shandong Peninsula is above 7 m/s, and that in the coastal areas of Jiangsu Province is 7~8 m/s. The wind speed in the areas farther away from the coastline is greater. The average wind speed in the coastal areas of Fujian and Zhejiang is above 9 m/s, which is rich in wind energy resources (Tab. 3-9).

[Tab. 3-8] Onshore wind energy at 70 m in Chinese provinces, regions, and cities (≥300 W/m^2), Source Chinese Renewable Energy Development Report

Province	Exploitable wind energy (10 MW)	Province	Exploitable wind energy (10 MW)
Inner Mongolia	145,967	Guizhou	456
Xinjiang	43,555	Henan	389
Gansu	23,634	Jiangsu	370
Heilongjiang	9651	Sichuan	340
Jilin	6284	Jiangxi	310
Liaoning	5981	Zhejiang	209
Hebei	4188	Hainan	206
Shandong	3018	Chongqing	138
Yunnan	2066	Hubei	126
Qinghai	2008	Taiwan	119
Shanxi	1598	Hunan	113
Ningxia	1555	Anhui	77

[Tab. 3-9] China's offshore wind energy potential in different provinces

Province (region, city)	Total (10 MW)
Zhejiang	3899
Guangdong	3863
Shandong	3603
Tianjin	213
Liaoning	3248
Fujian	2501
Jiangsu	2243
Guangxi	112
Hebei	841
Hainan	800
Shanghai	406

3.2.2 Present Status of China's Wind Power Industry

(1) Continuous expansion

Since 2010, with China's continuous support and investment in the wind power industry, wind power installed capacity continues to increase. At present, China has grown into the largest and fastest-growing wind power market in the world. According to a survey by the Chinese Wind Energy Association (CWEA), in 2017, the newly installed capacity of the whole country (except Taiwan) was 19.66 GW with a year-on-year decrease of 15.9%; the cumulative installed capacity reached 188 GW (see Fig. 3-10).

As China continues to optimize its wind power market, new installations are declining in the "Three North Regions" yet increasing in Mid-east and South China. In 2017, Mid-south China increased its new installations by 44% from last year, accounting for 23% of the country's total new capacity. Major growth contributors include Hunan, Henan, Guangxi, and Guangdong. By contrast, a year-on-year decrease in installed capacity was seen in China's northwest (>40%), southwest (>40%), northeast (32%), north (9%), and east (5%) (Fig.3-11).

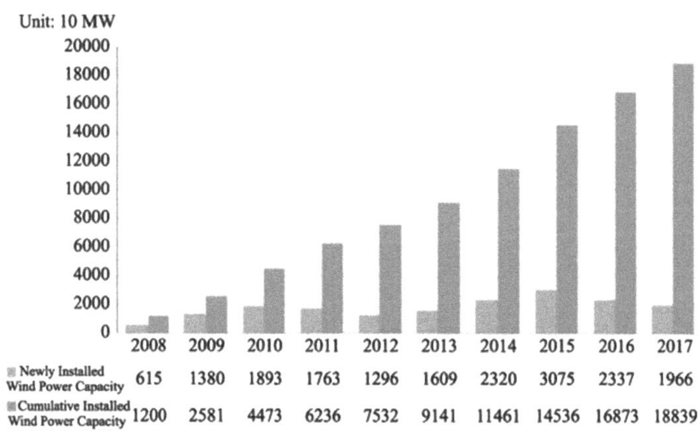

[Fig. 3-10] Newly/cumulative installed wind power capacity of China over the years. *Source* CWEA

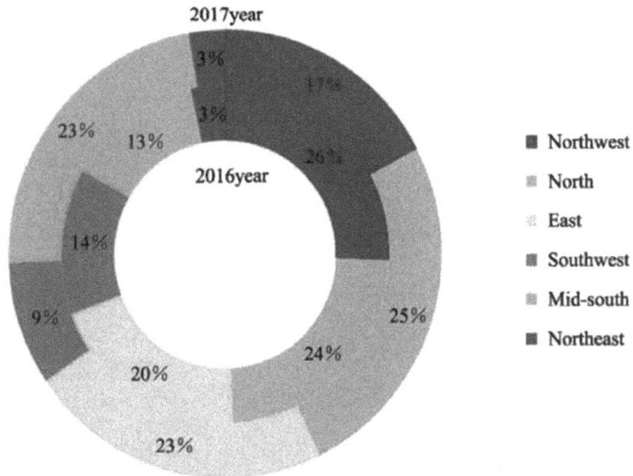

Source: CWEA

[Fig. 3-11] Shares of newly installed wind power capacity by regions of China in 2016 and 2017. *Source* CWEA

In 2017, China's top five regions in newly installed capacity include Hebei, Shan-dong, Jiangsu, Inner Mongolia, and Qinghai, with a combined share of 40.4%. With respect to cumulative installed capacity, Inner Mongolia, Xinjiang, Hebei, Gansu, and Shandong rank top five, contributing 47.6% to China's total. In 2017, China achieved breakthroughs in offshore wind power generation, adding 1.16 GW of wind capacity from 319 wind turbines with a year-on- year growth of 97%. The cumulative installed capacity came to 2.79 GW(Fig. 3-12).

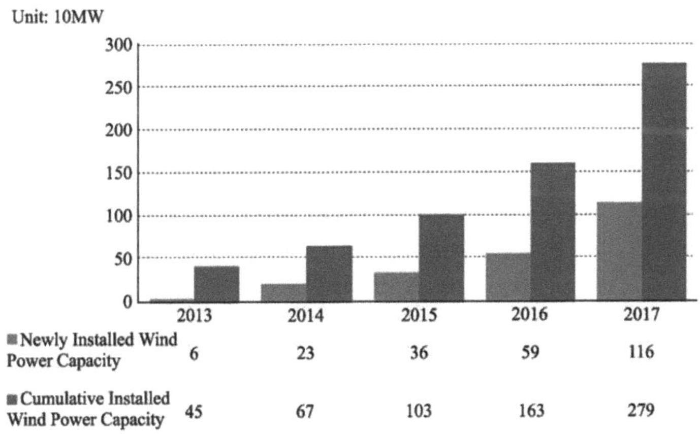

[Fig. 3-12] Newly and cumulative installed offshore wind power capacity in 2017. *Source* CWEA

(2) A well-established industry system

1) Manufacturers

CWEA data shows that 22 Chinese manufacturers were expanding their capacity by 19.66 GW in 2017, and Goldwind occupied the biggest share of 26.6% with a newly installed capacity of 5.23 GW, followed by Envision Energy, Mingyang Smart Energy, United Power, and CSIC Haizhuang Windpower (Fig. 3-13) ranking from top 2 to top 5. By the end of 2017, there were seven manufacturers whose cumulative installed capacity broke 10 GW, and their combined market share reached 67%. Goldwind took the first spot with a domestic share of 22.7% and a cumulative installed capacity of over 40 GW (Fig. 3-14).

In recent five years, the market share of wind turbine manufacturers has appeared as a defining concentrating trend. In China, the combined market share of the top five wind turbine manufacturers in new installations increased by 13 percentage points from 54.1% in 2013 to 67.1% in 2017 and that of the top ten suppliers grew by 12 percentage points from 77.8% in 2013 to 89.5% in 2017 (Fig. 3-15). In the 2017 list of the world's top 15 largest wind turbine suppliers, eight are from China.

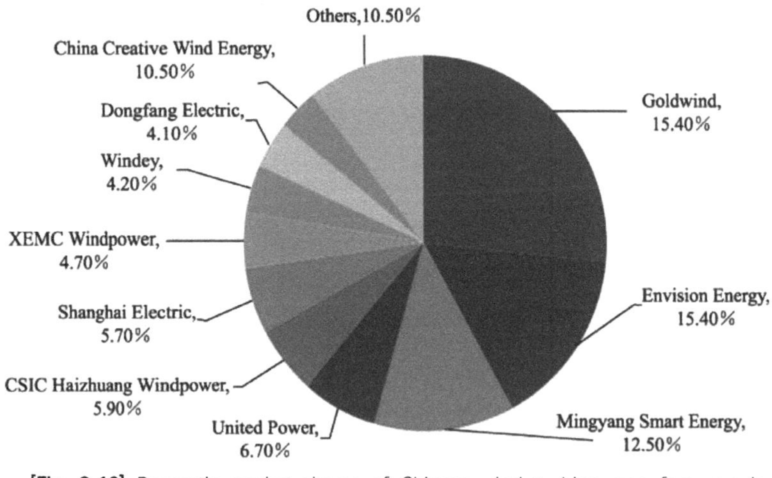

[Fig. 3-13] Domestic market shares of Chinese wind turbine manufacturers in 2017. *Source* CWEA

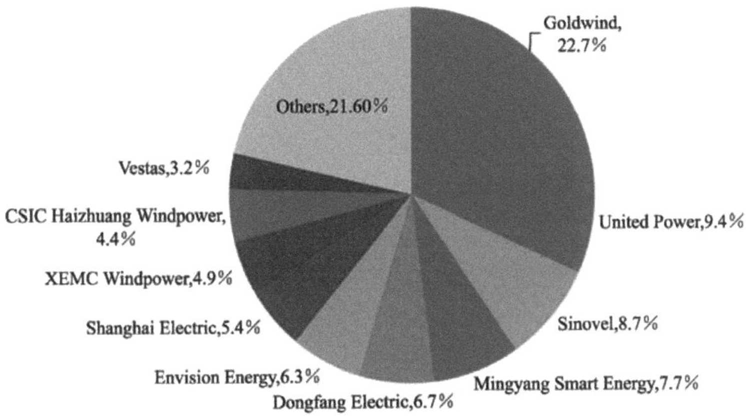

[Fig. 3-14] Cumulative market shares of Chinese wind turbine manufacturers in 2017. *Source* CWEA

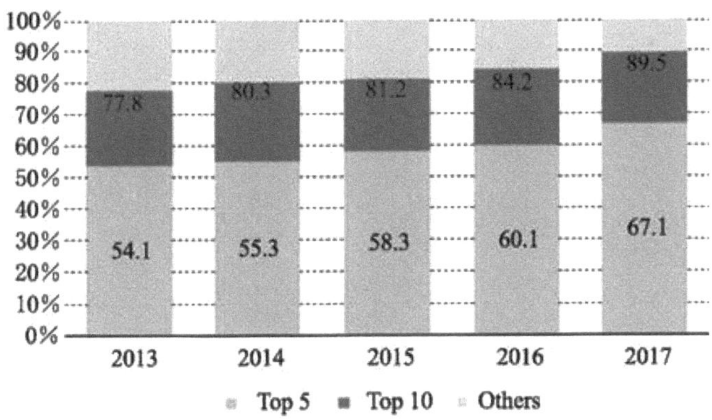

[Fig. 3-15] Market shares of top 5/10 Chinese wind turbine manufacturers from 2013 to 2017. Source CWEA

2) Developers

In 2017, there were over 80 Chinese wind power developers expanding their capacity in China, and top 15 of them added combined installed capacity of nearly 15 GW, accounting for 76% of the total increment (Fig. 3-16). By the end of 2017, the top 10 Chinese wind power developers combined realized a cumulative installed capacity of over 130 GW, accounting for 71% of the total (Fig.3-17).

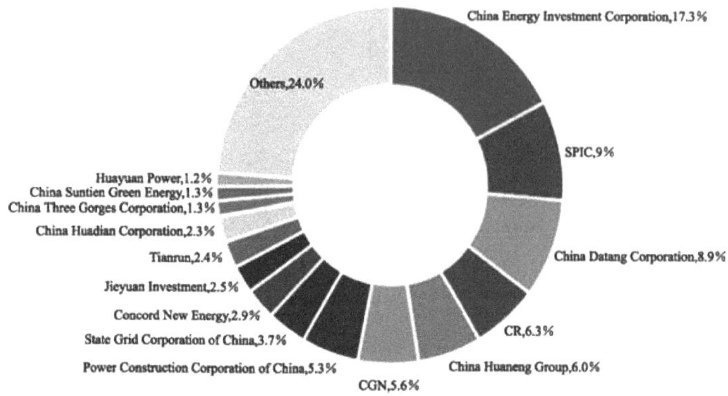

[Fig. 3-16] 2017 market share of Chinese wind power developers in new installations. *Source* CWEA

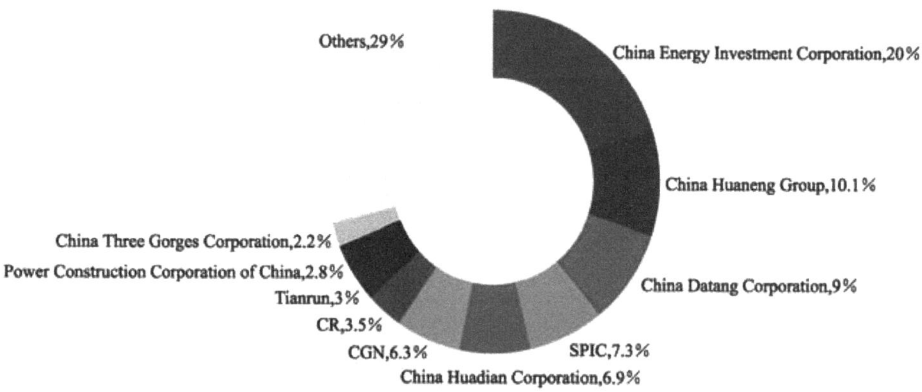

Source: CWEA

[Fig. 3-17] 2017 market shares of Chinese wind power developers in cumulative installations. Source CWEA

3) Operators

In recent years, wind capacity has claimed an ever-growing share of China's installed electricity capacity amid growing interest in wind farm O&M. The rapid growth in wind installations is also reflected in wind farm O&M. For long-term benefits of wind farms, developers have put more attention on keeping the wind turbine quality and high efficiency of wind farms out of warranty.

With a fully fledged wind power industry and cumulative rise in the number of warranty-expired wind turbines, an expanding market for wind farm O&M appeared gradually. Leading wind turbine suppliers, in particular, have gained prominent edges in this respect. The first batch of these wind turbines for aftermarket O&M comes from Sinovel, Goldwind, and Mingyang Smart Energy. Sinovel, for example, has over 8000 units of 1.5-MW wind turbines, representing a turnover of 300 million yuan. With the wide use of emerging information technologies, such as big data and cloud computing, using emerging Internet technology to improve the stability of wind turbine operation and wind power generation benefits of the wind farm has become the new trend in the wind power industry. Smart O&M is dominating the sector. Yet in terms of industry development, third-party companies are the better choices to make the sector specialized and efficient.

So far, there are three kinds of companies engaged in wind power aftermarket O&M services in China: wind turbine manufacturers, wind farm developers, and third-party professional O&M providers.

(3) Advancing wind energy technologies

In recent years, China's wind turbine technology research and development capacity has been continuously improved, from the early introduction of technology, digestion, absorption, and re-innovation, to independent research and development, gradually formed with independent intellectual property rights of multi-MW large wind turbine research and development capacity. The design of series and platform units has effectively shortened the response -ability of R&D design to meet the market demand. In 2017, the average power of newly installed wind turbines in China reached 2.1 MW with a year-on-year growth of 8%, and that of cumulative installed wind turbines increased by 2.6% to 1.7 MW.

At present, the mainstream wind farms in China have developed from 1.5- to 2~2.5-MW wind turbines. The 3~4-MW wind turbines have been produced in batches, and the 5- and 6-MW wind turbines have been connected to the grid. China is catching up with the international advanced level in the research, production, and application of large-scale wind turbines. Since 2010, the market share of 2-MW wind turbines has been rising in the new wind power installation market, exceeding that of 1.5-MW units for the first time in 2015. According to the latest statistics of CWEA, in 2017, the market share of newly installed wind turbines under 2 MW (excluding 2 MW) reached 7.3%, 2-MW wind turbine assembly machines accounted for 59% of the new installed capacity, and 2~3-MW (excluding 3 MW) accounted for 85%. The proportion of newly installed units for 3~4-MW (excluding 4 MW) units is 2.9%, and that for 4 MW and above units is 4.7% (Tab.3-10). In 2017, 1.5-MW wind turbines still dominated the total installed capacity, accounting for 45.8% of the total installed capacity, down about 5 percentage points year-on-year; 2-MW wind turbine market share rose to 35%, up about 3 percentage points year-on-year; 2~3 MW (excluding 3 MW) cumulative installed accounted for 8.9%. The cumulative installed ratio of 3~4-MW (excluding 4 MW) units is 2.1%, and that of 4 MW and above units is 1% (Tab. 3-11).

In addition, with the optimization and adjustment of the development layout, the wind turbine design technology adapting to the local complex terrain such as high-altitude and low-wind-speed areas obtained rapid development, which has promoted the development of wind farms in inland and mountainous areas, and the installed wind power capacity in high-altitude and low-wind-speed areas has also achieved a breakthrough growth. Cumulative wind power installed capacity presents a growing trend as well. China's high-altitude areas are mainly distributed in Yunnan, Guizhou, Sichuan, Qinghai, and Tibet, and low-wind-speed areas are mainly distributed in Henan, Anhui, Jiangxi, Hunan, Hubei, and other provinces.

With the continuous development of low-speed wind turbine technology, the diameter of the wind turbine is gradually increasing. The diameter of a 2.0-MW wind turbine increases obviously, and the diameter of a wind turbine above 110 m has increased significantly in recent two years. The development

[Tab. 3-10] Mix of the newly installed wind power capacity by power in 2017, Source CWEA

The power distribution of wind turbines	Newly installed wind power capacity (MW)	Share (%)
2.0 MW	1160.2	59.0
1.5 MW	121.5	6.2
2.1~2.9 MW	512.7	26.1
3~ 3.9 MW	57.7	2.9
4 MW and above	92.9	4.7
1.6~1.9 MW	20.9	1.1
<1.5 MW	0	0

[Tab. 3-11] Mix of the cumulative installed wind power capacity by power in 2017, Source CWEA

The power distribution of wind turbines	Cumulative installed wind power capacity (MW)	Share (%)
1.5 MW	8620	45.8
2.0 MW	6594	35.0
2.1~2.9 MW	1673	8.9
<1.5 MW	1112	5.9
3~3.9 MW	395	2.1
1.6~1.9 MW	259	1.4
4 MW and above	184	1.0

and application of low-speed wind power technology with a diameter of more than 120 m and a hub height of more than 100 m make the wind energy resources in high-altitude and low-wind-speed areas more effectively utilized. The annual utilization hours of 2000 h were once possible in the areas with an average wind speed of 6.5 m/s but are now realized with an average wind speed of 5 m/s. The wind power density in wind power development areas has also been reduced from 300 to 150 W/m^2, thus further improving the technical and economic efficiency of wind power development in complex terrain areas such as low wind speed.

(4) Stronger support measures

At present, China has initially established a relatively perfect industry management and policy system to promote the development of the wind power industry, and the safeguard measures have been gradually improved. The management regulations and technical requirements of key links such as wind power project development, construction, grid connection, operation management, and information supervision have been promulgated, the management process of wind power development and construction has been simplified, the technical standard system of wind power has been improved, the type certification of wind power equipment and key components has been carried out, and the information monitoring and the evaluation system of the wind power industry have been established, which formed a relatively standardized, fair, and perfect policy environment for the wind power industry and ensured the sustainable and healthy development of it.

1) Targets

To meet the 15% non-fossil energy consumption target by 2020, the National Energy Administration ("NEA") published Opinions on Establishing a Renewable Energy Target Guiding System in early 2016. Besides an overall goal for wind power development, it sets out that non-hydro renewables produce 9% of consumed electricity by 2020 as per the Renewable Energy Law. The target and industry scale are two equally important indicators to learn about China's energy transformation. As President Xi proposed to raise the share of non-fossil fuel in primary energy to about 20% by 2030, China will integrate the target into its specific plans and force market entities and governments at all levels to actively exploit and utilize new energy, thus taking the renewable industry to the next level.

2) Project management

On the one hand, wind power projects are not approved but managed based on annual development and management plans. A rolling execution plan will be prepared each year according to the national development plan. NEA will no longer issue the approval plan that includes specific projects but delegate the power to local governments. On the other hand, a monitoring and warning mechanism is put in place, which clearly defines the warning level for each region. It offers effective guidance for investment and facilitates better capital flow to curtailment-free Central and East China that offers better policies.

3) Guaranteed buyout

In March 2016, the National Development and Reform Commission ("NDRC") issued the Measures for the Administration of the Guaranteed Buyout of Electricity Generated by Renewable Energy Resources (F.G.N.Y. [2016] No. 625), which covers a clear definition, subject of responsibility, buyout range, and compensation measures.

It divides the feed-in electricity output from renewable energy resources into guaranteed electricity and marketable electricity, and the former will be purchased first. In May 2016, NDRC and NEA jointly published the Notice on the Administration of the Guaranteed Buyout of Electricity Generated by Wind Energy and Solar Energy. It sets out the minimum wind power and solar power availability subject to a guaranteed buyout in areas facing curtailment. Specific responsibilities and execution requirements are also clearly defined. Areas that fail to meet the requirements on the guaranteed buyout of availability are not allowed to start new projects. In addition, power grid enterprises and trading agencies should properly compensate for any curtailment.

4) Electricity price management

By the end of December 2016, NDRC published a Notice on Adjusting On-Grid Benchmark Prices for Solar Power and Onshore Wind Power. The benchmark feed-in tariff for offshore wind power generated by new farms located in Class I to IV resource areas is set to 0.4, 0.45, 0.49, and 0.57 yuan/kWh from January 1, 2018, down 0.07, 0.05, 0.05, and 0.03 yuan/kWh from that of 2016~2017, respectively. The benchmark feed-in tariff for offshore wind power remains unchanged. A new price mechanism can force wind power grid parity.

5) Plans

In November 2016, NEA released the *13th Five-Year Plan for the Development of Wind Power*. As a guide on the industry development over the next five years and beyond, it offers detailed plans for maintaining market scale, optimizing industry structure, eliminating curtailment, and improving market mechanisms. Firstly, China will raise the total wind generation capacity to 210 GW by 2020. Secondly, China will accelerate scale development in Mid-east and South China for a better industry structure. Thirdly, it proposes effective cures to power abandonment, including prioritized operation of wind farms included in the plan, and compulsory execution of minimum guaranteed buyout. Lastly, it provides such measures to improve the market environment as streamlining administration and delegating power to lower levels and building a comprehensive mechanism for monitoring and information disclosure. In July 2017, NEA released the Opinions *on Guiding the Execution of the 13th Five-Year Plan for Developing Renewable Energy* ("the Opinions") to solve issues including the curtailment of wind, solar, and hydropower, and subsidy deficit. The Opinions guide goal-directed planning and operation, power transmission and consumption, and scale management mechanism. In addition, NEA announced the plan for annual solar and wind power additions for 2017~2020. The Opinions sets the future direction and reflect China's ambitious renewables push.

6) Renewable portfolio standard

In March 2018, NEA published *Renewable Power Quota and Assessment Methods (Draft for Comments)*. The energy authority under the State Council should set annual RPS for each provincial administrative region and keep a hold on monitoring, evaluation, and assessment. At the same time, minimum supply-side RPS requirements will be set to boost consumption, which will in turn boost renewables expansion. All these efforts will help fulfill the targets to raise the share of non-fossil fuel in primary energy to about 15% and 20% by 2020 and 2030.

7) Distributed wind farms

In June 2017, NEA published the *Notice on Boosting the Construction of Distributed Wind Farms* to boost distributed wind farms. The energy authorities at the provincial lev-els should timely revise their planning based on facts, and distributed wind farms are not counted in the annual recommended capacity size. The authorities are encouraged to simplify approval procedures for distributed wind farms in the plan. In April 2018, NEA released *Interim Measures for Administering the Development and Construction of Distributed Wind Farms.* It further improves policies and mechanisms for distributed wind farms in such five aspects as planning guidance, project construction and management, grid connection, operation management, and financial and investment mode. The strong institutional support will power the expansion of distributed wind farms in China.

(5) Constraints in wind energy development

1) The issue of the increasing wind power curtailment

With the rapid and large-scale development of the wind power industry in China, the bottleneck problem of grid connection is becoming increasingly prominent. In 2016, China's wind curtailment amount was 41.9 TWh, with a year-on-year decrease of

7.8 TWh; the national average wind curtailment rate reached 12%, with a year-on-year decrease of 5 percentage points (Fig.3-18).

Power shedding has been one of the biggest constraints to China's wind power industry. A 50-MW offshore wind farm, for example, will realize an after-tax IRR of 8.04% when it operates at full availability (2500 h per year) based on an average construction cost of 8000 yuan/kW and a 2017 feed-in tariff of 0.5 yuan/kWh (tax included) for wind power generated in Class II

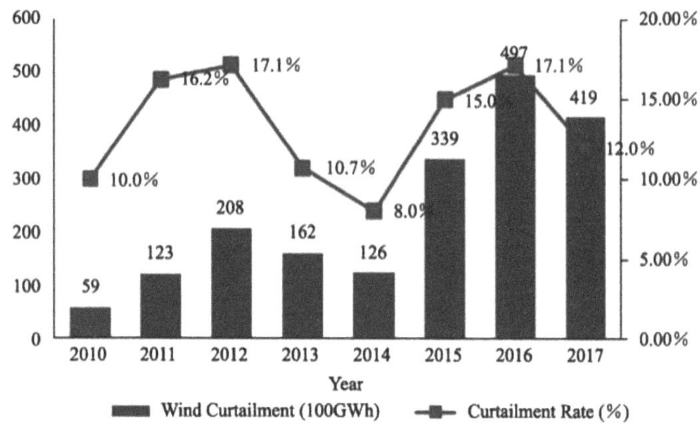

[Fig. 3-18] China's wind curtailment records over the years. Source National Energy Administration

area as per the No. 3044 document released by NDRC in 2015. A curtailment rate of 10% will result in a plunge in after-tax IRR to 6.34%, seriously restricting financial returns.

The problem, however, will not be solved in the short term of a combination of challenges: the slowdown in electricity demand under the new normal, insufficient local consumption and underdeveloped wind power grids in some regions, a monotonous mix of peak-shaving power sources, and a low load proportion; coal project investments remain high.

2) The issue of a growing subsidy deficit

Another constraint is delayed subsidy payment. China registered over 110 billion yuan subsidy shortfall by the end of 2017, according to NEA. The shortfall continues to widen by China's ambitious renewables push. A growing number of new energy projects demand more money support, while old projects are still unpaid.

As a result, wind power operators struggle with a tight cash flow and have to incur extra expenses from interest on overdue bank loans. Some underfunded companies are facing the risk of capital chain rupture. Estimation shows that if the year-end cumulative subsidy unsettled is 1.7

billion yuan, financial expenditure will increase by 77 million yuan based on a benchmark one-year lending rate of 4.6%. This means a profit drop of 0.012 yuan/kWh for wind farms.

3) The issue of the limited cost reduction potential

These years since 2015 have witnessed no significant decrease in wind farm cost, about 8300 yuan/kW. For onshore wind farms, more than 60% of investments are paid to wind turbines whose price, however, has remained stable at around 4200 yuan/kW since 2013. It is proved that it is the bottom price to ensure the reliability of wind turbines, and there is no room for a decrease in the short run. As China implements its 13th Five-Year Plan, the wind power industry is being transferred from the curtailment-plagued "Three North Regions" to Mideast and south regions. Yet new problems-great development difficulties and higher unit construction costs—will emerge in these populated destinations given a complicated terrain and high land cost. In other words, some regions will see a rise in wind power development costs though partly offset by technological progress and integrated development. With inferior wind resources, these regions can only sustain projects at an average annual wind speed of around 5 m/s. Considering the current technical conditions, these projects are economically fragile. Soaring land and forest acquisition costs have resulted in higher wind farm costs, and at the same time, wind speed continues to slow. In this light, there is limited room for cost reduction in these areas.

3.2.3 Perspective

(1) Steady market expansion

To promote the sustainable exploitation of wind power and realize the 15% non-fossil energy consumption target by 2020, it is essential to set objectives and develop annual plans accordingly. A series of new guidelines including the 13th Five-Year Plan for Developing Electricity, the 13th

Five-Year Plan for Developing Renewable Energy, and the 13th Five-Year Plan for the Development of Wind Power have been issued one after another, by which the wind power development is laid out and guided from the strategic level.

1) Goal in the 13th Five-Year Plan

According to the *13th Five-Year Plan for the Development of Wind Power*, it is estimated that the cumulative installed wind capacity will exceed 210 GW in 2020 with an average annual increase of 20 GW from the year-end capacity of 140 GW in 2015. The national guidelines have injected confidence and vitality into the wind power industry for larger-scale development.

① Overall capacity: By the end of 2020, the total installed on-grid wind capacity will exceed 210 GW, including over 5 GW from offshore wind farms.

② Goal on wind power consumption: The problem of curtailment will be effectively solved by 2020, and the minimum guaranteed buyout will be fully realized in the Three North Regions.

③ Industrial development goals: Wind power equipment manufacturing level and R&D capabilities continue to improve, 3~5 equipment manufacturing enterprises have fully reached the international advanced level, and the market share has significantly increased.

2) Industrial distribution in the 13th Five-Year Plan

① First, speed up the development of land-based wind energy resources in the central and southern regions: By 2020, the newly installed capacity of land-based on-grid wind power in the central and southern regions will be over 42 GW, and the cumulative installed wind power capacity will be over 70 GW.

② Second, orderly promote the local consumption and utilization of wind power in the "Three North Regions": By 2020, based on basically solving the problem of wind curtailment, by promoting local consumption and utilizing exist-ing channels for outward transmission, the newly installed on-grid wind power capacity will be about 35 GW, and the cumulative on-grid wind power will be about 135 GW.

③ Third, optimize the resource allocation through cross-provincial and inter-regional transmission channels: During the 13th Five-Year Plan period, 40 GW (including stock items) of wind power in the "Three North Regions" will be consumed in an orderly manner.

④ Actively and steadily promote offshore wind power construction: Focus on pro-moving offshore wind power construction in Jiangsu, Zhejiang, Fujian, Guang-dong, and other provinces. By 2020, the scale of offshore wind power construction in the four provinces has reached more than 1 GW. We will actively promote offshore wind power construction in provinces such as Tianjin, Hebei, Shanghai, and Hainan. Explore and promote offshore wind power projects in Liaoning, Shandong, Guangxi, and other provinces.

(2) Industry transformation and upgrade

In the early days of wind power development, manufacturing has always been the backbone of the development of China's wind power industry. With the increasing proportion of wind power installed in China total power installed in recent years, the market capacity of wind power operation and maintenance is expected to reach about 30 billion yuan, and the development prospect is broad. As the construction of wind power service system and service personnel in China are relatively lagging, wind power investors also rely more on the services of the main equipment manufacturers, the operation and maintenance services after the warranty period has not yet formed a complete industrial chain. The focus of the future wind power market will gradually shift from the front end of the industrial chain to the back end. In addition, with the continuous integration of large data and the wind power industry, the digital transformation project is carried out. Starting from large data of wind power, the whole life cycle data of wind power is integrated by using the Internet, cloud computing, and large data technology, and the design and construction of wind farm based on large data, intelligent manufacturing of wind power equipment and intelligent tube of the wind farm are carried out, which has also become an important way and direction for industrial transformation and upgrading in the future.

In September 2016, the Ministry of Industry and Information Technology, the Development and Reform Commission, the Ministry of Finance, and the Ministry of Science and Technology jointly issued the "Guidelines for the Implementation of Green Manufacturing Projects (2016~2020)", which calls for vigorous implementation of green manufacturing to achieve the transformation and upgrading of China's traditional manufacturing industry. In the field of wind power, green design, green technology and process, green production, green management, green supply chain, and green recycling can be carried out throughout the product life cycle through technical innovation and system optimization, to achieve the minimum environmental impact and the best utilization of resources and energy in the whole industrial chain and realized the transformation and upgrading to green manufacturing as the final goal. "Made in China 2025" proposed to achieve the strategic goal of manufacturing power through "three steps": In the first step, by 2025 China will become the manufacturing power; in the second step, by 2035, China's manufacturing industry as a whole reaches the middle level of the world's manufacturing powers; in the third step, till the time of New China's founding for 100 years, the position of China as manufacturing power is more consolidated and the comprehensive strength will be entered into the forefront of the world's manufacturing powers.

(3) Stronger technological innovation capability

China's wind energy technology has experienced the development process of technology introduction, digestion and absorption, and joint R&D and independent R&D, which has played a very important role in the development of China's wind energy. However, due to the differences between China's national conditions and those of Europe and the USA, we should learn from foreign advanced technology and at the same time improve our independent innovation and international competitiveness. Technological innovation in the future is mainly reflected in the following aspects:

1) Technology for large-scale and intelligent wind turbines

According to the "Energy Technology Revolution and Innovation Action Plan 2016~ 2030" issued by China, in the medium and long term, China will focus on the integrated optimization design and lightweight design technology of high-power onshore wind turbines and research integrating the load reduction and optimization technology for blades, loads and advanced sensor control of high-power units, the intelligent diagnosis of the electric control system of a powerful wind turbine, fault self-recovery, and maintenance-free technology and R&D of green manufacturing technology for high-power onshore wind turbine and key components, etc. In addition, in the design and manufacture of offshore wind turbines and key components, the emphasis will be laid on the overall design technology of offshore wind turbines of 10-MW class and above, including the integrated design technology of wind turbines, towers, and foundations, as well as the design optimization technology considering the limit load, fatigue load, and reliability of the whole unit. Developing wind turbines of 10 MW and above with independent intellectual property rights and key components such as bearings and generators, etc. In addition, with the increasing proportion of installed wind power capacity in the future and the higher requirements for intelligent operation control and maintenance of wind farms, the R&D of intelligent wind turbines has become an important technical direction for future development.

2) Low-wind-speed turbines

To make full use of wind energy resources in low-wind-speed areas, low-wind-speed wind turbines and low-wind-speed wind farms have developed rapidly in recent years. The diameter and tower height of wind turbines with low wind speed have exceeded 100 m. In addition, the complex terrain and wind turbulence characteristics in low-wind-speed areas will bring many new challenges to wind turbine design and wind farm construction; especially, the reliability design of low-wind-speed wind turbines and the layout optimization of low-wind-speed wind farms in complex mountainous areas need more comprehensive research.

3) Complete key technologies for offshore wind farm design and construction

With the continuous accumulation of experience in offshore wind power development, the breakthroughs of the complete key technologies in the design and construction of offshore wind farms will be achieved in the next five years. With the development of offshore wind power development, China will also start the development of wind farms in deep and offshore waters. According to the Energy Technology Revolution and Innovation Action Plan (2016~2030), China will focus on the site selection technology of offshs developed to meet the requirements of construction and manufacture under the characteristics of offshore marine environment load.

4) Lean design and O&M technology of wind farms

With the increasing installed capacity of wind farms and the increasing proportion of it in the power grid, the lean design, development, operation, and maintenance management of wind farms have gradually become a new technical direction. With the rapid development of "The Internet ", China will focus on the comprehensive intelligent sensor technology of wind turbines and wind farms, large data collection, and analysis technology of wind power, study on optimal design method of wind farms and large-scale grid-connected wind farms under complex terrain and special environmental conditions, and the optimization technology of wind farms based on big data. The key technologies of intelligent operation and maintenance of different types of wind farms based on the integrated application of the

Internet of things, cloud computing, and large data, as well as the key technologies such as optimal coordinated control, real-time monitoring, and grid adaptability suitable for access to distribution networks, will become the important development directions in the future. In the aspect of monitoring and operation and maintenance of offshore wind farms, it is necessary to analyze the factors that affect the safety and cost of operation and maintenance of offshore wind farms, study the operation and maintenance technology of offshore wind farms, develop a dynamic intelligent operation and maintenance management system based on life assessment, and develop special inspection and maintenance equipment and health model for offshore wind farms. The technology of condition assessment, operation risk assessment, residual life prediction, and operation and maintenance decision support are studied. The equipment technology of the new state detection system for offshore units, the maintenance technology of intelligent fault prediction, the remote network monitoring, and the intelligent diagnosis technology of key components are studied.

5) Operational control technology for wind farm cluster

Wind energy is hard to control because of its randomness, fluctuation, and intermittence. Large-scale wind power centralized grid connection requires to be more and more strict on the security and stability of power system operation, dispatch, and control. Wind farm cluster coordinated control is to integrate and centralize the wind farm group to make it the ability to flexibly respond to grid dispatch and control, and to form a detailed power supply in the scale and external control characteristics of Pitfish conventional power plant to improve the utilization of wind power. The coordinated control of the wind farm clusters can ensure the stable operation of the wind farm cluster according to the expected power generated under grid dispatch. The fluctuation of wind farm output power is gentle during the change process, which restrains the fluctuation of wind farm output power. Compared with the traditional wind farm control strategy alone, it has a great deal of stability voltage. The advantages show that wind power is not only controllable but also can be controlled quickly and accurately, which can provide effective support for the large-scale development of wind power.

6) Diversified wind power application

With the continuous development and improvement of wind power technology, the application fields of wind energy have been expanding, gradually showing a trend of diversified applications. Firstly, the adaptability of wind turbines in different scenarios has been improved; second, the development efficiency of wind energy resources has been realized through fine evaluation of wind resources and micro-location layout, and the market competitiveness of wind farm development has been improved; thirdly, the development of wind power heating and hydrogen production technology has been promoted; Fourthly, relying on the technologies such as pumped storage, energy storage devices, and smart grid technology to achieve multi-energy complementary integrated optimization and improve wind power consumption; Fifthly, make full use of new energy and multi-energy complementation to provide safe, economic, and environment-friendly power supply to islands, remote areas, specific industrial loads, communication base stations, municipal construction, and residential life. Through in-depth development of wind power diversification applications, break through the technical bottlenecks and policy constraints of wind power development, and then realize the healthy and sustainable development of the wind power industry.

(4) Further optimized policies

The renewable energy quota system is one of the important policy options for some countries in the world to develop renewable energy. According to statistics, the renewable energy quota system has been widely implemented in more than 20 countries, such as the USA, Britain, Italy, and the Netherlands, and many countries have achieved a leap in the proportion of renewable energy generation. In addition, the green power certificate (green card) trading system is widely adopted internationally. Green certificate is a kind of negotiable and cash-for-money certificate, which is an indicator to confirm the renewable energy power generation mode. It can be used not only as a measurement tool for renewable energy generation but also as a trading tool for transferring positive external ownership, such as a renewable energy environment. Promoting green power certificate trading is one of the effective ways to promote the sustainable and healthy development of the the renewable energy industry. On April 22, 2016, the General Department of the State Energy Administration issued the Notice on the Requirements for Establishing a Quota System for Non-water Renewable Energy Generation of Coal-fired Thermal Power Units (Draft for Comments), making it clear that the country is formulating and establishing a Quota System for Non-water Renewable Energy Generation of Coal-fired Thermal Power Units, and putting forward a clear assessment method. In 2020, each coal-fired power generation enterprise shall undertake a renewable energy generation quota of more than 15% of the thermal power generation, and the coal-fired power generation enterprise that has not fulfilled the requirements of the non-water renewable energy quota shall cancel its power generation business license. On January 18, 2017, the National Development and Reform Commission and the Ministry of Finance and National Energy Administration jointly issued the Notice on the Trial of the Renewable Energy Green Power Certificate Issuance and Voluntary Subscription and Transaction System (Development and Reform of Energy 2017, 132), and tried to issue and subscribe for renewable energy green power certificates nationwide. In March 2018, the National Energy Administration issued the "Renewable Energy Electricity Quota and Assessment Measures (Draft for Comments)", proposing that the competent department of energy under the State Council will formulate the quota indicators of renewable energy power for each provincial administrative region annually, and monitor, evaluate, and assess them.

It is expected that the quota policy and green card trading system will gradually mature and be widely used in the future, which will be of great benefit to the consumption of renewable energy. It will also make up for the deficiency of the existing subsidy system to a certain extent and promote the market-oriented development of renewable energy industries such as wind power.

Environmental loads on offshore wind turbine infrastructure

4

The offshore wind turbine works at sea all year round, and the environmental loads it receives are more complex than those on land. It mainly includes the loads generated by natural factors such as wind, waves, currents, earthquakes, and sea ice on the platform, as well as the changes in hydrostatic pressure and buoyancy of components caused by water level changes caused by waves and tides. Based on the characteristics of offshore wind power infrastructure, such as the high center of gravity and large wind area of blades, the environmental loads that play a major role can be summarized as aerodynamic loads on blades, hydrodynamic loads on the foundation below the water surface and ocean current loads.

4.1 Wind load

Because the offshore wind sites are subject to complex and variable wind forces with high strength, the blades, and towers of offshore wind turbines will exert a great force on their substructure under the action of wind. In addition, the persistence of wind force, and wind load becomes one of the main reasons affecting the fatigue life of offshore wind power foundations.

The wind load on offshore wind turbines mainly includes two parts. One is the aerodynamic load on the upper blade, which can be calculated based on momentum theory, blade element theory, and other relevant formulas; The other part is the wind load on the lower tower structure.

4.1.1 Basic characteristics of wind

The wind is a two-dimensional random process, assuming that the wind direction is only distributed on the horizontal plane. At this time, the instantaneous wind speed $u(z,t)$ can be decomposed into the average wind speed $\bar{u}(z)$ and the fluctuating wind speed $\tilde{u}(z,t)$. Then, the wind speed expression of t at any time can be written as:

$$u(z,\ t)=\bar{u}(z)+\tilde{u}\ (z,\ t)$$

(1) Average wind

The average wind means that the direction and size of the wind do not change with time in a period, but because of the influence of the roughness of sea level, the average wind speed will change along the height direction, and its change law can be approximated by the exponential law or logarithmic law.

- Exponential law

$$\bar{u}(z)=\overline{u_r}\left(\frac{z}{z_r}\right)^{\alpha}$$

Where :

z —Height above sea level ;

z_r—Reference height above sea level, in general, the value is 10 m ;

$\bar{u}(z)$ —Average wind speed at height , m/s ;

$\overline{u_r}$ —Average wind speed at reference height , m/s ;

α —The profile index of wind, which indicates the roughness of the sea surface. The larger the value, the rougher the sea surface.

- Law of logarithm

$$\bar{u}(z)=\overline{u_r}\frac{\ln z-\ln z_0}{\ln z_r-\ln z_0}$$

[Table. 4-1] Parameter value range of different geomorphic classifications

landforms	A	B	C	D
α	0.10~0.30	0.13~0.18	0.18~0.28	0.24~0.44
z_0 m	0.001~0.01	0.01~0.10	0.10~0.50	0.20~2.00

Where : z_0 — When the average wind speed is equal to zero, the higher the height, the rougher the sea surface.

According to the Chinese load specification, landforms are divided into four categories according to roughness: Category A refers to the desert, lakeshore, coast, island, and sea surface; Category B refers to fields, jungles, villages, and suburbs; Category C refers to small and medium-sized cities; Category D refers to the center of a large city. Their respective parameter values are shown in [Table 4-1].

(2) Pulsating wind

The fluctuating wind is random, and its wind direction and size will change with time. A large number of measured and analyzed results show that fluctuating wind speed is a stationary random process with zero mean value, and its probability distribution follows a normal distribution (RUAN Shengfu, 2010). Therefore, the power spectral density of wind speed can be used to express the distribution of fluctuating wind energy in the frequency domain. The common gust spectrum can be divided into the following two types according to whether the turbulence integral scale changes with height:

One is the wind spectrum that does not change with height, that is, the fluctuating wind spectrum at the standard height of 10 m. The disadvantage of this type of wind spectrum is that it will produce certain errors when solving the correlation function. The most common expression is the Davenport spectrum (WANG Xiuqiong & CUI Jianfeng, 2002) :

$$S(\omega)=\frac{4k\bar{u}(z_r)^2 r^2}{\omega(1+r^2)^{4/3}}$$

$$r=\omega L_r/2\pi\bar{u}(z_r)$$

where , $S(\omega)$ is the power spectral density of wind speed ; ω is angular frequency ; L_r is the turbulence integral scale, and the recommended value is 1200~1800 m; k is the surface resistance coefficient, which is related to the roughness of the sea surface. It is generally recommended to take 0.001~0.003, and 0.003 for near sea surface; $\bar{u}(z_r)$ is the average wind speed at the reference height z_r, generally taking the average wind speed at 10 m above sea level.

The second is the wind spectrum that varies with height, and the calculation results of the such spectrum can ensure high accuracy and reliability. The most common expression is NDP spectrum (WANG Xiuqiong & CUI Jianfeng, 2002):

$$S(z,f)=\frac{320\left[\frac{\bar{u}(z_r)}{10}\right]^2\left(\frac{z}{10}\right)^{0.45}}{\left(1+f_m^n\right)^{5/(3n)}}$$

$$f_m=172f\left(\frac{z}{10}\right)^{2/3}\left[\frac{\bar{u}(z_r)}{10}\right]^{-3/4}$$

where, $S(z,f)$ is the power spectral density of wind speed; f is the frequency, its value between 1/600 Hz and 1/2 Hz; $\bar{u}(z_r)$ is the average wind speed at the reference height z_r, generally taking the average wind speed at 10 m above sea level; z is the height above sea level; n is the parameters, its value is 0.46.

4.1.2 Momentum theory

The momentum theory originated from the study of marine propellers in the 19th century. At the beginning of the 20th century, Betz first applied the momentum theory to aircraft propellers and started its research on airflow (Wilson et al., 1976). This method is to apply the basic laws of fluid mechanics to study the effect of propeller disks on airflow. Its main principle is to estimate the force characteristics of fan blades through the relationship between the force of wind load acting on the wind wheel and the wind incoming velocity (ZHAO Jing, 2009). In the momentum theory, the method of simplifying the flow field near the wind turbine is adopted, assuming that it is an ideal flow pipe passing through the plane of the wind turbine as shown in [Fig. 4-1], and the following assumptions are adopted:

(1) The airflow is incompressible, uniform, and steady;

(2) The wind turbine is regarded as a plane propeller disk without thickness, and the influence of friction is not considered;

(3) Assume that the wind turbine is a flow pipe through which the air flows, the air pressures at both ends of the wind turbine at infinity on both sides are equal, and the axial thrust is evenly distributed on the plane of the rotor disc (ZHAO Jing, 2009)

In [Fig. 4-1], let the air flow from left to right along the horizontal direction (x axis), and the initial velocity at the inlet is V_1 , the initial cross-sectional area is A_1 ; The flow rate at the disk is V, the upstream area is A ; Air flow velocity is V_2 ; The sectional area at the outlet is A_2.

Assume that the induced velocity in the horizontal direction at the panel is:

$$u = \bar{u} \cdot V_1$$

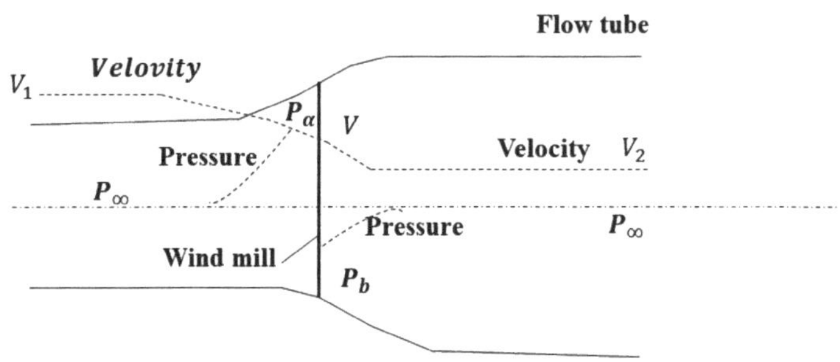

[Fig. 4-1] Betz Schematic Diagram of Momentum Theorem

where, is inflow factor, $0 < \bar{u} < 1$.

The air velocity at the panel is:

$$V = V_1 - u = (1 - \bar{u})V_1$$

According to the continuity equation of the flow tube:

$$V_1 \cdot A_1 = V \cdot A = V_2 \cdot A_2$$

Therefore, the momentum change rate of incoming flow can be expressed as:

$$\rho \cdot A_2 V_2 \cdot V_2 - \rho \cdot A_1 V_1 \cdot V_1 = \rho \cdot A \cdot V(V_2 - V_1) = \rho \cdot A(1-) \cdot V_1(V_2 - V_1)$$

In unit time, the force exerted by the disk on the fluid in the horizontal direction is:

$$f_x = (p^- - p^+)A$$

Bernoulli equation is applied to both sides of the disk in the flow tube:

$$\frac{1}{2}\rho V_1^2 + p_\infty = \frac{1}{2}\rho V^2 + p^+$$

$$\frac{1}{2}\rho V_2^2 + p_\infty = \frac{1}{2}\rho V^2 + p^-$$

Subtract the two equations:

$$p^- - p^+ = \frac{1}{2}\rho(V_2^2 - V_1^2)$$

According to the momentum theorem, the equation is equal to the equation :

$$\rho \cdot A_2 V_2 \cdot V_2 - \rho \cdot A_1 V_1 \cdot V_1 = \rho \cdot A \cdot V(V_2 - V_1) = \rho \cdot A(1 - \bar{u}) \cdot V_1(V_2 - V_1)$$

is equal to the equation $f_x = (p^- - p^+)A$:

then

$$f = (p^- - p^+)A = \frac{1}{2}\rho(V_2^2 - V_1^2)A = \rho \cdot A(1 - \bar{u}) \cdot V_1(V_2 - V_1)$$

Simplified:

$$V_1 + V_2 = 2(1-\bar{u}) \cdot V_1$$

So:

$$V_2 = (1-2\bar{u}) \cdot V_1$$

According to Newton's third law, the thrust of air flow acting on the disk is:

$$t = -f = -\rho \cdot A(1-\bar{u}) \cdot V_1(V_2 - V_1) = 2\rho \cdot AV_1^2 \cdot \bar{u}(1-\bar{u})$$

The power absorbed by the wind turbine is:

$$P = t \cdot V = 2\rho \cdot AV_1^3 \cdot \bar{u}(1-\bar{u})^2$$

From this, the thrust coefficient is:

$$C_t = \frac{t}{\frac{1}{2}\rho V_1^2 A} = 4\bar{u}(1-\bar{u})$$

Similarly, the power coefficient of the wind turbine is:

$$C_P = \frac{P}{\frac{1}{2}\rho V_1^3 A} = 4\bar{u}(1-\bar{u})^2$$

Therefore, the maximum thrust coefficient, power coefficient, and the corresponding inflow factor are as follows (Betz theory)

When $\bar{u} = \frac{1}{2}$, $C_{tmax} = 1$;

When $\bar{u} = \frac{1}{3}$, $C_{tmax} = \frac{16}{27} \approx 0.593$.

4.1.3 Leaf element theory

In the calculation of the aerodynamic load of a fan, another commonly used theory is the blade element theory. Based on the idea of calculus, the theory decomposes the wind turbine blade along its length into several elements called blade elements and assumes that the fluid motion on each blade element is independent of each other. Each blade element is regarded as a plane airfoil, and the force and moment received by it are integrated along the blade direction to obtain the force and moment received by the whole fan blade (Wilson et al., 1976). The blade element theory starts with the flowing gas around the blade micro-segment to describe the action of the force on the wind turbine and the energy transformation. Compared with momentum theory, blade elements can more accurately describe the influence of gas flow on blades, so it is widely used in fan design and aerodynamic load calculation (Wilson et al., 1976).

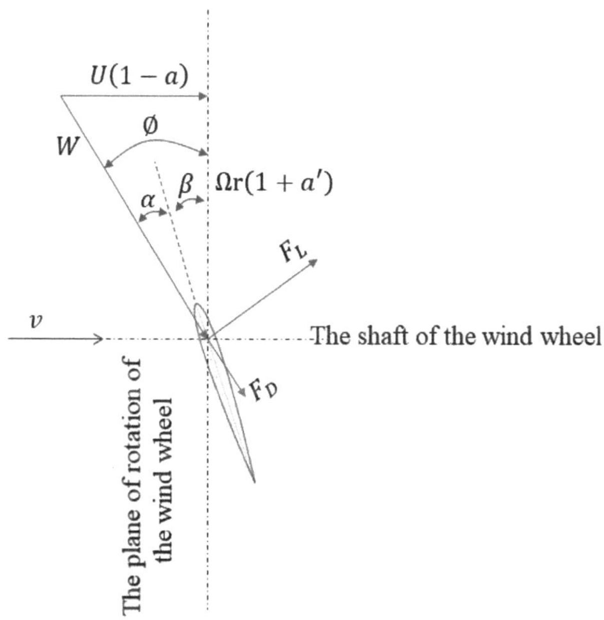

[Fig. 4-2] Wind speed component on the single leaf element

Taking the single leaf element in [Fig. 4-2] as an example, take a leaf element with the length of dr at the radius r of the wind turbine. U in the figure is the incoming wind speed at infinity; Ω is the circumferential velocity ; W is the relative velocity ; a is the axial induction factor ; a' is the circumferential induction factor ; α is the attact angle ; β is the pitch angle ; ϕ is the inflow angle.

Then the speed of the blade element in two directions is respectively the speed component in the direction perpendicular to the rotation plane of the impeller is $U(1-a)$ and the velocity component in the direction parallel to the rotation plane of the impeller is $\Omega r(1+a')$, the relative speed is obtained by combining the vertical speed with the horizontal speed W, The included angle ϕ between the velocity direction and the vertical direction can be expressed as :

$$\tan\phi = \frac{(1-a)U}{(1+a')\Omega r}$$

If the chord length of leaf element is l, the lift and resistance can be expressed as:

$$F_L = \frac{1}{2}\rho C_L W^2 l dr$$

$$F_D = \frac{1}{2}\rho C_D W^2 l dr$$

where, C_L is the lift coefficient of the element wing type; C_D is the drag coefficient of the blade element airfoil. Assume that the number of known blades is B, then the thrust of blade element at radius r is:

$$dT = B(F_L\cos\phi + F_D\sin\phi) = \frac{1}{2}\rho W^2 Bl(C_L\cos\phi + C_D\sin\phi)$$

Torque is:

$$dQ = Br(F_L\sin\phi - F_D\cos\phi) = \frac{1}{2}\rho W^2 Brl(C_L\sin\phi - C_D\cos\phi)$$

4.1.4 Calculation of aerodynamic load of fan

The above momentum theorem can be applied to the calculation of aerodynamic load on fan blades. Assuming that the fan is in normal working condition, the average pressure p_H acting on the swept area of the wind turbine can be determined by the following formula:

$$p_H = \frac{1}{2}\rho C_{FB} V_r^2$$

where, C_{FB} is the correlation coefficient, according to Bezt's theory in the momentum theorem $C_{FB} = \frac{8}{9}$; ρ is the destiny, $\rho = 1.297kg/m^3$; V_r is the rated wind speed.

The above equation can be simplified into:

$$p_H = \frac{V_r^2}{1800}$$

When the wind turbine works normally, the force on the top of the tower is:

$$F_{XH} = p_H A = p_H \frac{\pi D^4}{4}$$

where, A is the swept area of the wind turbine, that is, the projected area of the wind turbine on its rotating plane; D is the impeller diameter.

Therefore, the moment generated by the wind turbine can be calculated according to the basic bending moment calculation method:

$$M_{XH} = F_{XH} H$$

4.2 Wave load

Compared with onshore wind turbines, offshore wind turbines are more complex, and the main factor is the impact of wave loads. Wave theory can be divided into deterministic and stochastic wave theory. The wave height and period of the so-called deterministic wave theory are fixed and change uniformly with time, which can be expressed by a specific formula; However, the wave height and period described by the random wave theory are irregular and have strong randomness (HAN Chaoshuai, 2014).

4.2.1 Deterministic wave theory

Deterministic wave theory, also known as regular wave theory, is a variety of wave models based on certain assumptions and simplifications, including linear wave theory, Stokes wave theory, elliptical cosine wave theory, etc. During analysis, appropriate wave theory can be selected according to the operating water depth, wave period, and wave height data of the platform.

(1) Linear wave theory

Linear wave theory, also known as small amplitude wave theory, is the simplest form of wave, and also the most basic and widely used wave theory. The linear wave theory assumes that the water quality point vibrates harmonically with a fixed circular frequency, and the waveform propagates forward at a certain speed. This theory is the basis for studying complex finite amplitude waves and random waves.

Let the wave surface equation of a two-dimensional small amplitude advancing wave with constant depth be:

$$\eta = a\cos(kx - \omega t)$$

where a is wave amplitude , $a = \frac{H}{2}$, H is wave height ; $kx - \omega t = \theta$ is wave phase; ω is circular frequency of waves, $\omega=\frac{2\pi}{T}$; T is wave period; k is wave number; $k=\frac{2\pi}{L}$; L is wave length, $L=\frac{gT^2}{2\pi}th\frac{2\pi d}{L}$; d is water high, in deep water, when $\frac{d}{L}\geq 0.5$, $th\frac{2\pi d}{L}\propto 1$, $L=\frac{gT^2}{2\pi}$.

The velocity potential expression of the propulsion wave is:

$$\phi=\frac{TgH}{2\omega}\frac{chk(z+d)}{chkd}\sin(kx-\omega t)$$

Where x is the horizontal position of the water quality point; z is the vertical position of the water quality point.

From the above formula, the velocity and acceleration expressions in the horizontal and vertical directions of the water quality point can be obtained.

Horizontal speed:

$$u(t)=\frac{\partial\phi}{\partial x}=\frac{\pi H}{T}\frac{chk(z+d)}{shkd}\cos(kx-\omega t)$$

Vertical speed:

$$v(t)=\frac{\partial\phi}{\partial z}=\frac{\pi H}{T}\frac{shk(z+d)}{sh\square kd}\sin(kx-\omega t)$$

Horizontal acceleration:

$$\frac{\partial u(t)}{\partial t}=\frac{2\pi^2 H}{T^2}\frac{chk(z+d)}{shkd}\sin(kx-\omega t)$$

Vertical acceleration:

$$\frac{\partial v(t)}{\partial t}=-\frac{2\pi^2 H}{T^2}\frac{sh\,k(z+d)}{sh\,kd}\cos(kx-\omega t)$$

In deep water, when $\frac{d}{L} \geq 0.5$, it can be reduced that:

$$u(t) = \frac{\pi H}{T} e^{kx} \cos(kx - \omega t)$$

$$v(t) = \frac{\pi H}{T} e^{kx} \sin(kx - \omega t)$$

$$\frac{\partial u(t)}{\partial t} = \frac{2\pi^2 H}{T^2} e^{kx} \sin(kx - \omega t)$$

$$\frac{\partial v(t)}{\partial t} = -\frac{2\pi^2 H}{T^2} e^{kx} \cos(kx - \omega t)$$

For the above small amplitude wave theory, the wave pressure at any point in the wave is:

$$p = -\rho g z - \rho \frac{\partial \phi}{\partial t}$$

Where the first part is the hydrostatic pressure; The second part is the static wave pressure, which is caused by fluctuation. ρ is the density of seawater.

In the condition of low water:

$$p = -\rho g z + \frac{pgH}{2} \frac{chk(z+d)}{chkd} \cos(kx - \omega t)$$

In the condition of deep water:

$$p = -\rho g z + \frac{pgH}{2} e^{kz} \cos(kx - \omega t)$$

(2) Stokes wave theory

In the linear wave theory, because the ratio of wave height to wavelength (i.e., wave steepness H/L) and the ratio of wave height to water depth (i.e., relative wave height H/d) are infinitely small, the nonlinear effect caused by the fluctuating free water surface can be ignored. However, in practice, the wave steepness and relative wave height of ocean waves are generally limited,

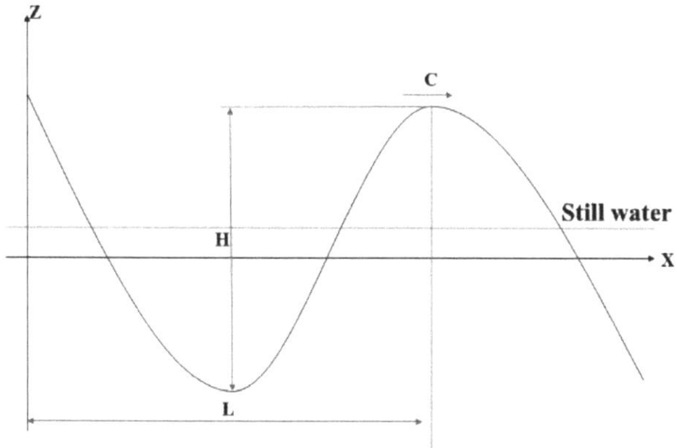

[Fig. 4-3] Stokes wave profile

even reaching a large value. In such a finite amplitude wave, the nonlinear effect of the fluctuating free water surface must be considered, so it is also called a nonlinear wave (ZHU Yanrong, 1991).

Stokes wave theory is one of the most commonly used finite amplitude wave theories in ocean engineering. Unlike linear waves, it does not regard wave steepness H/L as infinitesimal, but as the main factor determining wave properties. It proves that the wave surface is no longer a simple cosine shape, but a similar cycloidal shape with a narrow wave crest and wide wave trough, as shown in [Fig. 4-3] (ZHU Yanrong, 1991).

Stokes's theory assumes that the basic equation of wave motion can be expressed by a power series expansion of parameter ϵ. Parameter ϵ is a dimensionless constant related to the wave characteristic value. The most effective wave characteristic value can be expressed as steep H/L(large water depth) or relative wave height H/d(small water depth). Stocks fifth order wave has been widely used because of its remarkable nonlinear characteristics. The main calculation results are as follows:

Velocity potential :

$$\phi = \frac{c}{k}\sum_{n=1}^{5} x_n chnk(z+d)\sin n(kx-\omega t)$$

Height of wave surface :

$$\eta = \frac{1}{k}\sum_{n=1}^{5} y_n \cos n(kx - \omega t)$$

Wave speed :

$$c^2 = \frac{C_0^2}{k}\left(1 + \lambda^2 C_1 + \lambda^4 C_2\right)$$

Waveform factor :

$$x_1 = \lambda A_{11} + \lambda^3 A_{13} + \lambda^5 A_{15}$$
$$x_2 = \lambda^2 A_{22} + \lambda^4 A_{24}$$
$$x_3 = \lambda^3 A_{33} + \lambda^5 A_{35}$$
$$x_4 = \lambda^4 A_{44}$$
$$x_5 = \lambda^5 A_{55}$$
$$y_1 = \lambda$$
$$y_2 = \lambda^2 B_{22} + \lambda^4 B_{24}$$
$$y_3 = \lambda^3 B_{33} + \lambda^5 B_{35}$$
$$y_4 = \lambda^4 B_{44}$$
$$y_5 = \lambda^5 B_{55}$$

Wavelength L and coefficient λ can be obtained from transcendental equations:

$$\begin{cases} \dfrac{\pi H}{d} = \dfrac{1}{\left(\dfrac{d}{L}\right)}\left[\lambda + \lambda^3 B_{33} + \lambda^5 (B_{35} + B_{55})\right] \\ \dfrac{d}{L_O} = \dfrac{d}{L} thkd \left[1 + \lambda^2 C_1 + \lambda^4 C_2\right] \end{cases}$$

The corresponding horizontal and vertical velocities and accelerations can also be obtained by derivation.

where, The meaning of k, d, x, z, H, T, ω is the same as that in linear wave; $L_0 = \dfrac{gT^2}{2\pi}$; The 18 coefficients of A_{jj}, B_{ij} and C_{ij} are all function of relative water depth d/L, and has their corresponding expression.

(3) Elliptic cosine wave theory (Cnoidal)

For shallow water depth, an elliptical cosine wave is used to describe a finite amplitude periodic wave. When the waves enter the nearshore shallow water area $\left(0.05 < \frac{d}{L} \leq 0.1\right)$. The correlation between the water quality point movement characteristics at the wave crest and the wave well H/L becomes smaller, and the correlation between the water quality point movement characteristics and the relative wave H/d becomes larger. Both become the common factors determining the wave properties. The wavelength equation:

Two expressions of wave period are:

$$T\sqrt{\frac{g}{d}} = \sqrt{\frac{16d}{3H}}\left\{\frac{kK(k)}{\sqrt{1+\frac{H}{d}\left[-1+\frac{1}{k^2}\left(2-3\frac{E(k)}{K(k)}\right)\right]}}\right\}$$

$$T\sqrt{\frac{g}{d}} = \sqrt{\frac{16d}{3H}}\left\{\frac{kK(k)}{1+\frac{H}{d}\frac{1}{k^2}\left(\frac{1}{2}-\frac{E(k)}{K(k)}\right)}\right\}$$

where, The symbolics meaning of d, H, g is the same as that specified above ; k is elliptic integral modulus; are respectively the first and second complete circle planting integrals of module , namely:

$$K(k) = \int_0^{\frac{\pi}{2}} \frac{d\theta}{\sqrt{1-k^2\sin^2\theta}}$$

$$E(k) = \int_0^{\frac{\pi}{2}} \sqrt{1-k^2\sin^2\theta}\, d\theta$$

Wave surface equation:

$$\eta = \frac{16d^3}{3L^2}\{K(k)[K(k)-E(k)]\} - Hcn^2\left[2K(k)\left(\frac{x}{L}-\frac{t}{T}\right), k\right]$$

where, $cn^2\left[2K(k)\left(\frac{x}{L}-\frac{t}{T}\right), k\right]$ is Jacobi elliptic cosine function, function

cn^2's periodic is $2K(k)$. k value between $0\sim1$. when $k=0$, the wave surface equation is in the form of a linear wave, and when $k=1$, is the form of a solitary wave.

(4) Solitary wave theory

The trajectory of water particle motion described by linear wave theory is closed without net displacement; However, the motion track of wave water quality points described by nonlinear wave theory is not closed but has a small amount of net displacement. The wave generates mass transfer in the propagation direction. In nature, when the wave propagates to the near-shore shallow water area and is about to break, its wave surface shape and motion characteristics are close to solitary waves.

A solitary wave is a special form of elliptical cosine wave with modulus $k=1$, namely $2(k)=1$, $K(k)=\infty$. So, wave length $L\rightarrow\infty$, wave speed $c=\sqrt{g(d+H)}$. The wave surface equation can be expressed as:

$$\eta = H\sec h^2\left[\sqrt{\frac{3H}{4d^3}}(x-ct)\right]$$

where the physical meaning of is the same as that specified above ; The starting point of is set at the wave crest.

4.2.2 Scope of application of various deterministic wave theories

The general application scope of the linear wave, Stokes fifth order wave, and elliptical cosine wave theory can be summarized as follows:

$T\sqrt{g/d}<6.0$, $(d/L>0.2)$, $H/d\leq 0.2$, using linear wave theory ;

$T\sqrt{g/d}<10.0$, $(d/L\geq 0.1)$, using Stokes fifth order wave theory ;

$T\sqrt{g/d}<10.0$, $\left(\frac{d}{L}<0.1\right)$, using the Elliptic Cosine Wave Theory.

The deep-water wave region can be completely calculated by linear wave and

Stokes wave theory; The very shallow water wave region is mainly calculated by the elliptic cosine wave and solitary wave theory; However, the situation in shallow water wave area is more complex, and many wave theories need to be used for calculation.

4.2.3 Random wave theory

The wave motion of offshore engineering structures in the marine environment is irregular and has obvious randomness. The study of random waves requires a large number of wave data obtained through long-term observation. Combined with the relevant theories of random processes, and using certain mathematical models, the random waves are represented in the form of a wave spectrum based on semi-theory and semi-experience. The general expression of irregular waves is:

$$S_\eta = \frac{A}{\omega^p}\exp\left[-\frac{B}{\omega^q}\right]$$

where. ω is wave frequency; A、B、p、q are four adjustable coefficients. The value of A、B is related to wind, wave, and other factors in the sea area, including wind speed, wave height, wave period, etc., which can be selected according to the actual observation data, $p = 4 \sim 6$, $q = 2 \sim 4$.

At present, common wave spectra include the following:

(1) Neumann spectrum

Neumann spectrum is a kind of wave spectrum first proposed based on fully developed wave data. Its formula is:

$$S_\eta(\omega) = \frac{2.40}{\omega^6}\exp\left[-\frac{145.8}{\omega^2 (H_s)^{4/5}}\right]$$

Spectrum peak frequency:

$$\omega_m = 6.97/(H_s)^{\frac{2}{5}}$$

where, H_s is the significant wave height.

(2) Pierson-Moskowitz spectrum (P-M spectrum)

P-M spectrum is the measured wave energy spectrum obtained from the integrated analysis of observation data, which is more full and true than the observation data of the Neumann spectrum. P-M spectrum was first proposed in 1964 and has been widely used in wave research and related engineering fields. At the same time, it was also defined as the standard wave spectrum in the 11th International Ship Model Pool Conference (ITTC), and its expression is:

$$S_\eta(\omega)=\frac{0.78}{\omega^5}\exp\left(-1.25\left(\frac{1.257}{\sqrt{H_s}\,\omega}\right)^4\right)$$

Spectral peak frequency:

$$\omega_m=1.257/\sqrt{H_s}$$

(3) JONSWAP spectrum

JONSWAP spectrum is the wave spectrum used in the limited wind area obtained by the Joint North Sea Wave Project jointly developed by the United Kingdom, Netherlands, Germany, the United States, and other countries in the 1960s through the observation of waves in the North Sea. Its specific expression is as follows:

$$S_\eta(\omega)=\frac{ag^2}{\omega^5}\exp\left[-\frac{5}{4}\left(\frac{\omega_m}{\omega}\right)^4\right]\gamma^a$$

where $a=0.076\left(\frac{X}{u^2}\right)^{-0.2}$

Spectral peak frequency:

$$\omega_m=7\pi\frac{g}{u}\left(\frac{gX}{u^2}\right)^{-0.33}$$

where X is the wind area length, u is the wind speed above the sea surface 10 m. Energy scale coefficient:

$$\alpha = \exp\left[-\frac{(\omega-\omega_{re})}{2\sigma^2\omega_r^2}\right]$$

Spectral peak raising factor:

$$\gamma = \frac{S_\eta(\omega_m)}{S_\eta(\omega_m)_{P-M}}$$

where σ is the peak type coefficient, $\sigma = \begin{cases} 0.07\,(\omega \le \omega_m) \\ 0.09\,(\omega > \omega_m) \end{cases}$ $S_\eta(\omega_m)$, $S_\eta(\omega_m)_{P-M}$ are peak values of the JONSWAP spectrum and P-M spectrum.

4.2.4 Calculation of wave forces on small-scale sediment

The effects of waves on fixed offshore engineering structures are mainly reflected in four aspects:

(1) Viscous effect caused by the viscosity of seawater;

(2) The added mass effect caused by the change of wave field velocity around the structure due to the inertia of the fluid;

(3) Scattering effect caused by the influence of the medium on the incident wave;

(4) The free surface effect is caused by the large relative height of the structure disturbing the free surface of the original wave field.

For small-scale structures ($D/L < 0.2$), it has no significant effect on wave motion itself, and the wave action on structures is mainly manifested as viscous action and additional mass action. At present, the Morrison equation, which is a semi-theoretical and semi-empirical formula based on diffraction theory, is mainly used to calculate the wave forces on small-scale structures. Its basic assumptions include: the fluid is an incompressible ideal fluid; the motion of the fluid is a potential motion; the boundary of the structure can be regarded as a part of the fluid boundary. For the Morrison equation, the water depth is assumed to be d, height H is the incident wave along the $p$ositive direction of the x-axis, Then the horizontal wave force f on the cylindrical structure with diameter D at any height z (seabed $z=0$) $f_H$can be considered as two parts: one is the horizontal drag force f_D caused by the horizontal velocity u_x of the wave water particle movement; The other is the horizontal acceleration of wave water particle motion frac $\frac{du_x}{dt}$ Horizontal inertial force f caused by f_I. The formula can be expressed as:

$$f_H = f_D + f_I = \frac{1}{2} C_D \rho D u_x |u_x| + C_M \rho \frac{\pi D^2}{4} \frac{\partial u_x}{\partial t}$$

where, C_D is drag coefficient; C_M is inertial force coefficient; ρ is sea water's destiny.

When using the Morrison equation to calculate wave forces on small-scale structures, two key issues should be paid attention to: first, according to the water depth of the sea area where the structure is located and the wave elements (wave height, period) of the selected design wave and other factors, select a suitable wave theory to calculate the wave surface equation of water quality points η, horizontal velocity u_x and horizontal acceleration frac $\frac{\sigma u_x}{\sigma t}$. The specific basis for selecting wave theory and the calculation method of horizontal velocity and acceleration have been described in detail in the previous section; The second point to note is to select the appropriate C_D, C_M value according to the test experience or relevant specifications.

4.3 Current load

4.3.1 Current speed

The current load is generated by the current action caused by wind, tide, and other factors. Therefore, the current speed is also related to tidal current and wind-driven current. Generally, only the horizontal velocity component of the current is considered, and its distribution along the vertical direction can be expressed as :

$$u_c(z) = u_t(0)\left(\frac{z}{d}\right)^{\frac{1}{7}} + u_w(0)\left(\frac{z}{d}\right)$$

where , $u_c(Z)$ is the current velocity at altitude z ; z is the vertical distance from the sea bed ; $u_t(0)$ is the tidal current velocity at mean sea level $z=d$; $u_w(0)$ is the wind-induced current velocity at mean sea level $z=d$; d is water depth.

4.3.2 Current load

The current load acting on small-scale structures is calculated in the same way as the drag force in Morison's formula, which is: (China Classification Society, 2005)

$$F_c = \frac{1}{2} C_D \rho A u_c^2$$

where , C_D is the drag coefficient same as the drag force coefficient used in wave load calculation ; ρ is the seawater density ; A is the projected area of the component on a vertical plane ; u_c is current speed.

In the case of wave current joint action in the sea area where the structure is located, it is necessary to calculate the current load on the small-scale structure after the vector superposition of the current velocity and the motion velocity of the wave water quality point.

4.4 Sea ice load

The applicability of the random dynamic ice load model on wind power foundation structure is studied through the field observation data of sea ice breaking characteristics of wind power foundation cone structure in cold regions; Based on field observation data, it is found that ice force period and ice force amplitude obeys normal distribution. (Wang Guojun et al., 2020).

Due to the randomness of the physical and mechanical properties of sea ice, the destruction process of the interaction between sea ice and cone structure is also random. The larger the width of the structure, the thinner the sea ice is, and the stronger the randomness of breaking. As shown in [Fig. 4-4], when the ratio of the diameter of the cone structure to the thickness of ice (diameter thickness ratio) is larger, the smaller the size of sea ice breaking is, and the greater the non simultaneity of sea ice breaking is. Therefore, to more accurately simulate the ice forces based on tapered wind power, it is necessary to analyze the randomness of the ice load.

(a) Diameter to Thickness Ratio:20

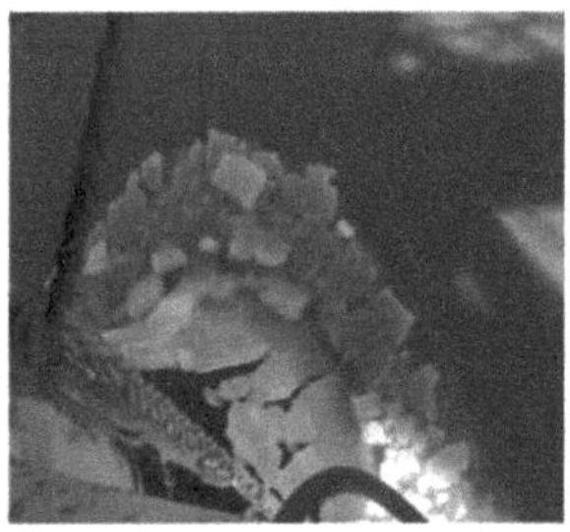

(b) Diameter to Thickness Ratio:45

(c) Diameter to Thickness Ratio:90

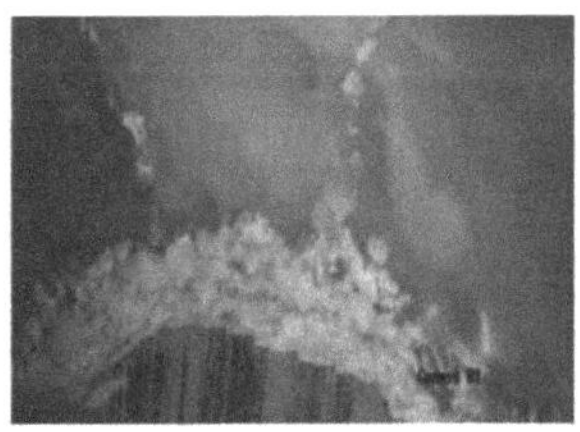

(d) Diameter to Thickness Ratio>180

[Fig. 4-4] Sea ice bending failure on cones of different scales, source: (Wang Guojun et al., 2020)

According to the field video data processing of the tapered wind power foundation, the bending and breaking of sea ice in front of the cone mainly goes through three stages: from the contact of the cone to the bending and breaking of sea ice (i.e. the ice force loading stage); During the climbing stage of broken ice along the cone driven by the subsequent ice plate, the thrust of the subsequent ice plate on the broken ice is getting smaller and smaller (ice force unloading stage). At the same time, the cone surface is relatively smooth and the friction force is small, so it can be considered that the ice force is unloaded to 0; The broken ice slides away from both sides of the cone, and the subsequent ice sheet has not contacted the cone surface. The ice forces during this period are 0. The breaking characteristics of sea ice in front of the wind power cone are the same as those of sea ice in front of the Bohai jacket plus the cone structure.

Qu et al.(Qu Yan et al., 2006) proposed a model of random ice force function, which simplified the cone ice load into an isosceles trigonometric function with randomly distributed peak and period of ice force. The time from loading to unloading of ice force is 1/3 of the whole period, as shown in [Fig. 4-5].

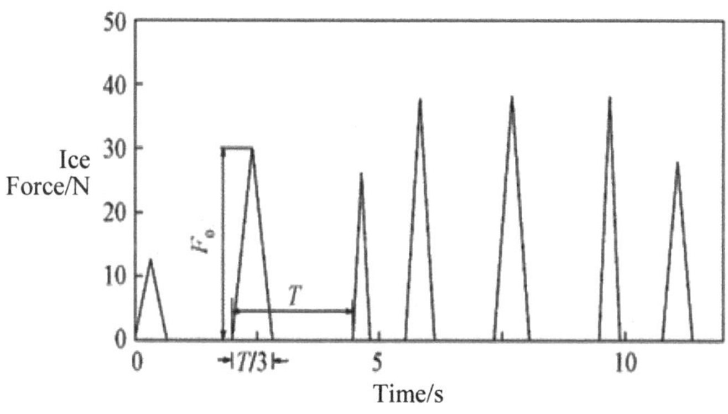

[Fig. 4-5] Random ice force simplified model on the conical structure, source: (Qu Yan et al., 2006)

The ice force function in a period can be defined as:

$$f_i(t) = \begin{cases} \frac{6F_{0i}}{T_i}t, & 0 < t < \frac{T_i}{6} \\ 2F_{0i} - \frac{6F_{0i}}{T_i}, & \frac{T_i}{6} < t < \frac{T_i}{3} \\ 0, & \frac{T_i}{3} < t < T_i \end{cases}$$

where: i is the period; is time; $f_i(t)$ is the ice force at time t; T_i is the size of the ith ice force cycle; F_{0i} is the amplitude of ice force in this period. The random ice force function of any length of time can be extended to:

$$F(t) = \sum_{i=1}^{N} f_i(t - t_i^0)$$

where: N represents the number of ice force cycles to be established; when $i = 0$, $t_1^0 = 0$; when $i > 1$, $t_i^0 = \sum_{j=1}^{i-1} T_i$. When the ice force cycles are expanded to N, the ice force amplitude F_{0i} and ice force cycle T_i of each cycle form two random series. To obtain the applicable random ice force function, it is necessary to make statistics on the ice force cycle and amplitude of the real ice force and determine the random distribution of the ice force cycle and amplitude sequence.

Wang Guojun analyzed the statistical characteristics of the ice force amplitude and ice force period of the pheasant body based on the measured data of the Bohai platform and found that the ice force period and ice force amplitude obey normal distribution, as shown in [Fig. 4-6] and [Fig. 4-7].

Ice force cycle sequence $T_i \sim N(\overline{T}, \sigma_T^2)$, Its probability density function p_T can be expressed as:

$$p_T = f(T, \overline{T}, \sigma_T) = \frac{1}{\sigma\sqrt{2\pi}} e^{-\frac{(T-\overline{T})2}{2\sigma_T^2}}$$

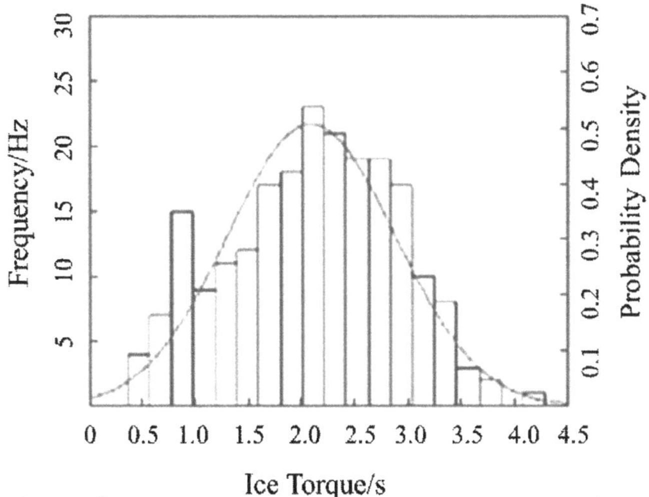

[Fig. 4-6] Normal distribution of ice force period, source: (Wang Guojun et al., 2020)

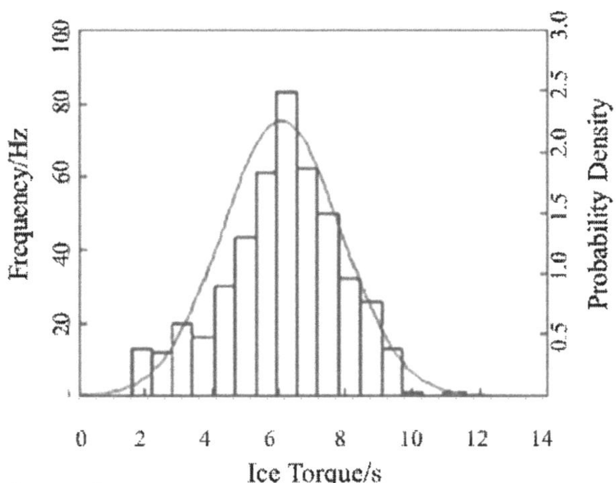

[Fig. 4-7] Normal distribution of ice force amplitude, source: (Wang Guojun et al., 2020)

Ice force amplitude $F_l \sim \left(F, \sigma_f^2\right)$, probability density function pg can be expressed as

$$p_F = f\left(F, \overline{F}, \sigma_F\right) = \frac{1}{\sigma\sqrt{2\pi}} e^{-\frac{(F-F)2}{2\sigma_F^2}}$$

Qu et al. (Qu Yan et al., 2006) obtained the standard deviation of ice force through analysis σ_τ is usually 40% sim70% of the average ice force period T. In application, the average value of 0.5 times can be taken as the value of its standard deviation.

$$\sigma_\gamma = 0.5$$

At the same time, the differential coefficient of ice force amplitude obeys random distribution at 0.2~0.5, and its average value is about 0.4. In application, the standard deviation σ_ρ of ice force, the amplitude can be taken as 0.4 times the average value $\vec{F}$ of ice force

$$\sigma_f = 0.4F$$

4.4.1 Analysis of ice force average period characteristics

The ice force period T can be calculated by the ratio of the ice-breaking length L before the structure to the ice velocity v (Yu Xiao, 2001):

$$T = \frac{L}{v}$$

Li Feng et al.(LI Feng & Yue Qianjin, 2000) found that the bending fracture length of sea ice is mainly controlled by the thickness of sea ice according to experimental analysis The two can be expressed as a linear relationship.

$$k = \frac{L}{h}$$

k is the ratio of ice-breaking length and ice thickness referred to as the "length thickness ratio coefficient". Based on the field measured data and indoor test data, through the analysis of the length thickness ratio k on cones of different scales, it can be found that the width thickness ratio of structures of different scales is also different, and generally increases with the increase of cone diameter (as shown in [Tab. 4-1] and [Fig. 4-8]). This is mainly because with the increase of cone diameter, the line width per unit thickness of the ice sheet increases, and the fracture length of sea ice increases. At the same time, the dispersion of fracture length increases (mainly for ice thickness of 5~40 cm).

[Tab. 4-1] Structural characteristic of different ice-resistant cones and the length-thickness ratio k, source: (Wang Guojun et al., 2020)

cone	fundamental frequency /Hz	Waterline stiffness /()	Maximum cone diameter /m	k (L/h)
JZ202 MUQ (Four leg cone)	0. 87	4. 5 × 107	4	7.16
JZ202 NW (Single leg cone)	1	2. 54 × 107	6	6.28
HSVA Test (Indoor test cone)	—	—	0.61	5.29
ZH WTF (Offshore wind power)	0.2~0.3	2. 66 × 107	9.2	9.19

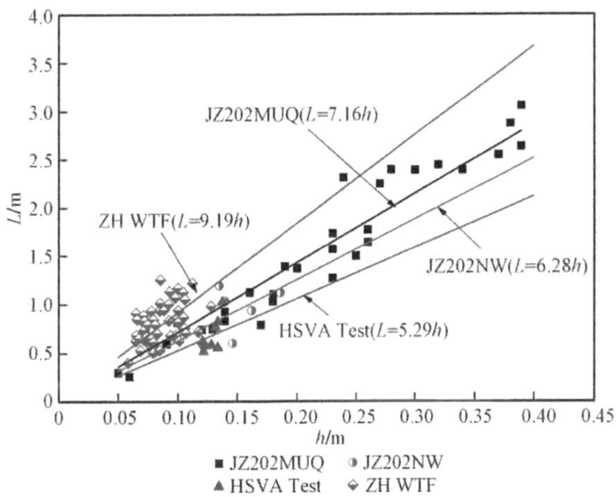

[Fig. 4-8] Length-thickness ratio of the frontal sea ice bending failure of different scales, source: (Wang Guojun et al., 2020)

Because the width of the cone influences the length-thickness ratio coefficient obtained by fitting, Xu Ning introduced a dimensionless parameter, the width-thickness ratio D/h (D is the diameter of the cone where the sea ice acts), into the analysis of the ice cone(XU Ning, 2011). Let the coefficient $q = D/h$;

Therefore, this coefficient is selected to analyze the variation law of width thickness ratio. Through video data analysis, it is found that when the sea ice acts on the wind reactance ice cone when the ice thickness is less than 5 cm, the dispersion of the sea ice breaking length is large, and multiple breaks occur, breaking into small pieces of ice, or splitting failure occurs. Therefore, the measured data with $q < 160$ is selected for analysis.

Through data analysis, it is found that the coefficient k increases rapidly with the width thickness ratio when the width thickness ratio q is small and increases slowly with the width thickness ratio when the width thickness ratio q is large. Therefore, the logarithmic equation is used for analysis:

$$k = a + b\ln q$$

From the analysis of measured data, it is found that the length thickness ratio coefficient k is greater than 4, ISO19906(ISO 19906, 2010), and the length thickness ratio coefficient k recommended is 4~10. Therefore, let $a = 4$ in the above equation, and fit the relationship between length thickness ratio coefficient k and width thickness ratio coefficient q according to the measured data, to obtain the relationship between the average period of sea ice breaking in front of the cone structure, ice thickness and ice speed:

$$k = 4 + 0.98\ln q$$

$$\overline{T} = \frac{kh}{v}$$

According to the data of ice thickness and ice speed measured on site, the above-average ice force cycle relationship is used to calculate the ice force cycle, which is compared with the measured ice force cycle. The results are shown in [Fig. 4-9] and [Fig. 4-10].

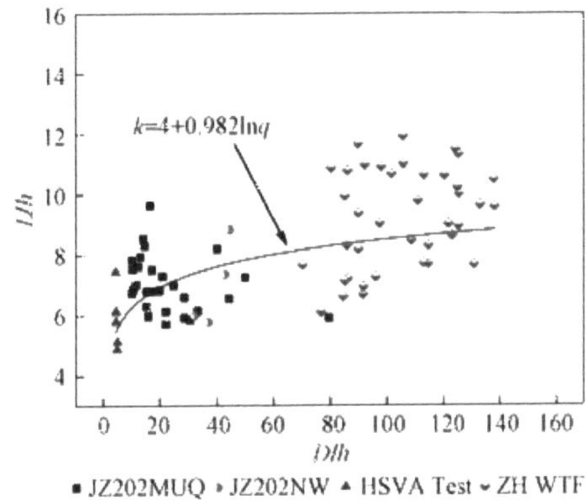

[Fig. 4-9] The length-thickness ratio versus the width-thickness ratio q, source: (Wang Guojun et al., 2020)

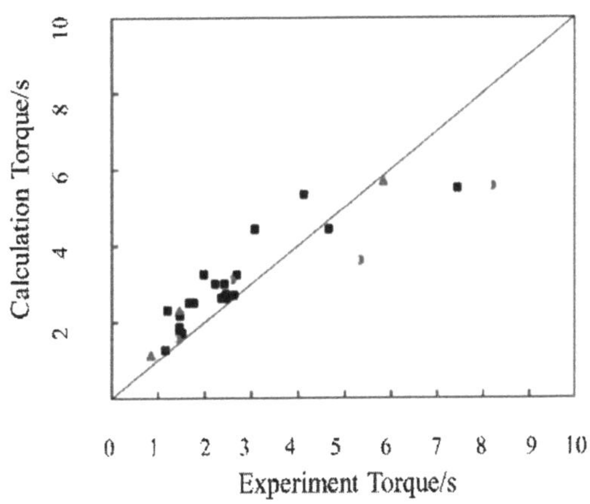

[Fig. 4-10] The calculated results versus the measured data of the ice period, source: (Wang Guojun et al., 2020)

The calculated average period of the whole is slightly larger than the measured ice force period. The measured ice force period is obtained based on the analysis of pressure box data. Due to the non-uniformity and secondary fracture of sea ice, the extracted ice force period is smaller. The above expression is used to calculate the average ice force period to meet the engineering requirements.

4.4.2 Analysis of ice force amplitude characteristics

The average ice force $\overline{F}$ of sea ice is related to the extreme ice force. Qu et al.(Qu Yan et al., 2006) obtained $\overline{F}= F_{max}/1.8$ through statistical analysis of measured data, where F_{max} is the extreme static ice force on the cone structure.Li Feng, et al(LI Feng & Yue Qianjin, 2001)、Yu Yonghai, et al(Yu Yonghai & Yue Qianjin, 2000)、Xu Ning (Xu Ning, 2001) studied the extreme static ice load model on the cone structure, analyzed the applicability of the existing static ice load model on the Bohai Sea plus cone jacket platform, and obtained the Hiryama Obara formula, which is suitable for the analysis of extreme static ice load on the narrow cone structure:

$$F_{max} = B\sigma_f h^2 \left(\frac{D}{L_c}\right)^2$$

where: σ_f is sea ice bending strength , by analyzing the physical and mechanical properties of sea ice in the wind farm sea area, the average bending strength is 750 kPa; is the empirical coefficient.

Floating system

5

5.1 Single-column foundation (Spar)

5.2 Semi-submersible foundation (Semi-sub)

5.3 Tension leg foundation (TLP)

To tackle climate change and promote global climate governance, China committed in June 2015 to reduce carbon dioxide emissions per unit of GDP by 60%~65% in 2030 compared with 2005(Dongfang Wind Power Network, 2022). The energy strategy has risen to the national strategic level. To achieve this strategic goal, China needs to vigorously develop clean renewable energy such as wind power.

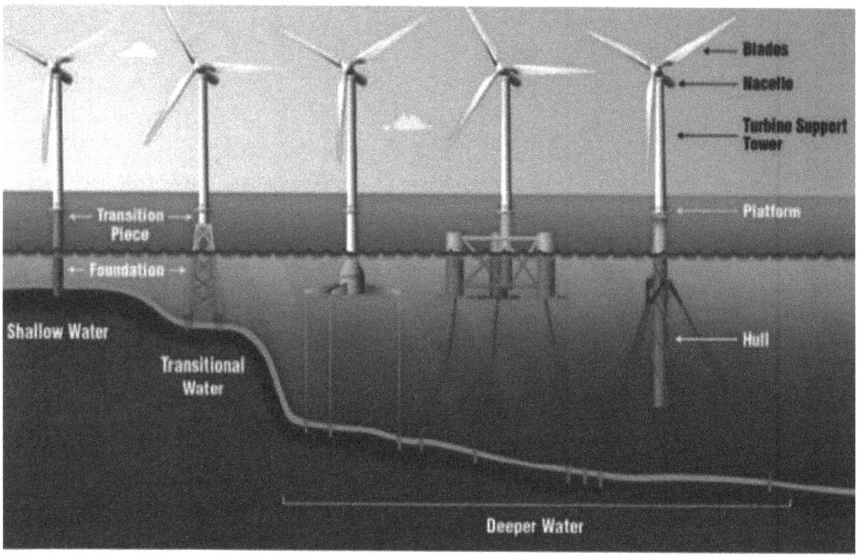

[Fig. 5-1] Source:(Dongfang Wind Power Network, 2022)

Wind power is a renewable and clean energy, which has the following advantages: (1) Environmental protection and energy conservation. It is pollution- free and renewable, which can reduce the consumption of nonrenewable resources such as coal and oil. (2) The construction technology of onshore wind farms is mature, the construction period is short, and small wind turbines can supply power flexibly. (3) Large-scale application technology has been mature, and the power grid has a strong ability to accept wind power. (4) Relatively low cost in renewable energy (ZHOU Xuhong et al., 2020).

Relevant research and engineering practice show that(XU Liangliang & HU Xiaozhen, 2018), (CHI Yongning et al., 2016), (WEN Feng, 2016b), (YANG Fang et al., n.d.), (LIN Heyun et al., 2011), compared with onshore wind power, offshore wind power has more obvious advantages: (1) wind turbine power generation is higher. The power generated by the wind turbine is proportional to the 3rd power of the wind speed, and the wind speed at sea is significantly higher than that on land. (2) The fan operates more stably. The terrain on land fluctuates, and the wind speed and direction change frequently, which is easy to cause damage to the drive system of the fan; The change frequency of wind speed and direction at sea is lower than that on land, and the wind is more stable. (3) The construction of offshore wind farms does not require land occupation.

China has a coastline of more than 18000 km and an ocean area of about 3 million km^2, which has the natural advantage of developing offshore wind power. The regions with the richest offshore wind resources in China are adjacent to economically developed regions with large power demand. Vigorously developing offshore wind power can realize the digestion of wind power energy nearby and effectively reduce power transmission costs.

The offshore wind turbine is mainly composed of fan blades, fan nacelle, fan tower, foundation structure, etc. For floating foundations, a mooring system should also be provided. It is the key to ensuring the safety, stability, reliable operation, and economy of offshore wind turbines to select different foundation structures according to the different hydrological environments, seabed soil conditions, and use requirements of offshore wind farms.

In shallow sea areas with a water depth of less than 50 m, fixed foundation structures are generally used. At present, the common fixed foundation structures in projects mainly include pile foundations, gravity foundations, and barrel foundations. In deep sea areas with a water depth of more than 50 m, if the fixed infrastructure is used, the cost will increase significantly, and it is difficult to achieve technically at present. Compared with fixed infrastructure, floating infrastructure has the advantages of good mobility, easy disassembly, and recycling after service.

Floating foundations can be divided into TLP, Spar, Semi, and Barge according to different water depths, weather, and installation requirements. They are box platforms floating on the sea and anchored to the seabed with an anchoring system. They apply to waters larger than 50 m. They are one of the trends in the development of wind turbine foundations in the deep sea in the future (JIA Xiaohui & GUAN Hui, 2017). In China, this type of infrastructure is still in the stage of learning and accumulating experience. At present, it has not been used in offshore wind power projects. In other countries outside China, this type of infrastructure is already in the development stage. For example, Blue H Technologies of the Netherlands developed the world's first floating wind power generator (80 kW) with offshore oil well technology in 2008, which is installed in the wind farm off Puglia, southern Italy; Hywind in Karmoy Sea Area, Norway in 2009 (using Spar foundation); Wind Float of Portugal Price Power (2011); The University of Maine (2013) built a 1:8 large-scale prototype model and carried out tests on the sea. The foundation is equipped with three anchors and three buoyancy tanks; In 2013, Japan built two prototypes: a semi-submersible and spar foundation; In 2017, Norske Petroleum installed five 6 MW wind turbines in Pilot Park floating wind farm in Scotland. The foundation was spar foundation and fixed to the seabed by an anchor chain.

In deep sea areas, a wind turbine foundation is called floating structure support, including a floating structure and anchoring system. When the water depth is greater than 50 m and it is not economical to use other forms of foundation, floating structure support can be considered. The floating structure is a closed box floating on the sea, on which the fan tower is fixed. Different shapes are adopted according to different anchoring systems, generally rectangular, triangular, or circular. The anchoring system mainly includes a submarine bucket foundation and connecting equipment. The connecting equipment can be generally divided into anchor rods and anchor chains. The anchoring system is correspondingly divided into a fixed anchoring system and a catenary anchoring system. In comparison, the fixed anchorage system is better than the catenary anchorage system, because the former can make most of the floating structure sink below sea level, minimize the wave load, and enhance the stability of the floating structure. The floating structure support also includes tension leg type and ship type. The tension leg is fixed on the sea floor by chain tension and becomes a floating structure support that can not be moved or moved. The floating body is in a semi-submersible state by operating the tension connection equipment, and its stability is better than the ship-type support. These platform structures are commonly used for oil production platforms and have not been applied in offshore wind farms so far. However, when the shallow sea area is developed and the wind farm is developed into the deep sea, the floating support mechanism will have broad application prospects(XU Rongbin, 2011).

Scope of application of offshore wind power foundation: In the face of complex and rapid development, a variety of foundation design schemes can be described as endless. Many factors should be considered when selecting the foundation of the offshore wind turbine. First, the basic investment in offshore wind turbines accounts for 20%~30% of the total investment. Secondly, the marine environment is quite different from the land environment, and construction is relatively difficult. The floating foundation is composed of a floating body and a tension tendon system. The upper part of the floating body is connected with the fan tower, and the lower part extends several cantilever pontoon structures to provide buoyancy and connect with the tension tendon. The tension tendon is a polyester cable, and the lower part is connected with the vertical stressed anchor. At present, an experimental unit floating foundation is put into operation 10 km away from the southwest coast of Norway. According to the company that developed the Hywind project, the Hywind wind turbine can be used in the sea area with a water depth of 120 m ~700 m.(SUN Xuting et al., 2019).

The typical foundation structures of a floating fan mainly include a Spar structure, Semi-submersible structure, and TLP structure, as shown in the figure below (CCS, 2021).

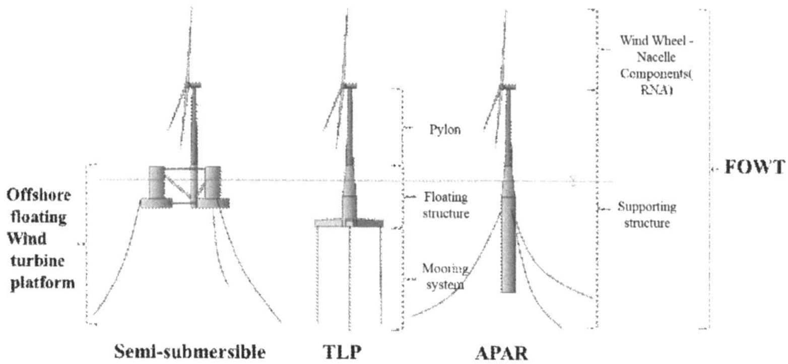

[Fig. 5-2] Schematic diagram of offshore floating fan facilities

(a) Spar

(b) TLP

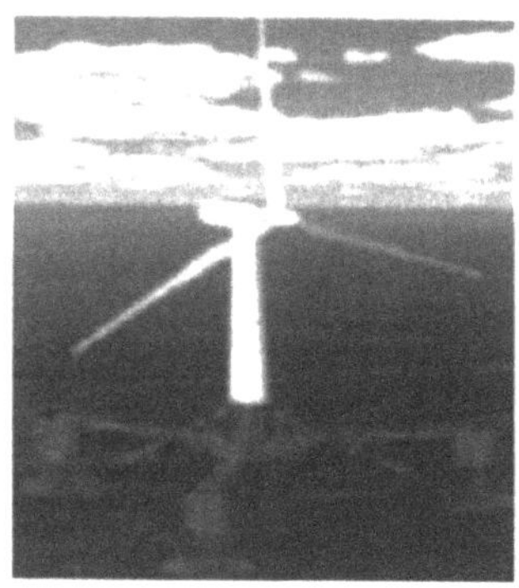

(c) Semi

[Fig. 5-3] Floating offshore wind turbine infrastructure

5.1 Single-column foundation (Spar)

Single column platform is a typical floating platform used in deep water (Dinh & Basu, 2013), (Karimirad & Moan, 2012). With the development of the ocean, many new floating offshore platforms have emerged, including the Spar platform. Spar platform also belongs to the category of compliant platform. Since the 1980s, the Spar platform has been widely used in the cause of human development of the deep sea, undertaking various tasks such as drilling, production, offshore crude oil processing, oil storage, loading, and unloading. It is regarded as the development direction of the next-generation deep-water platform by many oil companies.

The spar platform technology has been applied to human ocean development for more than 30 years, but before 1987, in the work of human ocean development, the spar platform was always used as an auxiliary system rather than a direct production system. They are used as buoys, marine scientific research stations, or marine communication transfer stations, and sometimes as offshore loading and unloading and storage centers. The structure of the early-built spar platform is different from that of the current spar platform used in deep-sea oil and gas development. However, through the observation of these early spar platforms, researchers in various countries have collected a lot of data, laying a solid foundation for the birth and development of modern spar platforms.

In 1987, Edward E. Horton designed a special spar platform for deep-water drilling and oil production, and its structure is particularly suitable for deep-water operation environments. The spar platform designed by Horton is recognized as the ancestor of the modern spar production platform.

A spar platform is superior to other floating platforms in terms of economy and stability. In the past two decades of development, three generations of spar platforms have been developed, including Classic Spar, Truss Spar, and Cell Spar. According to Offshore's statistics, as of May 2010, there were 17 Spar platforms in the world, 16 of which were operated in the Gulf of Mexico in the United States, and only one (Kikeh) was in Malaysia.

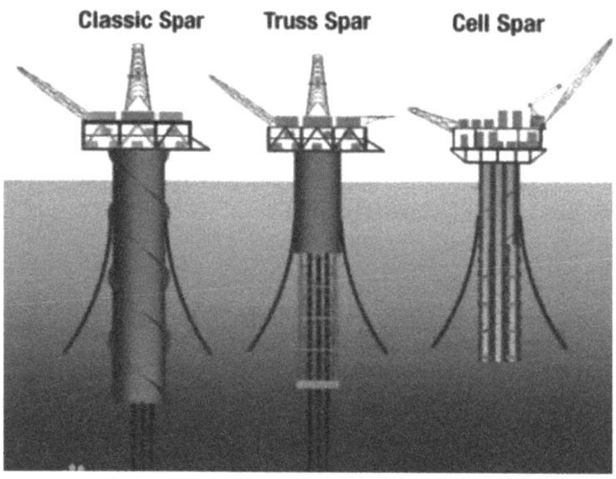

(a) single column foundation

(b) Single column foundation

[Fig. 5-4] Single-column foundation

Spar structure is a slender single-column structure, and the bottom can be ballasted. The mooring system is a combination of catenary and wire rope. The wind turbine is installed at the top of the floating body. Due to the huge thrust and torque in the horizontal direction of the fan, a large static angle will be generated during normal operation(Fig. 5-5). As the waterline area is small and the draft is large, the stability of the spar-type floating fan is generally good, and it is conducive to improving the overall performance of the floating fan in deep water. However, the main body of the spar structure is too long, which will lead to difficulties in construction and installation.

As shown in [Fig. 5-5(a)], the single-column floating foundation mainly consists of long columns and mooring systems. The basic stress principle is that the center of gravity of the whole structure is far lower than the center of buoyancy. When the structure inclines under the action of horizontal load, the buoyancy will produce a large restoring moment to the center of gravity, as shown in [Fig. 5-5(b)]. Single-column floating foundation structure has the advantages of simple structure, small vertical wave excitation force, good stability, etc., but its column length is generally long, which makes it difficult to install offshore. Single-column floating foundation structures can be used in ultra-deepwater areas above 100 m.

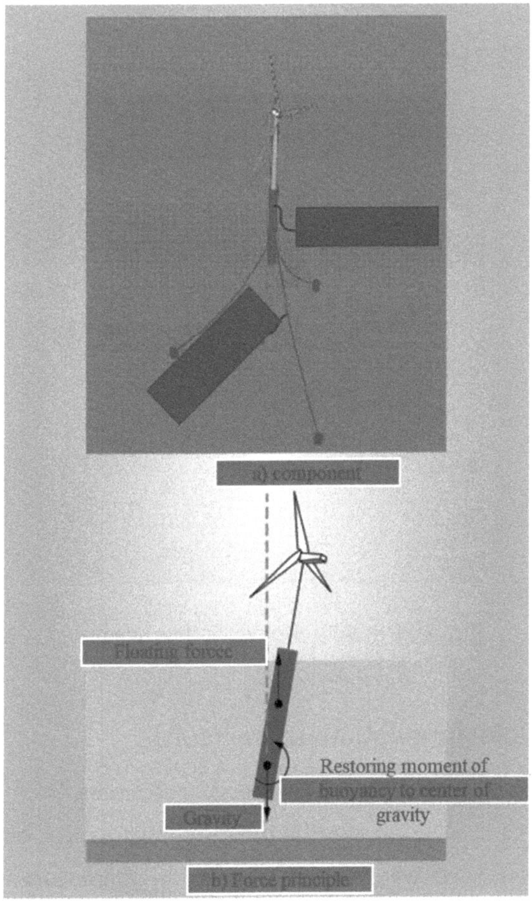

[Fig. 5-5] Spar floating foundation structure, source: (ZHOU Xuhong et al., 2020)

In 2017, the British Hywind offshore floating wind farm was officially put into operation, and the wind farm adopted a single-column floating foundation structure. [Fig. 5-6] shows the process of factory fabrication, on-site installation, and construction completion. The site is located 25 km away from the northeast coast of Scotland. It has five fans with a unit capacity of 6 MW. The fan blade diameter reaches 154 m. The total investment of the project is 210 million pounds. The water depth of the site is 95~120 m.

(a) Made in factory (b) Construction and installation on site

(c) Construction completed

[Fig. 5-6] Hywind offshore floating wind farm in the UK, source: (ZHOU Xuhong et al., 2020)

5.2 Semi-submersible foundation (Semi-sub)

There are abundant offshore wind energy resources, and the development of offshore wind power is of great significance for improving China's energy structure. At present, most offshore wind farms use fixed foundations, but in the sea area with a water depth greater than 50 m, fixed foundations are no longer applicable due to high cost. Therefore, in recent years, researchers have proposed a variety of conceptual design schemes for floating fan foundations (Musial, 2007), among them, the semi-submersible floating fan platform has attracted much attention due to its good stability, a wide range of applicable water depths and convenient construction and transportation. Henderson et al. (Henderson & Witcher, 2010) and Thiagarajan(Thiagarajan & Dagher, 2014) summarized the development of different types of floating fans, and Gao Wei introduced the research status of floating fan projects in different countries. However, there is no systematic summary and analysis of the conceptual design and performance research of semi-submersible floating fan platforms.

(1) Brief introduction

At present, most of the Semi structures studied are buoy combination structures, and the mooring system is fixed in a way similar to offshore oil platforms (Fig.5-7) (ZHOU Xuhong et al., 2020) , (LIU Zhenya, 2015).

[Fig. 5-7] Fixing method of the mooring system

The semi-type floating fan has strong wind wave resistance and a wide range of applicable water depth, but its stability is lower than that of spar type floating fan. Semi-type floating fans can be built on the wharf as a whole, so their offshore installation cost is low but their manufacturing cost is high. The key to the design of the integral structure is to reduce the motion response and reasonably select the mooring system to ensure the normal operation of the fan.

The concept of the offshore floating fan was proposed by Heronemus (Heronemus, 1972) as early as 1972. However, due to the constraints of technology and cost, it began to develop gradually in recent years. All countries are actively developing offshore floating fan foundations, and have made some progress in conceptual design, numerical simulation, and model testing. The report of the European Wind Energy Association (Arapogianni et al., 2013) pointed out that, as of the beginning of 2013, there were 2 full-scale and 3 experimental floating fans in the world that had been connected to the grid for power generation, and 35 floating fan projects were in the development stage, including at least 12 semi-submersible floating fan projects, as shown in [Tab. 5-1].

[Fig. 5-8] Semi-type floating wind turbine

[Tab. 5-1] Semi-submersible floating fan project

Project name	Country/area	Progress
WindFloat	USA	2010, Full-scale model test
Mitsui Semi-sub	Japan	2013, Full-scale model test
DeepCwind	USA	2013, 1/8 model test
HiPR Wind	EU	2013, 1/10 model test
Winflo	France	2014 Full-scale model test
Vertiwind	France	2014 Full-scale model test
Mitsubishi	Japan	2015 Full-scale model test
DIWET Semi-sub	France	Conceptual design stage
Tri-Floater	Netherlands	Water tank test stage
Hitachi Semi-Sub	Japan	Conceptual design stage
Shimizu Semi-sub	Japan	Model test stage
Windsea	Norway	Model test stage

(2) Structure

As shown in [Fig. 5-9 (a)], a semi-submersible floating foundation structure mainly consists of a central short column, edge short column, beam, diagonal brace, water pressure plate, and mooring system. Its basic stress principle is: when the structure inclines under the horizontal load, the buoyancy of the short column at one side edge increases, and the buoyancy of the short column at the other side edge decreases, and the buoyancy difference produces a restoring moment to the center of gravity, as shown in [Fig. 5-9 (b)]. Semi-submersible floating foundation structure has the advantages of a wide range of applicable water depth and small installation difficulty, but its stability is lower than that of single column foundation structure.

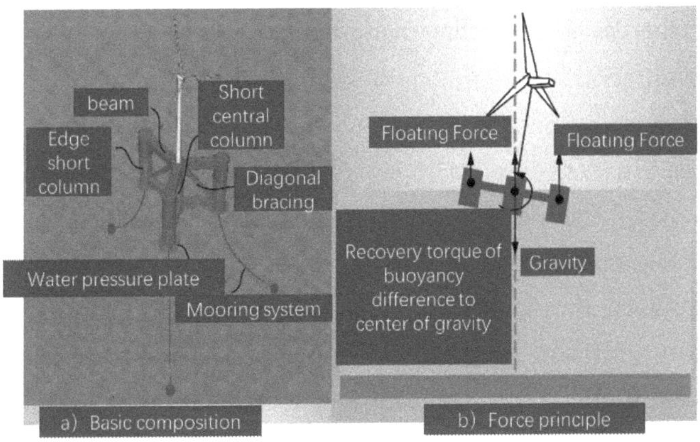

[Fig. 5-9] Semi-submersible floating foundation structure, source: (ZHOU Xuhong et al., 2020)

In 2011, the Portuguese Wind Float floating wind turbine test platform was put into operation, and the foundation structure was semi-submersible. The test platform has a wind turbine with a capacity of 2 MW(All Daily, 2017). In 2013, Fukushima Offshore Floating Wind Farm in Japan was put into operation, as shown in (ZHOU Xuhong et al., 2020). The foundation structure is semi-submersible. The wind farm is located about 20 km offshore in Fukushima County and has a fan with a unit capacity of 2 MW (WU Shaobo et al., 2017), (People's Daily Online, 2013). As shown in [Fig. 5-10, the semi-submersible foundation can be installed on the shore with the wind turbine superstructure in advance, and then towed to the designated place as a whole, reducing the construction cost and difficulty.

[Fig. 5-10] Offshore floating wind farm in Fukushima, Japan, source: (ZHOU Xuhong et al., 2020)

(3) Code for the design of floating platforms

The development of floating fans can not be separated from the guidance of professional design specifications, but the structural design specifications in the field of floating fans are still very lacking at present, so many scholars are actively studying the design specifications and guidelines of floating fans while exploring new conceptual design schemes. Lefranc pointed out that there is manual operation during normal operation of the oil platform, and oil is usually stored in the platform, so the overturning of the oil platform may cause serious consequences such as oil leakage and even casualties, while the floating fan platform is not. Therefore, he proposed that the partial factor of environmental load should be reduced from 1.3 to 1.15, and the partial factor of material should be reduced from 1.15 to 1.1 in the design of the floating fan platform. Ronald 等(Knut O. Ronold et al., 2011) put forward the key problems to be solved in the development of the specification from the aspects of design criteria and safety level, site conditions, loads and load models, stability, and mooring systems. Cordle et al. (Cordle & Jonkman, 2011) pointed out that the current design standards and guidelines related to floating fans refer to the experience of fixed fans and floating oil platforms, evaluate the applicability of these specifications, and propose the key technologies to be solved in the development of floating fan specifications. Sirnivas et al. (Sirnivas et al., 2014) proposed that

IEC series standards can be used for structures above sea level such as fans and towers, ISO series standards and American API can be used for structures below sea level, and the design guidelines of classification societies such as ABS, BV, DNV, and GL can be used as a supplement to the specifications.

5.3 Tension leg foundation (TLP)

The main body of the TLP structure is connected to the seabed through the tension tendons with large vertical stiffness (as shown in [Fig. 5-11]). The TLP type comprises a floating foundation (platform) to carry the wind turbine as shown in [Fig. 5-11]. Unlike the spar type which needs to be assembled offshore, this TLP wind turbine may be assembled and commissioned onshore, thereby avoiding the logistical difficulties of offshore assembly. The floating platform is held in position by vertical tendons (also called tethers) which are anchored either by suction piles, driven piles, or a template foundation. The pretensioned tethers provide the righting stability.

[Fig. 5-11] TLP structure, Source:(Ma et al., 2021)

Since the overall buoyancy is greater than its gravity, the tension tendons need to provide greater pretension. Therefore, the TLP floating fan can fully resist the effects of various marine environmental loads such as wind, waves, and current. Its biggest feature is that the motion response of rolling, pitching, and heaving is very small, and it can effectively reduce the motion response of the fan caused by the movement of the floating foundation, thus improving the performance of the fan. However, the setting of top pretension and the design of tension tendons are particularly important for TLP floating fans. The higher the top pretension is, the better. However, excessive pretension will increase the requirements for the tensile tendon materials and the mechanical properties of seabed anchorage, and will also increase the cost. At the same time, the construction cost will also increase with the increase of water depth, which does not apply to deep water areas. Its natural period is small, so the influence of the local frequency vibration of the foundation on fan performance should be fully considered in fan design.

The following is from Wikipedia: https://en.wikipedia.org/wiki/Tension-leg_platform

A tension-leg platform (TLP) or extended tension leg platform (ETLP) is a vertically moored floating structure normally used for the offshore production of oil or gas and is particularly suited for water depths greater than 300 meters (about 1000 ft) and less than 1500 meters (about 4900 ft). The use of tension-leg platforms has also been proposed for offshore wind turbines.

The platform is permanently moored using tethers or tendons grouped at each of the structure's corners. A group of tethers is called a tension leg. A feature of the design of the tethers is that they have relatively high axial stiffness (low elasticity), such that virtually all vertical motion of the platform is eliminated. This allows the platform to have the production wellheads on deck (connected directly to the subsea wells by rigid risers), instead of on the seafloor. This allows a simpler well completion and gives better control over the production from the oil or gas reservoir, and easier access for downhole intervention operations.

TLPs have been in use since the early 1980s. The first tension leg platform was built for Conoco's Hutton field in the North Sea in the early 1980s. The hull was built in the dry-dock at Highland Fabricator's Nigg yard in the north of Scotland, with the deck section built nearby at McDermott's yard at Ardersier. The two parts were mated in the Moray Firth in 1984.

The Hutton TLP was originally designed for a service life of 25 years in Nord Sea depth of 100 to 1000 meters. It had 16 tension legs. Its weight varied between 46,500 and 55,000 tons when moored to the seabed, but up to 61,580 tons when floating freely. The total area of its living quarters was about 3,500 square meters and accommodated over 100 cabins though only 40 people were necessary to maintain the structure in place.

The hull of the Hutton TLP has been separated from the topsides. Topsides have been redeployed to the Prirazlomnoye field in the Barents Sea, while the hull was reportedly sold to a project in the Gulf of Mexico (although the hull has been moored in Cromarty Firth since 2009).

Larger TLPs will normally have a full drilling rig on the platform with which to drill and intervene in the wells. The smaller TLPs may have a workover rig, or with most recent TLPs, production wellheads located at remote drill centers subsea.

The deepest (E)TLPs measured from the sea floor to the surface are:

5,185 ft (1,580 m) Big Foot ETLP

4,674 ft (1,425 m) Magnolia ETLP. Its total height is some 5,000 feet (1,500 m).

4,300 ft (1,300 m) Marco Polo TLP

4,250 ft (1,300 m) Neptune TLP

3,863 ft (1,177 m) Kizomba A TLP

3,800 ft (1,200 m) Ursa TLP. Its height above the surface is 485 ft (148 m) making a total height of 4,285 ft (1,306 m).

3,350 ft (1,020 m) Allegheny TLP

3,300 ft (1,000 m) W. Seno A TLP

Use for wind turbines

Although the Massachusetts Institute of Technology and the National Renewable Energy Laboratory explored the concept of TLPs for offshore wind turbines in September 2006, architects had studied the idea as early as 2003. Earlier offshore wind turbines cost more to produce, stood on towers dug deep into the ocean floor, were only possible in depths of at most 50 feet (15 m), and generated 1.5 megawatts for onshore units and 3.5 megawatts for conventional offshore setups. In contrast, TLP installation was calculated to cost a third as much. TLPs float, and researchers estimate they can operate in depths between 100 and 650 feet (200 m) and farther away from land, and they can generate 5.0 megawatts.

Computer simulations project that in a hurricane TLPs would shift 0.9 m to 1.8 m and the turbine blades would cycle above wave peaks. MIT and NREL researchers say dampers could be used to reduce motion in the event of a natural disaster.

MIT and NREL researchers plan to install a half-scale prototype south of Cape Cod. Sclavounos said, "We'd have a little unit sitting out there to show that this thing can float and behave the way we're saying it will."

Blue H Technologies of the Netherlands deployed the world's first floating wind turbine on a tension-leg platform, 21.3 kilometers (13.2 mi) off the coast of Apulia, Italy in December 2007. The prototype was installed in waters 113 meters (371 ft) deep to gather test data on wind and sea conditions and was decommissioned at the end of 2008. The turbine utilized a tension-leg platform design and a two-bladed turbine. Seawind Ocean Technology B.V., which was established by Martin Jakubowski and Silvestro Caruso -the founders of Blue H Technologies, acquired the proprietary rights to the two-bladed floating turbine technology developed by Blue H Technologies.

6

Pile supported the foundation

6.1 Single pile foundation(Monopile)

6.1.1 Brief introduction and structure diagram of single pile foundation

The single pile foundation, namely the single steel pipe pile foundation, is composed of a steel pipe pile with a diameter of 3~7 m. It applies to water areas less than 25 m. Its structural characteristics are simple structure and clear stress. For soft soil foundations, the hammering pile sinking method can be adopted, for rock foundations, the drilling method can be adopted, and large-diameter bored piles can also be formed in rock foundations. Due to its simple production process, low construction cost, easy control of the construction process, and rich experience of the construction unit, the foundation has become the mainstream infrastructure of offshore wind turbines. At present, this type of foundation structure is widely used at home and abroad, such as China's Yancheng Dongtai Project (2016) and Dalian Zhuanghe Project (rock drilling method) (2017); Sheringham Shoal, UK (2012); Denmark's homes Rev (2009); Arklow Bank of Ireland (2003); WP Q7 (2005) and other projects in the Netherlands (JIA Xiaohui & GUAN Hui, 2017).

A single pile foundation is the simplest foundation structure, as shown in [Fig. 6-1]. It is composed of welded steel pipes. The connection between the pile and the tower can be either welded flange connection or a casing flange connection. The single pile foundation transmits the fan load through the pressure of the side soil. The insertion depth depends on the geological type of the seabed. It is generally 10~20 m deep under the seabed. The diameter depends on the load, generally 3~5 m, and the wall thickness is about 1% of the pile diameter (XU Rongbin, 2011).

Single pile foundation (see (XU Rongbin, 2011)) is now widely used in built offshore wind farms. Single pile foundation is particularly suitable for shallow and medium water depths with good bearing stratum. The advantages of a single pile foundation are simple structure, simple and fast construction, low cost, and strong adaptability of the foundation; The disadvantage is that when the seabed is relatively hard, the pile foundation should be installed on the seabed mud surface using the drilling technology, which has a high cost. The foundation is

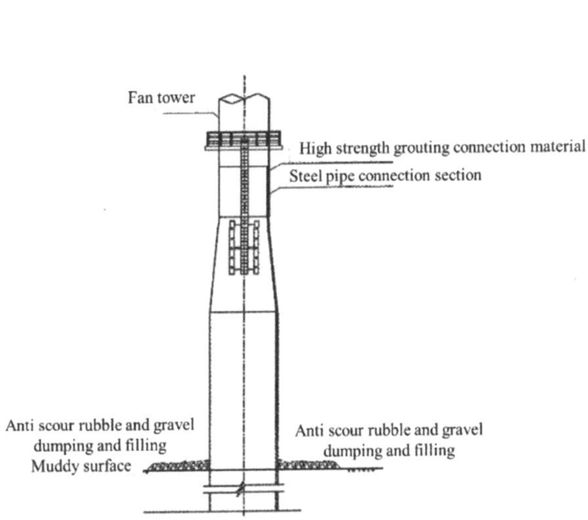

[Fig. 6-1] Schematic Diagram of Single Pile Foundation, Source: (XU Rongbin, 2011)

suitable for use in the sea area with a shallow depth of 30 m and a good bearing stratum. In the offshore waters of some European countries, the existing wind farms such as HomsRev, Samso, Utglnden, Aiklow Bank, Seroby Sands, and Kentish Flats all adopt the form of single pile foundation; Single pile foundation is also used in domestic wind farms in Jiangsu, Fujian and other places. Single pile foundation is the most widely used offshore wind turbine foundation in the world. As of the end of 2017, the cumulative data shows that there are 3270 wind turbines with a single-pile foundation in Europe, accounting for 81.7% of the total.

Physical structure diagram, as shown in [Fig. 6-2](HUANG Shen et al., 2018)

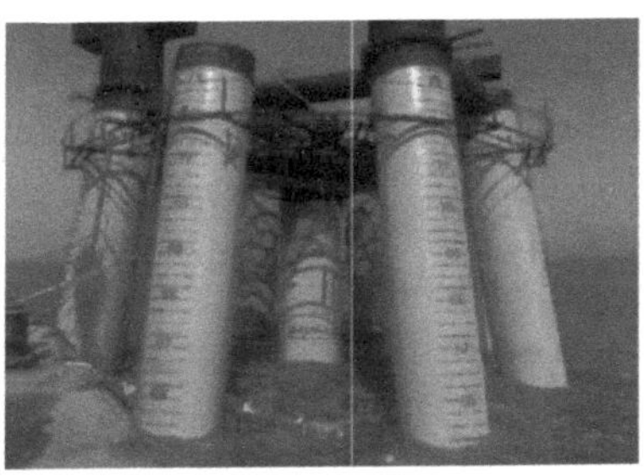

(a) The test platform of pile ZK01 and ZK 28

(b)The test platform of S1 and S2

[Fig. 6-2] Physical structure diagram Source: (HUANG Shen et al., 2018)

6.1.2 Scope of application of single pile foundation

The pile foundation is mainly used for most soils containing deep soft soil layers and applies to soft soil areas with a water depth of 10.0 m~25.0 m. As offshore wind farms in China are in offshore shallow water areas, single-pile foundations are more suitable in theory, but in practice, single-pile foundations are not widely used in wind power in China. The main reason is the lack of single-pile construction equipment, and the shallow water depth in the intertidal zone will also affect the construction ship.

6.2 Tripod foundation(Tripod)

Three steel pipe piles with a diameter of 1~3 m are located on the seabed and buried 10~20 m below the seabed. The tripod truss is a prefabricated component that bears the load of the upper tower and transmits the stress and moment to the three steel piles. The three piles are uniformly arranged as equilateral triangles, and the pile top supports the upper tripod truss structure through steel casing to form a composite foundation. The foundation has light deadweight and good overall structural stability and is suitable for water areas of 15~30 m. At present, some projects at home and abroad adopt a tripod foundation, such as the Nantong Rudong Project (intertidal zone) in China; 6 units of Multigrid in AlphaVentus (2010) of Germany; Nogersund (1990) and other projects in Sweden(JIA Xiaohui & GUAN Hui, 2017).

The tripod pile foundation (see Fig. 6-3) adopts a standard three-leg supporting structure, which is composed of a central column, three cylindrical steel pipes inserted into the seabed at a certain depth, and a diagonal bracing structure. The steel pipe pile is connected to the upper structure through special grouting or pile formwork. Vertical or inclined pipe sleeves can be used. The central column provides the basic support for the fan tower, similar to the single pile foundation. This kind of foundation evolved from the simplification of a single tower structure, and at the same time, it enhances the stiffness and strength of surrounding structures, which is common in the offshore oil and gas industry (XU Rongbin, 2011).

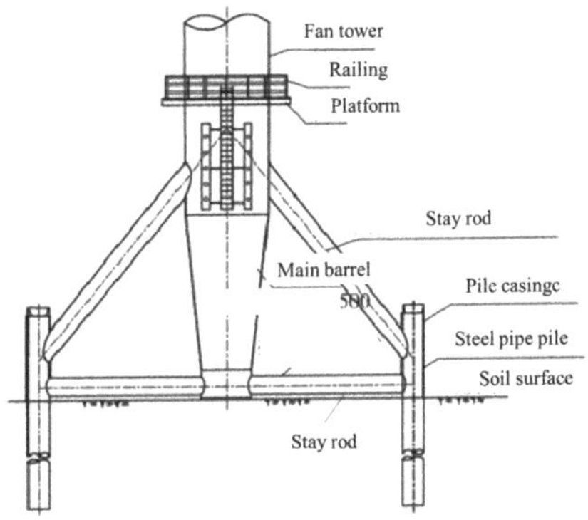

[Fig. 6-3] Schematic Diagram of Tripod Pile Foundation, source: (XU Rongbin, 2011)

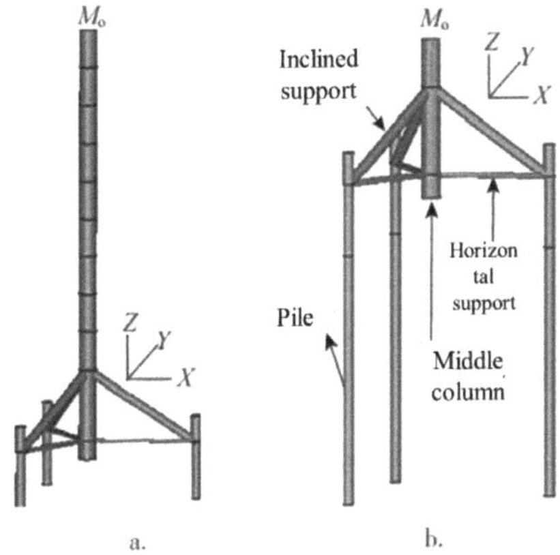

[Fig. 6-4] (a)Sketch of tripod-type and (b)tripod foundation of offshore wind turbine, source: (LI Wei et al., 2015)

According to the naming method determined by the Offshore Wind Working Group of the International Electrotechnical Commission (IEC) (IEC 61400-3 : Wind Turbines - Part 3: Design Requirements for Offshore Wind Turbines, International Electrotechnical Commission, 2005), an offshore wind power generation structure consists of two parts: a wind turbine generator (RNA) and a support structure (LI Wei et al., 2015). Supporting structure refers to the whole structure below the wind turbine, including the tower, substructure, and foundation.

The tripod foundation structure is similar to the simple platform for marginal oilfield development. The three piles are connected to the central column through a triangular rigid frame, and the wind turbine tower is connected to the column to form a structural whole, as shown in [Fig. 6-4a]. In this paper, the tripod foundation structure is roughly divided into four parts, namely, middle column, diagonal brace, horizontal brace, and pile, as shown in [Fig. 6-4b].

6.3 Jacket foundation(Jacket)

The jacket foundation is shown in [Fig. 6-5]. It is a steel cone platform-shaped space frame with steel pipes as ribs. The foundation is a three-leg or four-leg structure and is composed of cylindrical steel pipes. The jacket foundation form has been matured and applied in the construction of deep-sea oil production platforms, and can be popularized and applied to offshore wind power (XU Rongbin, 2011).

The jacket foundation is a steel cone platform-shaped space frame, with steel pipes as the skeleton, prefabricated on land, transported to the installation point, and driven into the seabed from the jacket. The foundation has high strength, lightweight, and large bearing capacity, and can effectively solve the underwater connection problem. It is suitable for water areas within the range of 5~50 m, but it is expensive, requires a lot of steel, and is vulnerable to weather and waves. A jacket foundation is one of the future development trends of deep-water wind farms. This type of foundation in China is still in the test stage, such as the Guishan project in Zhuhai, and it is also in the development stage abroad, such as the six Repower units in the German Alpha Ventus (2010) offshore wind farm; Two 5 MW units in the British Beatrice (2006) demonstration offshore wind farm, Ormonde (2012) and other projects(JIA Xiaohui & GUAN Hui, 2017).

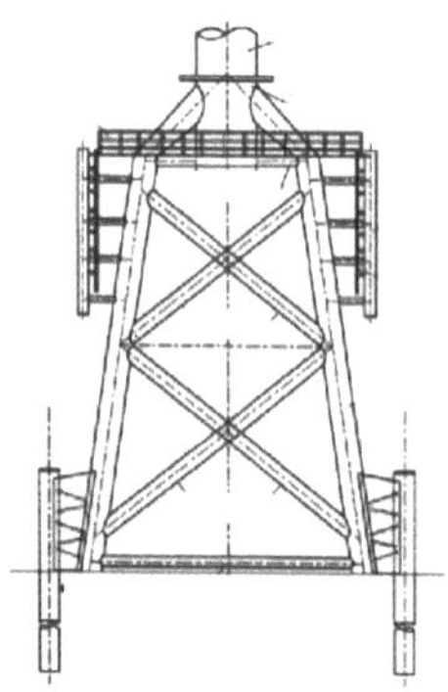

[Fig. 6-5] Jacket foundation, source: (XU Rongbin, 2011)

[Fig. 6-6] (Daily wind power, 2019)Source: http://www.chinawindnews.com/6313.html

The jacket foundation belongs to the "grid triangle foundation". The load of the jacket is borne by the piles driven into the foundation. The jacket foundation form has been quite mature in the construction of offshore oil drilling and production platforms and can be widely used in offshore wind power. The utility model has the advantages of high foundation strength, mature installation technology, and lightweight, and is suitable for large fans and deep-sea fields; The disadvantage is that a large amount of steel is required, the manufacturing time is long, the cost is relatively high, and the installation is seriously affected by the weather. It applies to the shallow sea area with a water depth of 300 m and does not apply to the situation where there are large areas of rocks on the seabed. It is one of the future development trends of wind farms in the deep sea (YUAN Ruhua et al., 2018).

6.4 Pile group cap foundation(Group pile cap foundation)

A Pile group foundation is a pile foundation composed of multiple piles. The pile group foundation can be regarded as a space rigid frame with a rigid cap. In design, the pile group is usually arranged with a vertical symmetry plane in both horizontal and vertical directions, and the load acts on the symmetry plane. At this time, each pile can be projected onto the symmetry plane and simplified as a plane rigid frame.

After the foundation is subjected to vertical load, its pile side resistance, pile tip resistance, settlement, and other properties change due to the interaction of pile cap, pile, and soil, which is different from that of a single pile. The bearing capacity is often not equal to the sum of the bearing capacity of every single pile, which is called the pile group effect. The size of the pile group effect is affected by soil property, pile spacing, pile number, pile forming method, and other factors.

It has the following characteristics:

1. The load transfer mode is not uniform.
2. The stress state of foundation soil is different.
3. The calculation of the vertical bearing capacity of pile groups has different meanings.

The most widely used pile group foundation is the high pile cap pile group foundation (with embedded foundation ring and embedded anchor bolt), which is composed of foundation pile and upper pile cap (including concrete pile cap and steel pile cap) and applies to a water area of 5~20 m. The inclined pile foundation piles are arranged along the circumference of a certain radius, which is favorable for resisting horizontal displacement and structural stress, but the pile foundation is relatively long, and the overall structure is thick and heavy. Since the wave has a large jacking force on the pile cap, it is necessary to take reinforcement measures for the connection of the pile cap. At present, this type of foundation is relatively common in China. For example, the Yancheng Xiangshui project uses the embedded foundation ring and embedded anchor bolt (test pile) high pile cushion cap foundation type, and the Shanghai East Bridge wind farm project uses eight steel pipe piles with a diameter of 1~3 m as foundation piles, which are uniformly arranged in a circular shape at the bottom of the cushion cap(JIA Xiaohui & GUAN Hui, 2017).

Gravity foundation(Gravity Based)

7

The gravity foundation makes the wind turbine stand on the sea surface through its mass. Compared with other foundation forms, gravity type foundation has a simple structure, little influence from the seabed, and good stability (SUN Xuting et al., 2019).

Gravity foundation is one of the main forms of offshore wind turbine infrastructure. It has been successfully applied in many offshore wind farms in Europe, showing great advantages in construction convenience and project cost control (YANG Wei et al., 2018). Gravity foundation mainly relies on its weight to make the fan stand vertically on the sea. It is generally a reinforced concrete caisson structure with a small bearing capacity and a simple manufacturing process. It is suitable for rock soil foundation and hard clay and sand. The foundation must have sufficient bearing capacity to support the self-weight of the foundation structure and environmental loads, and it should be used in water areas of 0-10 m. The early offshore wind power foundations are all this type. The early offshore wind farms in the world are all of the gravity types. Thornton Bank offshore wind farm is the first commercial offshore wind farm in the world that uses a gravity base. The water depth of the wind farm is about 12.0 m ~27.5 m. The wind farm uses a gravity base, reinforced concrete structure, and hollow, and its construction and transportation weight is about 1200 t. Fill with fine sand or crushed stone after installation, with a total weight exceeding 6000 t (SUN Xuting et al., 2019). China is still testing this type of foundation. For example, the foundation of two fans (composite barrel structure) in the Xiangshui project in Yancheng uses a combination of gravity foundation and suction foundation; LillgrundOresund (2008) in Sweden; Sprogo, Denmark (2009); Belwind (2011) and other projects in Belgium (JIA Xiaohui & GUAN Hui, 2017).

The gravity foundation (see Fig.7-1) is generally a reinforced concrete caisson structure. The fan is kept vertical by its gravity. To avoid the buoyancy between the foundation and the seabed, it needs to have enough weight. Therefore, the gravity foundation is the largest in volume and mass among all foundation types. To reduce the impact of ice load on the part in contact with the sea level, it can be designed as a cone (XU Rongbin, 2011).

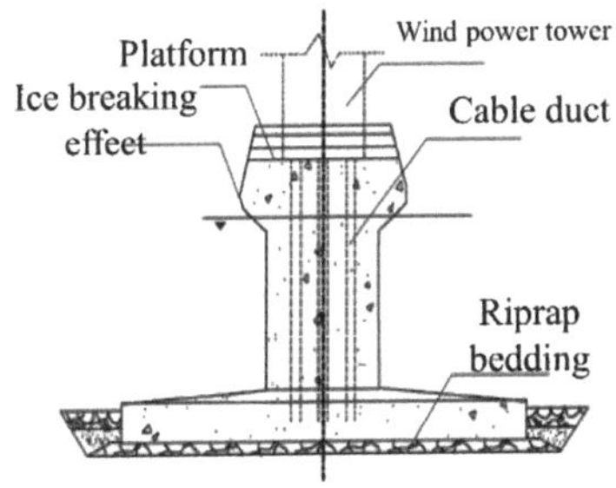

[Fig. 7-1] Schematic Diagram of Gravity Foundation, source: (XU Rongbin, 2011)

Gravity foundation is generally divided into concrete caisson type and gravity pedestal type (LU Jinliang et al., 2012). It is a foundation type suitable for shallow seas and with good seabed surface geology. It relies on its weight to stabilize horizontal loads such as wind load and wave load. Therefore, the gravity foundation has the largest volume and mass among all foundation types. It is generally prefabricated on land and then transported to the construction site by barge before being hoisted in place. This foundation is simple to install and requires less investment, but it has certain requirements for water depth, which is generally not suitable for wind farms with a water depth of more than 10m; There are also some restrictions on the geological conditions of the seabed surface, which is not suitable for the muddy seabed. At present, gravity foundations are applied to Middelgrunden, and Ronland wind farms in Denmark and Lillgrund wind farms in Sweden (YUAN Ruhua et al., 2018).

Scope of application. Gravity foundation applies to hard clay, sand, and rock foundation. The foundation must have sufficient bearing capacity to support the dead weight, service load, wave, and water load of the foundation structure. The foundation size is determined according to the bearing capacity of the foundation and the resistance required to resist sliding and overturning. The circular structure with a bottom is a circular caisson structure, and the bottomless structure is cylindrical. Before the circular structure is placed, a riprap bed with a certain thickness must be laid on the foundation. On the one hand, it plays a leveling role, and on the other hand, it can diffuse the stress of the structure on the foundation, reducing the foundation stress and reducing uneven settlement. The foundation can be prefabricated in the prefabrication yard and installed by a crane barge or semi-submersible barge. After being placed in place, sand or gravel is filled in, and then the concrete top plate is poured, and the tower foundation is installed on the upper part(LU Jinliang et al., 2012). In the area where the bottom foundation of the foundation can be exposed when the average tide level is near the shore, the installation vessel can not be in place due to the restriction of the draft, so the cast-in-place structure can be used. The circular structure bears less wave and current forces than the square structure. For bays with a small wave and current velocities, the foundation can also be square (LU Jinliang et al., 2012). It is mainly applicable to areas with good natural foundation, not suitable for soft foundation and scouring seabed; It applies to the water depth within 10 m and can be used within 20 m under the condition that the foundation is particularly good (LU Jinliang et al., 2012).

7.1 Caisson foundation

The caisson foundation, also known as the pneumatic caisson foundation, is the foundation of bridge piers and abutments or other structures built with pneumatic caissons.

The suction anchor foundation is divided into single buckets, three buckets, and four-bucket suction caisson structures. The suction anchor foundation pumps out the water in the steel skirt caisson to form a negative pressure to generate suction. This infrastructure was first used on offshore marginal oil platforms. The advantage of this foundation is that it can save a lot of steel consumption and offshore construction time, and has a good application prospect, which can be applied in both shallow sea and deep sea. At present, the foundation of the wind tower project of China Power is the suction anchor structure (YUAN Ruhua et al., 2018).

7.1.1 Introduction

A pneumatic caisson is a bottomless box structure. Because compressed air is required to provide working conditions, it is called pneumatic caisson or simply caisson.

The caisson foundation is the foundation for building bridge piers and abutments or other structures with pneumatic caissons. Caisson is composed of a top cover and a side wall, and its side wall is also called a blade foot. Holes are reserved on the top cover to install upward-connected gas cylinders (well pipes) and various pipelines. The upper end of the air cylinder is connected to an air brake. The airlock is composed of a central airlock, a human air transformer, and a material·air transformer (or feed barrel and outlet barrel). A cofferdam shall be installed on the top cover of the caisson or a permanent outer wall shall be built. The space under the roof is called the studio. Due to poor operation conditions of caissons, which are harmful to people's health, low work efficiency and high cost, and the human body can not withstand excessive atmospheric pressure, the caisson's water entry depth is generally controlled within 35m, which limits the buried depth of the foundation. Therefore, the caisson foundation is seldom used except in special cases(ZE Yankang, 1981).

7.1.2 Features and functions

Caisson is composed of a top cover and a side wall (Fig.7-2), and its side wall is also called a blade foot. Holes are reserved on the top cover to install upward-connected gas cylinders (well pipes) and various pipelines. The upper end of the air cylinder is connected to an air brake. The airlock is composed of a central airlock, a human air transformer, and a material air transformer (or feed barrel and outlet barrel). A cofferdam shall be installed on the top cover of the caisson or a permanent outer wall shall be built. The space under the roof is called the studio.

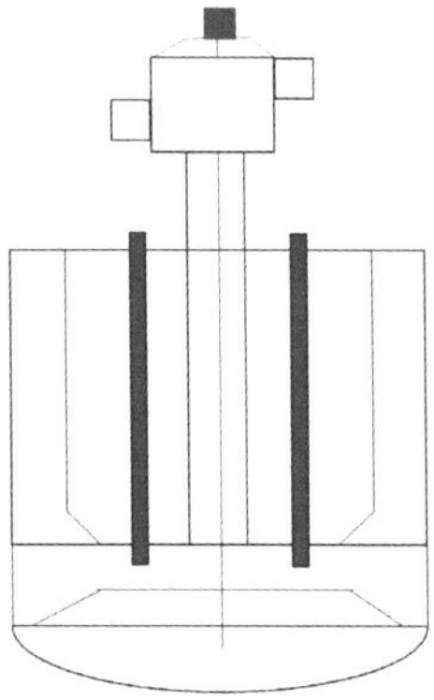

[Fig. 7-2] Structural Diagram of Caisson

When sinking the caisson underwater, use an air compressor outside the caisson to send compressed air to the airlock and caisson working chamber through the air reservoir and the oil separator through the air transmission pipe, and then press the water in the working chamber out of the room. The staff can enter the working room from the central airlock and the escalator in the air cylinder to work through the human air transformer. The function of the human air transformer is to gradually change the air pressure in the gate so that the staff can adapt to the air pressure difference between indoors and outdoors, and at the same time, prevent the high-pressure air from overflowing due to personnel entering and leaving the workshop.

In the caisson working room, the workers use excavation machines, hydraulic machines (including hydraulic mud flushes and suction dredgers), and other machines and tools to remove the earth and rock under the caisson, remove various obstacles, and make the caisson sink under its weight and the gradually increasing soil or other weight, overcoming the friction around it and the reaction of compressed air. After the caisson is lowered to the design elevation and the foundation is inspected and treated, the workshop is filled with soil, and the airlock cylinder is removed. Currently, the caisson becomes an integral part of the foundation. The necessary buildings can be built on it under the protection of cofferdams, such as bridge piers, underwater tunnels, subways, and other hydraulic and port structures(ZE Yankang, 1981).

7.2 Bucket foundation

Given the stress characteristics of offshore wind turbines bearing huge bending moments, Lian Jijian et al.(Lian et al., 2011),(Lian et al., 2012), (Ding et al., 2013) proposed a wide shallow composite cylindrical foundation with good bending resistance and suitable for soft clay foundation in China's coastal areas. Compared with the traditional tube foundation (small tube structure), the large-scale tube top-bearing tube foundation with prefabricated reinforced concrete as the main material is different in form and stress mode. In terms of size, the basic barrel of the general small barrel structure is about 10 meters high, and the outer diameter of the barrel is generally about 5 meters. Under the same tube height, the outer diameter of a large-scale tube foundation can reach more than 30 m, which is a typical wide shallow foundation (as shown in Fig.7-3). This new type of shallow and wide foundation determines its different bearing mode and failure mechanism from the traditional tube foundation: the bearing capacity of the traditional tube foundation mainly depends on the friction of the soil at the tube side, and the soil at the tube bottom does not work in the whole bearing system, even under the action of bending moment load, the traditional tube foundation is subject to the reaction of some passive soil; The large-scale tube top bearing type tube foundation, through certain foundation treatment methods, consolidates the soil inside the tube. The ultimate goal is to make the consolidated soil inside the tube and the tube foundation form a whole and bear the upper load together. The soil reaction under the "whole" plays a major role in the entire bearing capacity system, to achieve "tube top bearing".(YU Hongyan et al., 2013). This form of stress greatly improves its bearing capacity. In the selection of the force transfer transition section between the tower tube and the tube foundation, the new foundation selects the arc transition section, which is a new structure, and applies to prestress to it. The transition section effectively converts the huge bending moment of the fan tower into the limited tension and compression stress in the foundation structure through the arc structure, which not only enables the structure to have a good force transmission performance, but also avoids the problem of serious stress concentration in the traditional barrel foundation, and its crack resistance

[Fig. 7-3] Wind power structure system with composite bucket foundation, source: (YU Tongshun, 2016)

performance is extremely excellent, and its anti-corrosion performance is excellent. At present, the composite cylinder foundation has been successfully applied to the wind power development of Qidong Wind Farm in the Jiangsu sea area (YU Tongshun, 2016).

As an anchor and foundation type, bucket foundation has great application value in marine and port engineering, such as (1) offshore structure mooring system, such as single point mooring, towing platform, ship mooring, pipeline fixation, and towing; (2) Suction anchor; (3) Anti-slide pile of traction cushion platform; (4) Breakwater; (5) Platform foundations, such as jacket platforms and tension leg platforms; (6) Offshore wind power foundation, which is a new application. Compared with other offshore foundations, the bucket foundation mainly uses the pressure difference generated by the air/water pumped out of the cylinder to form suction (also called negative pressure when the pressure is lower than one atmosphere) to sink. It has the advantages of simple installation, no noise pollution, high anti-overturning bearing capacity, saving steel, and can be reused. It is expected to become one of the main foundation types of offshore wind turbines in the future.

7.2.1 Application

In the construction of offshore wind farms, bucket foundations are mainly applied to the foundation of wind measuring towers and offshore wind turbine structures, which can be roughly divided into three types: single-cylinder type, multi-cylinder type (three or four cylinders), and single cylinder multi cabin type developed by Tianjin University (research and development team of Professor Lian Jijian and Professor Ding Hongyan) and Daoda Company. Typical application engineering cases are shown in [Tab. 7-1] and [Fig. 7-4] to [Fig. 7-8].

[Tab. 7-1] Types of suction foundation and its typical applications in offshore wind turbine, source: (ZHANG Puyang & HUANG Xuanxu, 2018)

Classification	Typical Applications	Others
Single cylinder type: Mono Bucket Foundation	In August 2002, Frederikshavn Wind Farm in Denmark: 3 MW wind turbine, with a diameter of 12 m and a skirt height of 6 m. Clay, water depth 1~4 m. In 2005, Wilhelmshawen Wind Farm in Germany: a 4.5 MW wind turbine, with a diameter of 16 m and a skirt height of 15 m, was hit by a barge during installation, and it failed to buckle under negative pressure after penetrating 6. 8 m. Sand, water depth 18 m. In March 2009, HornsRev 2 Wind Farm in Denmark: anemometer tower, with a diameter of 12 m and a skirt height of 6 m. In July 2015, it was successfully recycled after use. Stratified soil, with a water depth of 9~17 m. In September 2013, Dogger Bank Wind Farm in England: 2 anemometers, 14 m in diameter and 8 m in skirt height. The water depth is 18 m. In September 2014, North Sea, UK, installed two tubular foundations (diameter 8 m, skirt height 6 m and diameter 4 m, skirt height 6 m). Three offshore wind farms, 24 days and 29 experiments, installation, and recovery. Lake Erie Wind Farm in America: 3 MW wind turbine, 18 m diameter, 12 m skirt height. Layered soil, water depth of 20 m, considering ice load, no construction started.	The self-floating installation is available, and the jet system ensures that the perpendicularity of the foundation is less than two thousandths.

Multicylinder type: Suction Jacket Foundation	In October 2014, BorkumRiffgrund 1 wind farm in Germany: 4 MW wind turbine, three-cylinder type, 8 m in diameter and 8 m in skirt height. Sand, water depth 25 m. In May 2016, Dudgeon Wind Farm in the United Kingdom, with a water depth of 18~25 m, installed two typhoon flashlight-type foundations with a diameter of 8 m, a height of 6 m, and a diameter of 4 m, a height of 6 m. In March 2018, Aberdeen Bay Wind Farm in Britain: two 8. 8 MW and nine 8. 48 MW fans, three barrel type, with a diameter of 9. 5~10. 5 m and a skirt height of 9~13 m. The water depth is 19~32 m. The project to be adopted in 2018: 57 7 MW wind turbines in Hornsea 1 Wind Farm, UK; 20 8 MW wind turbines in BorkumRiffgrund 2 wind farm in Germany.	After installation, grouting under the top cover of the cylinder can be used to ensure uniform contact between the cylinder top and the soil.
Single barrel multi cabin composite barrel foundation: Composite Bucket Foundation	September 2010, Qidong Sea Area, Jiangsu Province: 2.5 MW experimental prototype, foundation diameter of 30 m, concrete skirt height of 6 m, arc transition section height of 18 m, single barrel 7 steel cabin structure. In 2017, the Three Gorges Xiangshui Wind Power Plant in Jiangsu Province: two 3 MW wind turbine foundations, with a diameter of 30 m, a steel skirt height of 12 m, an arc transition section height of 23 m, a dead weight of 2500 t, and a single barrel 7 steel cabin structure. Verify the one-step whole machine transportation and installation technology, and the structural perpendicularity is 3/10000 after installation. In 2018, the Three Gorges Xiangshui Wind Farm in Jiangsu Province is planned to install 11 sets of 3.3 MW and 2 sets of 6.45 MW wind turbine foundations, with a diameter of 32~36 m, a steel skirt height of less than 11 m, and a single cylinder 7 steel cabin structure.	The multi-cabin structure in the cylinder can realize the technology of the foundation tower head complete machine transportation and sinking leveling.

(a) (b) (c) (d) (e) (f)

(a) Denmark,Frederikshavn (single cylinder type) **(b)** UK, Dogger Bank, anemometer tower (single cylinder multi arc type) **(c)** American, Lake Erie, (Single cylinder anti ice type) **(d)**German, Borkum Riffgrund 1 (Three barrel type) **(e)**UK, Abderdeen Bay (Three barrel type) **(f)** UK, Dudgeon Booster station (four barrel type)

[Fig. 7-4] Suction bucket foundations in the offshore wind farm, source: (ZHANG Puyang & HUANG Xuanxu, 2018)

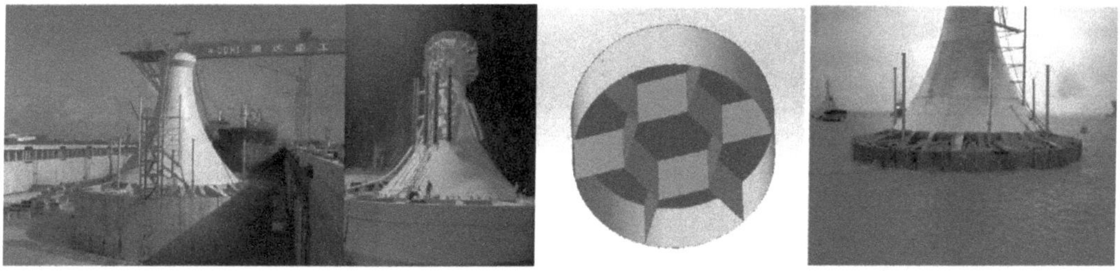

(a) (b) (c) (d)

(a) Concrete cylinder wall **(b)** Steel cylinder wall **(c)** Honeycomb like 7 compartment structure in the cylinder **(d)** Foundation self floating towing

[Fig. 7-5] Composite bucket foundation (CBF) with subdivisions, source: (ZHANG Puyang & HUANG Xuanxu, 2018)

(a) (b) (c) (d) (e)

(a)Batch prefabrication; (b)Foundation self floating;(c) Fan installation;(d) Integrated floating transportation;(e) Integral installation

[Fig. 7-6] One-step integrated installation technique with CBF, tower, and wind turbine, source: (ZHANG Puyang & HUANG Xuanxu, 2018)

[Fig. 7-7] Applications of suction foundations in UK Hywind Scotland floating wind farm, ource:(ZHANG Puyang & HUANG Xuanxu, 2018), (OVERVIEW -World's First Floating Wind Farm Moves towards Completion in Scotland, 2017)

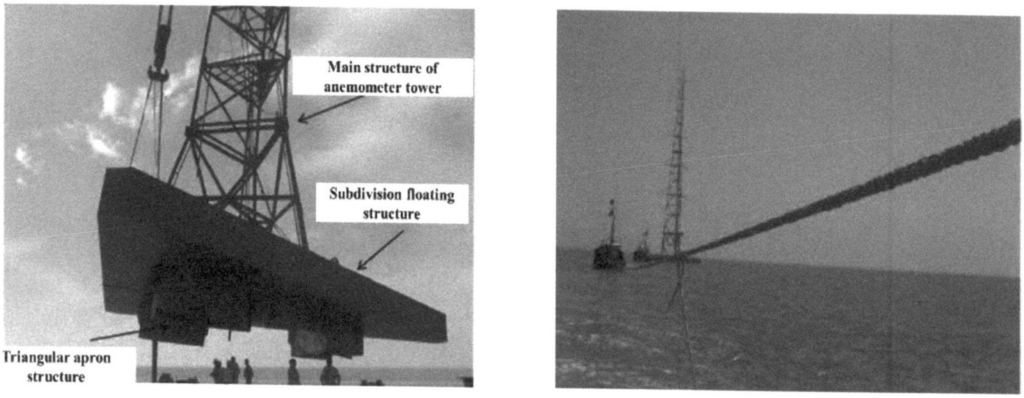

(a) Offshore anemometer tower bottom infrastructure; (b) Towing of offshore anemometer tower

[Fig. 7-8] One-step integrated installation technique of offshore anemometer tower, source: (ZHANG Puyang & HUANG Xuanxu, 2018)

The single barrel multi cabin composite barrel foundation is a wide and shallow foundation structure type of large-scale concrete steel plate reinforcement prestressed steel strand composite system (diameter: 25~40 m, height: 6~15 m), which organically combines the arc transition section and barrel foundation into a composite barrel foundation (weight: 2000~4000 t). The whole structural system effectively converts the huge bending moment of the fan tower into the limited tension and compression stress in the foundation structure through the prestressed concrete transition section. The composite structure of multiple materials effectively solves the deformation coordination and crack control of the steel-concrete structure, gives full play to the material advantages of the steel-concrete structure, and improves the safety and durability of the structure. The foundation cylinder wall can be divided into concrete and steel. The cellular subdivision structure in the cylinder can realize the self-floating towing and fine leveling of the foundation, as shown in Figure 2. This type of infrastructure can realize onshore batch prefabrication and offshore integrated installation, thus greatly saving offshore operation time, reducing production, transportation, and installation costs, and making it possible to develop offshore wind farms in an efficient, low-cost, and large-scale way. Its technological innovation ideas are shown in Figure 3. The foundation, tower, and 3 MW fan shown in the figure have been transported and installed in one step as a whole (overall towing for 72 h, 281 n mile, whole machine sinking for 8 h, levelness of 3/10000). At present, 11 sets of 3.3 MW and 2 sets of 6.45 MW composite tubular foundation wind turbines under construction will be installed in a one-step manner in Jiangsu Dafeng offshore wind farm of China Three Gorges Corporation this year.

As a suction anchor, it can also be used in the mooring system of the floating fan platform. For example, in July 2017, Norwegian National Petroleum Company UK built the world's first offshore floating wind farm, Hywind ScotlandPilot Park, which is composed of five 6 MW wind turbines. Each fan uses three suction foundations to provide the anchoring force, as shown in [Fig. 7-7].

The new integrated offshore wind power anemometer tower jointly developed by Tianjin University and DODA also uses a suction foundation for auxiliary sinking and leveling construction, as shown in [Fig. 7-8]. The combined structure system of offshore anemometers, which integrates the anemometer tower, floating structure, and suction skirt foundation structure, has been applied to the wind measurement projects of 8 offshore wind farms in Jiangsu and Hainan, with resource conservation, environmental friendliness, significant economic and social benefits, and the farthest towing distance of 350 n miles (Qidong Lianyungang).

In a word, the offshore wind turbine foundation can greatly improve the efficiency of offshore wind power construction and environmental friendliness. In particular, the single-cylinder multi-compartment offshore wind turbine composite cylinder foundation realizes the integrated transportation and installation of foundation tower wind turbine, which will significantly reduce the cost of offshore wind power construction and installation, and improve China's scientific and technological level and independent innovation ability in the field of offshore wind energy utilization. However, the soil conditions of offshore wind farms in China are very complex, and the surface soft soil, silt, sandy soil, and clay or layered soil are complex and diverse. How to effectively achieve the accurate control of the structural perpendicularity during the smooth sinking of the bucket foundation to the design depth and the sinking process, and avoid the installation risk caused by the soil plug in the bucket or the deformation or even buckling of the skirt and sub deck is bound to be the key problem in the application level of the bucket foundation.

7.3 Suction foundation

A suction foundation is also called a negative-pressure cylinder foundation. The foundation is divided into single-cylinder and multi-cylinder suction caisson foundations. It is a new type of offshore engineering foundation structure and applies to wind fields in various water depths with sandy soil or soft clay geological conditions. Because its material installation cost is lower than that of pile-supported foundations, and it is easy to transport and install at sea, it is favored by the offshore wind power industry. There are few application cases of suction foundations in offshore wind farm projects, such as Frederikshavn (2003) in Denmark. From this point of view, the development space of the suction foundation is large, but its reliability needs further verification (JIA Xiaohui & GUAN Hui, 2017).

The suction-type foundation can be used in shallow seas and deep seas. The suction-type foundation in the shallow sea is also called Suction Bucket, as shown in [Fig. 7-9]. This is a new concept of foundation structure, which is the combination of traditional pile foundation and gravity foundation. The so-called negative pressure refers to the method used to install the bucket foundation, with the purpose that the negative pressure effect can partially bear the dynamic peak load (XU Rongbin, 2011).

The suction cylinder foundation consists of a cylinder and an extension section. The cylinder is a cylinder with a bottom opening and a top seal, and the extension section is a gradual single pass whose diameter changes along a curve. The suction cylinder foundation commonly used in offshore wind power is generally in the form of reinforced concrete prestressed structure and steel structure. Compared with other foundation forms, the suction cylinder foundation has the following advantages: low cost, fast construction speed, recyclability, etc.

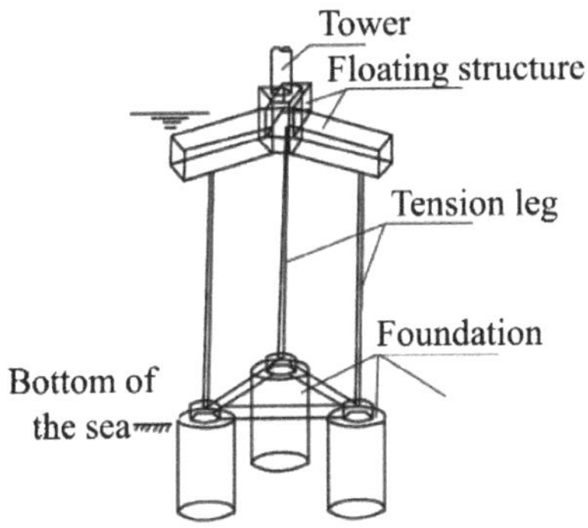

[Fig. 7-9] Schematic diagram of negative pressure bucket foundation, source: (XU Rongbin, 2011)

The development of suction tube foundation is inseparable from the development of skirt foundation, suction anchor, and suction pile (SHI Xiaochun et al., 2000),(HE Yanping & TAN Jiahua, 2002),(LU Xiaobing et al., 2003). In the 1970s, Norway's North Sea oil field began to use a skirt gravity foundation. The Ekofisk Doris oil storage tank installed by Phillips Oil Company in 1973 was the first skirt foundation structure. In 2002, Denmark installed a 12 m diameter bucket foundation for a V90 wind turbine in the construction of Frederikshavn offshore wind farm, which was first applied to large wind turbines(G. Houlsby et al., 2005).

With the vigorous development of offshore wind power, offshore wind turbines are gradually developing from shallow sea to deep sea. Compared with the offshore area, the deep sea area has the advantages of wider scope, richer wind resources, and no conflict with the channel. Therefore, many new basic forms have emerged. For example, the suction cylinder foundation is generally more suitable for the sea area with a water depth of 60.0 m. The self-weight and negative pressure penetration can greatly shorten the construction time and improve the construction economy.

8 Influence of wind load on offshore wind power foundation

8.1 Influence of wind load on single pile foundation

8.1.1 Numerical analysis of horizontal loading of monopile foundation of large-scale offshore wind turbine

Offshore wind energy offers a huge potential for the expansion of renewable energies in the world. As one foundation of offshore wind turbines, the monopile foundation has been the standard installation program of offshore wind turbines. To evaluate the horizontal ultimate bearing capacity of the monopile foundation, the behavior of large-diameter monopiles under horizontal loading induced by wind and wave loads is analyzed using a three-dimensional finite element model. The mechanics and stability of monopile foundations have been studied. It has been investigated the relationship between horizontal ultimate bearing capacity and horizontal displacement or angle of rotation. Those will be supplied important theoretical laws and references for designers and constructors(WU Ke et al., 2008).

(1) Finite element calculation model

According to the characteristics of marine soft soil, Mohr-Coulomb elastoplastic constitutive model was adopted, (WU Ke et al., 2008)Under saturated undrained conditions, soil parameters are $v = 0.49$, $r = \frac{16.0kN}{m^3}$, $S_u = 6$ OkPa ; Assuming that the deformation modulus is approximately proportional to its undrained shear strength, the fixed value $E = 500S_u$. For the single pile foundation structure, the linear elastic constitutive model is adopted, and the elastic constants are $E = 2.1 \times 10^5 MPa$, $v = 0.2$. For the calculation of the contact between the single pile foundation structure and the surrounding soil, the active-passive surface contact algorithm is adopted. The single pile foundation structure surface with high stiffness is selected as the active surface, the soil surface is the passive surface, and the interface sliding friction coefficient is taken as $u = 0.2$. According to the symmetry of the single pile foundation structure, half of the pile foundation and foundation are taken for analysis, and a finite element calculation model is established by using the

6-faceted 20-node element (C3D20R), as shown in [Fig.8-1]. In the example analysis, the diameter and height of a single pile are respectively $D = 7.5\,m$, the thickness of the single pile foundation structure $t = 9\,cm$, h is the distance from sea level to the seabed.

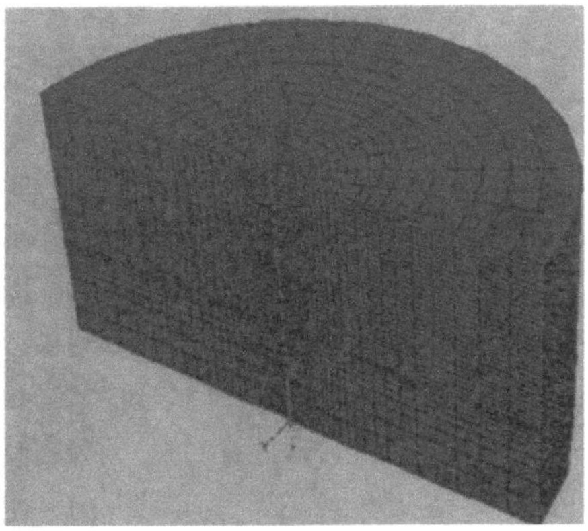

[Fig. 8-1] Finite element calculation model, source : (WU Ke et al., 2008)

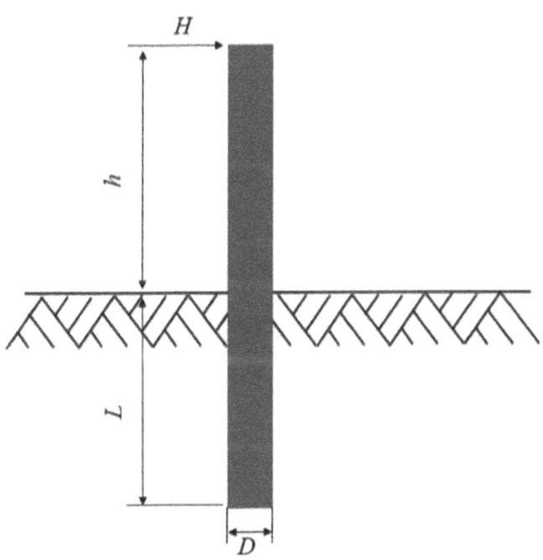

[Fig. 8-2] FEM method, source : (WU Ke et al., 2008)

(2) Study on bearing capacity characteristics of offshore wind turbine single pile foundation under wind load

For the horizontal bearing capacity characteristics of single pile foundations of offshore wind turbine generator units, based on two types of single pile foundations with lengths of $L = 30\ m$、 $L = 20\ m$ the finite element calculation of variable parameters is carried out for different cases of $h/L = 0.2$, $h/L = 0.5$, $h/L = 1.0$ respectively.

1) Working Mechanism and Failure Mode of Single Pile Tomb Foundation

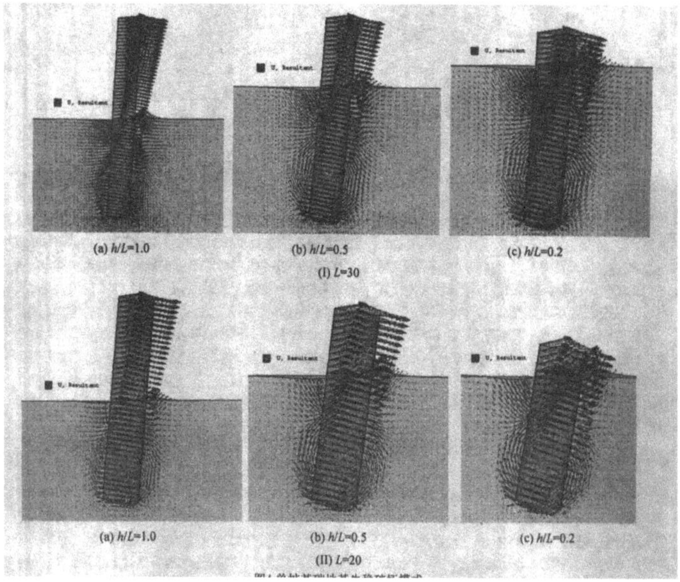

[Fig. 8-3] Failure mode of single pile foundation, source : (WU Ke et al., 2008)

It can be seen from [Fig. 8-3]: (1) Under the action of horizontal load, an obvious spherical rotation failure surface is formed inside the single pile foundation, and the rotation center is about 2/3 of the displacement of the single pile foundation. (2) A continuous spoon-shaped plastic zone is formed at the bottom of the single pile foundation; The molded plastic zone is formed in the soil on both sides of the foundation, and the passive zone is located outside the single pile foundation with the same direction as the horizontal load; The other side of the single pile foundation is the active area, and the separation between the single pile foundation and the foundation soil occurs. Therefore, the foundation instability failure mode of a single pile foundation under horizontal load such as wind is composed of a spherical rotation failure zone at the bottom of the foundation and active and passive wedge failure zones on both sides of the foundation, and the rotation center is about 2/3 of the displacement of the single pile foundation.

8.1.2 Influence of horizontal load on bearing characteristics of single pile foundation

8.1.2.1 Study on horizontal bearing characteristics of large-diameter offshore wind power single pile foundation

Based on the research on the bearing capacity of offshore wind turbines in China, this paper established the calculation mode of large-diameter offshore wind turbine foundation and seabed through theoretical analysis and numerical simulation by using the large-scale general finite element software COMSOL. Through this model, the lateral bearing capacity of large-diameter offshore wind turbine pile foundation under lateral loading was calculated, the influence of lateral loading, pile diameter, and other parameters on lateral displacement and a pile bending moment of offshore wind turbine pile foundation was analyzed, and the results can be used for similar projects(LIU Jianxiu et al., 2022).

(1) Calculation model

Liu Jianxu established the calculation model of offshore wind power large-diameter single pile foundation and seabed under horizontal load using the large-scale commercial finite element software COMSOL Physics. The seabed soil mass is mainly composed of muddy clay, silt, and silt. The main physical and mechanical parameters of the soil mass are shown in [Tab. 8-1]; The single pile foundation is made of steel pipe piles with a length of 66m and a depth of 50m. The detailed parameters of the foundation wall materials are shown in [Tab. 8-2]. The steel pipe pile adopts the isotropic linear elastic constitutive model, and the seabed soil adopts the Mohr-Coulomb model. In the calculation model, the contact between the single pile foundation and the surrounding seabed is rigid.

[Tab. 8-1] Parameter of soil around pile and soil at pile tip, source: (LIU Jianxiu et al., 2022)

layer	thickness/mm	Elastic modulus /MPa	Poisson's ratio	Effective gravity/ $kN \cdot m^{-3}$	Cohesion /kPa	internal friction angle /°
Muddy clay	28	6.6	0.3	7.5	17.6	12.8
silt	16	6.4	0.3	6.2	19.8	11.6
Silty sand	56	39.5	0.3	9.1	4.5	31.3

[Tab. 8-2] Physical and mechanical parameters of the pile, source: (LIU Jianxiu et al., 2022)

Pile name	Pile diameter /mm	thickness/mm	depth/mm	length/m	Elastic modulus /GPa	Poisson's ratio	Effective gravity $/kN \cdot m^{-3}$
Steel pipe pile	3	70	50	66	206	0.3	78.5

The calculation model of offshore wind power foundation and seabed under horizontal load is shown in [Fig. 8-4].

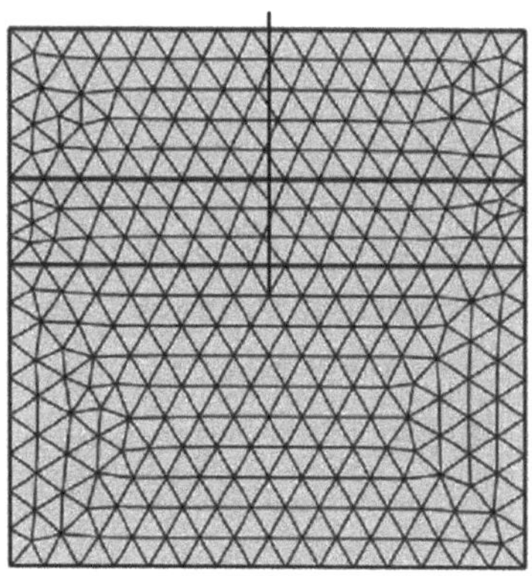

[Fig. 8-4] Offshore wind power foundation and seabed calculation model, source: (LIU Jianxiu et al., 2022)

(2) Analysis of numerical calculation results

Liu Jianxiu(LIU Jianxiu et al., 2022) calculated the offshore wind power foundation and seabed under horizontal load, and the results are as follows.

(LIU Jianxiu et al., 2022) shows the change curve of the horizontal displacement of the pile top with the horizontal load when the pile diameter is 3 m and the wall thickness of the steel pipe pile is 70 mm. It can be seen from(LIU Jianxiu et al., 2022) that the horizontal load horizontal displacement curve of the pile is a gradual type, and there is no obvious inflection point in the curve, which can be considered a progressive failure.

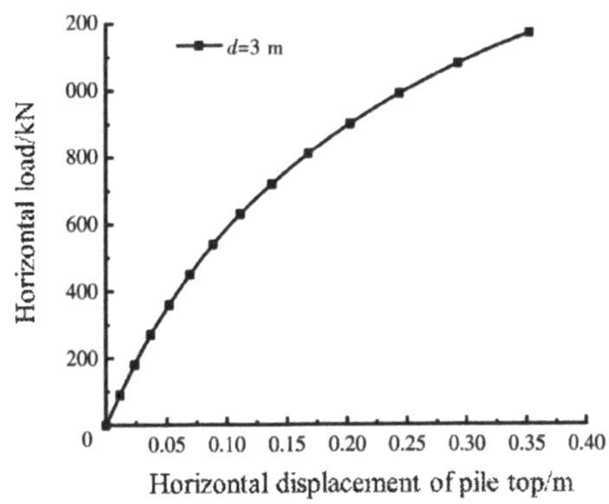

[Fig. 8-5] Pile top horizontal load — pile top horizontal displacement curve
Source: (LIU Jianxiu et al., 2022)

[Fig. 8-6] shows the curve of the horizontal displacement of the pile shaft along the buried depth of the pile under different horizontal loads when the pile diameter is 3 m and the wall thickness of the steel pipe pile is 70 mm. It can be seen from [Fig. 8-6] that when the horizontal load is constant, the horizontal displacement of the pile top is the largest and the pile bottom is the smallest, and the horizontal displacement of the pile shaft presents a nonlinear change law; When the pile diameter and the wall thickness of the steel pipe pile are constant, the horizontal displacement of the pile shaft increases with the increase of the horizontal load.

[Fig. 8-7] shows the curve of the bending moment of a single pile foundation along the buried depth under different horizontal loads when the pile diameter is 3 m and the wall thickness of the steel pipe pile is 70 mm.

It can be seen from [Fig. 8-7] that when the pile diameter is fixed, under the action of horizontal load, the bending moment of the pile shaft increases first and then decreases with the change of depth. At the same time, with the increase of the horizontal load, the position of the maximum bending moment of the pile shaft gradually moves down, and the maximum bending moment of the pile shaft gradually increases.

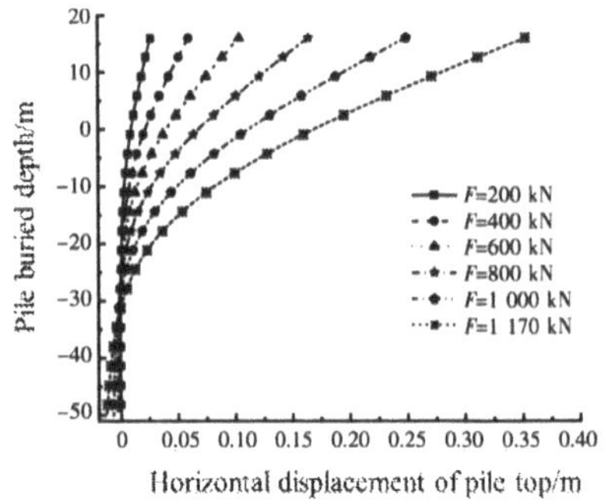

[Fig. 8-6] Curve of horizontal displacement of pile shaft changing along pile burial depth, Source: (LIU Jianxiu et al., 2022)

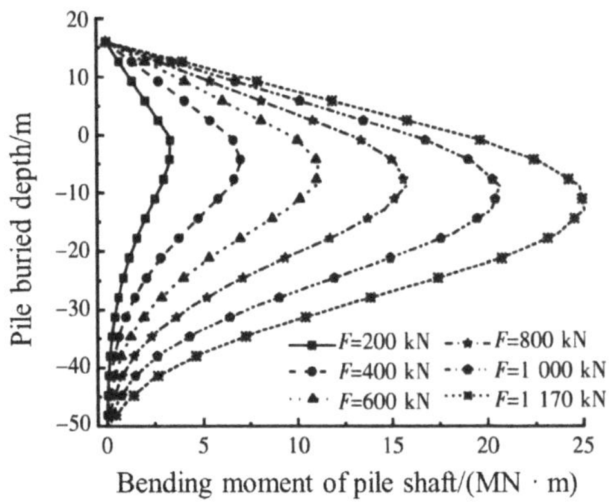

[Fig. 8-7] Curve of pile bending moment along pile burial depth, (LIU Jianxiu et al., 2022)

[Fig. 8-8] shows the change curve of pile bending moment with burial depth when the pile diameter is 1 m, 2 m, 3 m, 4 m, and 5 m, and the wall thickness of the steel pipe column is 70 mm. It can be seen from [Fig.

8-8] that within 5 m below the mud surface, changing the pile diameter has little effect on the bending moment of the pile shaft; At the depth of soil, the bending moment of the pile increases with the increase of pile diameter. When the pile diameter increases from 1m to 5 m, the maximum bending moment of the pile shaft increases from 10.91 to 11.60, and the location of the maximum bending moment of the pile shaft deepens from 4.625 m underground to 8.750 m.

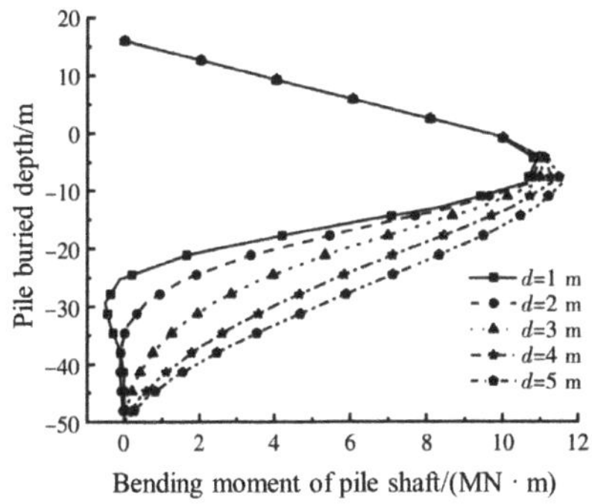

[Fig. 8-8] Curve of pile bending moment along pile burial depth under different pile diameters, Source: (LIU Jianxiu et al., 2022)

[Fig. 8-9] shows the curve of horizontal displacement along the pile shaft of a single pile foundation with different pile diameters under horizontal load when the wall thickness of the steel pipe pile is fixed. It can be seen from [Fig. 8-9] that when the applied horizontal load is unchanged, the outer diameter of the pile is changed. It is found that the horizontal displacement of the pile top of the single pile foundation decreases gradually with the increase of the pile diameter, indicating that the ability of the pile foundation to resist the horizontal load is gradually enhanced with the increase of the pile diameter. At the same time, when the pile diameter, steel pipe pile wall thickness, and horizontal load are fixed, the horizontal displacement of the single pile foundation along the pile shaft is still nonlinear.

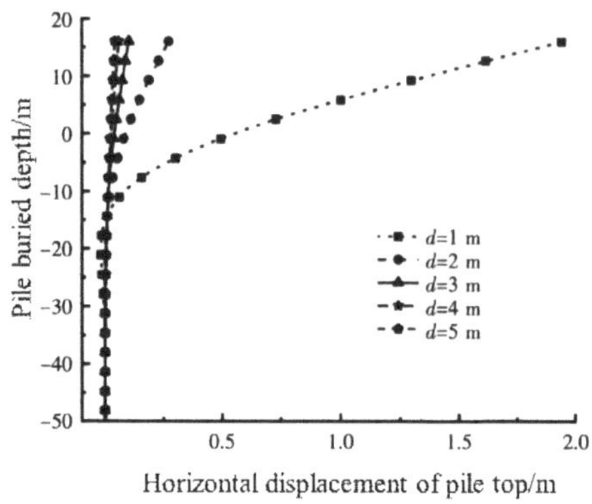

[Fig. 8-9] Curve of horizontal displacement of pile shaft along pile burial depth under different pile diameters, Source: (LIU Jianxiu et al., 2022)

From the above analysis, it can be seen that increasing the pile diameter can significantly reduce the horizontal displacement of the pile shaft, especially the horizontal displacement of the pile top, but it increases the bending moment of the pile shaft and increases the project cost. Therefore, in the actual process, it is necessary to obtain a suitable pile diameter to improve the bearing capacity of the pile shaft and reduce the cost.

8.1.2.2 Horizontal bearing capacity of offshore wind power single pile foundation and its improvement measures

Pan Chenchen (PAN Chenchen et al., 2019) used the large-scale general finite element software ABAQUS to study the bearing capacity of a single pile foundation under horizontal load, and studied the impact of different pile diameter conditions, filling materials in the pile and adding gravity structure on its bearing capacity, carried out indoor tests, proved the reliability of numerical simulation, and finally proposed a gravity single pile foundation with excellent horizontal bearing capacity, The mechanism of action was discussed.

(1) Finite element calculation model

1) Model parameters and dimensions

As the standard form of offshore wind power foundation, single pile foundation mainly consists of a 3.5~50 m steel pile. According to the existing pile driving technology and actual project needs, Liu Hongjun and Li Hongjiang determined that the steel pipe pile with a pile diameter of $D=5\,m$ should be studied as a typical pile; They have carried out systematic research on the Yellow River Delta and have abundant data. At the same time, the Yellow River Delta has small topographic relief and is close to the vast Bohai Sea. It has the advantages of strong wind, strong wind speed, and stable wind direction, which is particularly suitable for building offshore wind farms to develop the offshore wind power industry. The interbedding of silt and silty clay (LIU Hongjun & LI Hongjiang, 2014), is widely below the surface of the seabed in the Yellow River Delta. The most representative silt is selected as the soil model for the convenience of research.

ABAQUS is used to establish the finite element model. See [Tab. 8-3] for the parameters and dimensions of the pile and soil of the model.

[Tab. 8-3] Table 1 The physical and mechanical parameters and geometric dimensioning of foundation model, source: (PAN Chenchen et al., 2019)

Parts	Physical and mechanics parameters	Geometric dimensioning
Steel pile	Linear elastic material $\rho=7800\,kg/m^3$ $E=200\,GPa$, $v=0.3$	Length $L=60\,m$, out diameter $D=5\,m$, thickness $t=40\,mm$, Buried depth $h=50\,m$, Cantilever section $h'=10\,m$.
Soil	Elastic plastic M-C model, $\rho_{sat}=1800\,kg/m^3$, $E=3.5\,MPa$, $V=0.3$, $\varphi=20°$, $\psi=20°$. $c=15\,kPa$.	Out diameter $D'=100\,m$, height $H=65\,m$, The width of center is $t=40\,mm$, depth $d=50\,m$

Note: The calculated thickness of soil (the depth of foundation). H is defined as 65 m here, The radial length of the sail (the diameter of pile)=100 m.

2) Model Settings

To simulate the actual situation of the seabed, the floating weight is used for the calculation of the models in this paper. In the calculation process, the initial stress is balanced by assuming that the weight of each component is the same, and then the weight difference between the pile and soil is applied in the physical force balance analysis step to achieve the same state as the actual situation. The pile tip and soil are defined as binding constraints. The internal and external surfaces of the pile and the soil follow the Coulomb friction theory. The friction coefficient μ = 0.28 ; Small sliding and face-to-face contact modes are adopted between contact pairs. The bottom of the soil mass is fully constrained, and the side restricts the radial displacement. It can move freely in the vertical direction, and the top is a free surface without any constraint; All components are simulated by a three-dimensional 8-node reduced integral hexahedral element (C3D8R); The model after grid division is shown in [Fig.8-10].

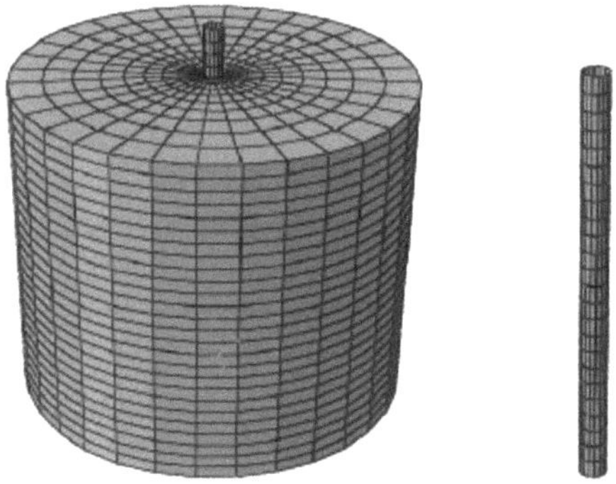

[Fig. 8-10] Mesh generation of the model, source: (PAN Chenchen et al., 2019)

The displacement control method is used for loading. A horizontal displacement is defined at the reference point of the pile top, and the load-displacement relationship is obtained by outputting the reaction at the reference point. In this paper, the horizontal displacement of the reference point of the pile top is used to reach μ=0.025 m. The load corresponding to 0.025 m is used as a reference to compare the bearing capacity.

(2) Stress and displacement analysis of single pile foundation

1) Model analysis under different pile diameters

First, to discuss the influence of pile diameter and wall thickness on pile bearing capacity, single pile foundations with different pile diameters and wall thicknesses are used for loading tests under the condition that the steel consumption is equal. See [Tab. 8-4] for specific parameters.

[Tab. 8-4] Table 2 The pile size of different pile diameters, source: (PAN Chenchen et al., 2019)

Pile number	1	2	3	4	5
Pile diameter D/m	1.5	2.0	3.3	4.0	5.0
Thickness t/m	0.07	0.05	0.03	0.025	0.02

The load-displacement curves obtained by defining the same horizontal displacement at the pile top reference point of each model and the curve of the change of the horizontal bearing capacity of the pile with the pile diameter are shown in [Fig. 8-11 and [Fig. 8-12]. It can be seen from [Fig. 8-11] that with the increase of displacement, the load is also increasing. At the initial stage, the load is distributed in a straight line, and then there are turning points in the curve. The change of the turning points of each curve is obtained by making the tangent line of each curve, as shown in the red dotted line in the figure, In the range of 0~5 m pile diameter, the load corresponding to the turning point increases with the increase of the pile diameter, which shows that the elastic deformation range of the pile increases gradually with the increase of the pile diameter in the range of 0~5 m pile diameter, that is, the bearing capacity of the pile increases. [Fig. 8-12] shows this relationship more clearly. Therefore, the horizontal bearing capacity of the single pile foundation can be improved by increasing the pile diameter in the range of 0~5m pile diameter.

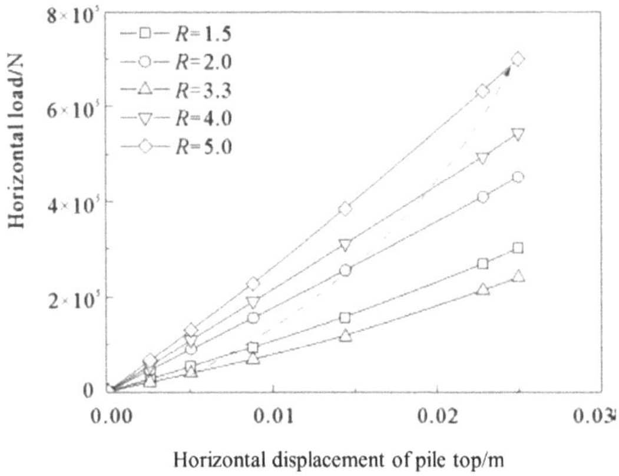

[Fig. 8-11] Load-displacement curve of different diameters, source: (PAN Chenchen et al., 2019)

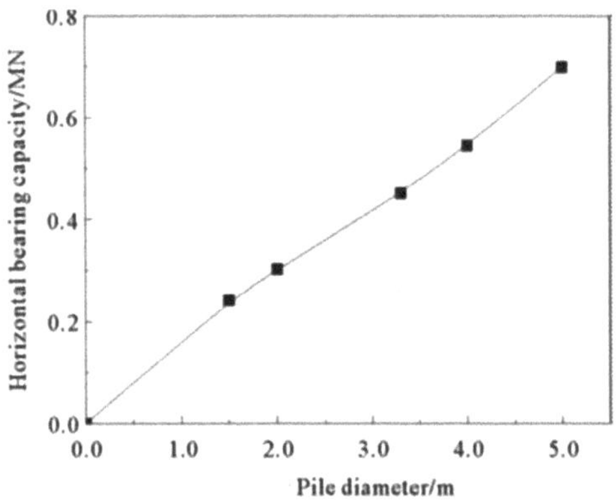

[Fig. 8-12] The horizontal bearing capacity change with diameters of monopole foundation, source: (PAN Chenchen et al., 2019)

2) Horizontal bearing capacity of single pile foundation under different structural forms

According to the existing pile driving technology and the needs of the actual project, the pile diameter of the single pile foundation can not be increased without limitation, generally 3.5~5.0 m, so the single pile foundation with the pile diameter D=5 m is selected as the typical pile for later research. To change the horizontal bearing capacity of single pile foundations, it is considered to improve their horizontal bearing capacity by combining single pile foundations with other structural forms. See [Tab. 8-5]for the measures taken for each model.

[Tab. 8-5] The combination of different structures and monopile foundation (D=4 m), source: (PAN Chenchen et al., 2019)

Pile No.	6	7	8	9
Type	a	a+b	a+c	a+b+c

The components represented by a, b, and c in [Tab. 8-5] are steel pipe piles, concrete filling materials, and gravity structures (see Fig.8-13), all of which are concrete materials. The material parameters are linear elastic materials, the density is $\rho = 2500kg/m^3$, Young's modulus $E = 20GPa$, and Poisson's ratio $v = 0.17$; Structure outer diameter $D_b = 4.92m$, height $H_b = 10m$; Structure outer diameter $D_c = 12m$, inner diameter $D'_c = 12m$, height $H_c = 2m$.

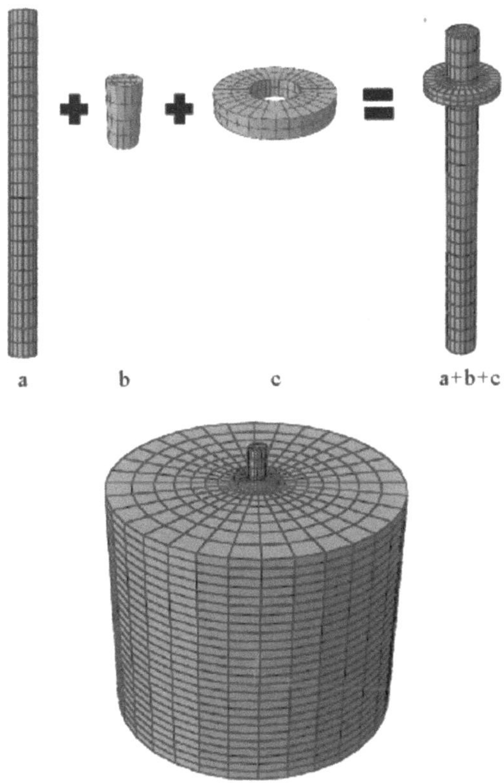

[Fig. 8-13] The combination of monopile foundation and aided structure, source: (PAN Chenchen et al., 2019)

The binding constraint is adopted between ab, and in fact, there will be 30 cm pores left between ac to connect them through grouting. After balancing the ground stress, the model directly sets ac as a binding constraint to facilitate research. Friction contact is adopted between the c structure and soil, and the friction coefficient is 0.28.

The load-displacement curves of each model can be obtained by defining the horizontal displacement at the reference point of the pile top, as shown in [Fig. 8-14].

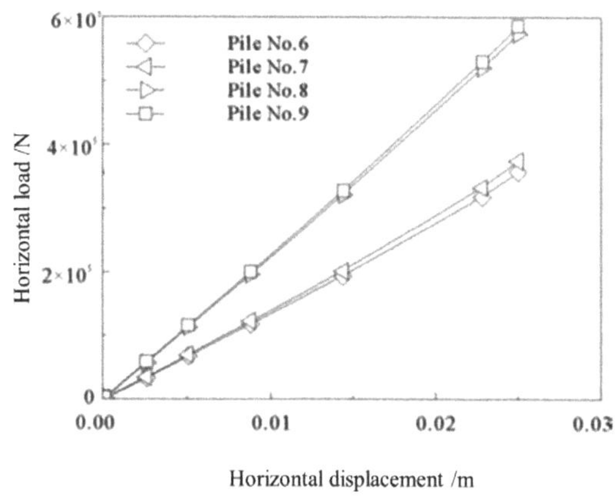

[Fig. 8-14] Load-displacement curve of different structure combinations, source: (PAN Chenchen et al., 2019)

It can be seen from the comparison results of piles 6 and 7 or piles 8 and 9 in [Fig. 8-14] that the capacity of filling concrete materials in steel pipe piles to improve the horizontal bearing capacity of piles is limited, while the comparison results of piles 6 and 8 show that the horizontal bearing capacity of single pile foundation is increased by about 50% due to gravity structure c. Another measure to improve the horizontal bearing capacity of single pile foundations can be set with an external gravity structure.

For a gravity structure single pile foundation, an important factor affecting its horizontal bearing capacity is the friction between the gravity structure and soil. The contact surface between the two is shown in the shaded part of [Fig. 8-15].

[Fig. 8-15] The contact surfaces between gravity structure and soil, source: (PAN Chenchen et al., 2019)

The friction coefficient adopted for pile 8 is $\mu=0.28$, to deeply study the influence of friction coefficient on the horizontal bearing capacity of a single pile foundation, the following takes 8 # pile as a typical pile and the friction coefficient as a research variable to set different friction coefficients between gravity structure and soil. See [Tab. 8-6] for the friction coefficient settings.

[Tab. 8-6] The gradient of friction coefficient, source: (PAN Chenchen et al., 2019)

Pile No.	8	10	11	12	13
Friction coeffi/μ	0.28	0.50	0.75	1.00	Rough

The horizontal displacement of the single pile foundation in [Tab. 8-6] is defined at the reference point of the pile top, and the corresponding load friction coefficient curve is obtained as shown in [Fig. 8-16].

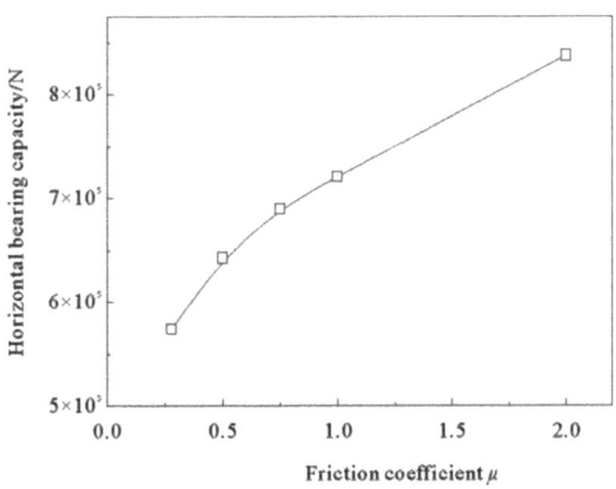

[Fig. 8-16] The relationship between friction coefficient and horizontal bearing capacity of monopile foundation, source: (PAN Chenchen et al., 2019)

It can be seen from [Fig. 8-16] that the friction coefficient between the gravity structure and the soil can significantly increase the horizontal bearing capacity of the composite single pile foundation. When the friction coefficient is 1, the horizontal bearing capacity of the composite single pile foundation is increased by 25%, while when the friction coefficient is set as Rough, the horizontal bearing capacity is increased by nearly 50%, Therefore, how to increase the friction coefficient of the contact surface between the two or how to connect the gravity structure and the soil is of great significance to improve its horizontal bearing capacity.

8.1.3 Experimental study on the effect of the typhoon on cumulative deformation of offshore wind power monopile foundation

To study the influence of typhoons on the cumulative deformation of offshore wind power monopile foundations, an indoor small-scale model test was carried out. Large-scale loads were set at different times during the loading process to simulate the situation when typhoons acted, and the lateral movement of the pile body was measured. To calculate the rotation angle of the pile at the mud surface, draw the curve of the rotation angle changing with the number of cyclic loadings, and analyze the influence of the typhoon on the cumulative rotation angle from the curve (LU Zhiyue et al., 2019).

8.1.3.1 Model Experiment Design

(1) Model test similarity ratio

For the offshore wind power monopile foundation with a diameter of 7.5 m, a buried depth of 37.5 m, and a pile length of 80 m, Lu Zhiyue selected the commonly used geometric similarity ratio of 1:100 to make a model. A similar, similar design is carried out according to Buckingham's Π theory, and the similarity ratio of related physical quantities in the test is shown in [Tab. 8-7](LU Zhiyue et al., 2019).

[Tab. 8-7] Model test similarity ratio, source: (LU Zhiyue et al., 2019)

Physical quantity	Similitude ratio	dimension
Length, L	$L = L/\lambda$	L [m]
Force, F	$F_M = F_P/\lambda^3$	F [N]
Stress, σ	$\sigma_M = \sigma_P/\lambda$	F/L^2 [Pa]
severe, γ	$\gamma_M = \gamma_P$	F/L^3 [N/m^3]
bending moment, M	$M_M = M_P/\lambda^4$	$F \times L$ [N m]
Bending stiffness, EI	$(EI)_M = (EI)_P/\lambda^5$	$F \times L^2$ [N m^2]
Time, t	$t_M = t_P/\lambda^{1/2}$	T [s]
Frequency, f	$t_M = t_P/\lambda^{-1/2}$	$1/T$ [Hz]

(2) Test conditions

The model tests were all carried out in a specially designed model tank with a length of 1.2 m, a width of 0.8 m, and a depth of 0.9 m, in which 0.85 m of dry sand was buried. Use a marker to mark the inner wall of the model groove from bottom to top at intervals of 0.17 m in depth, which are divided into 5 layers and filled in layer by layer. When filling the soil, the mass of dry sand at a depth of 0.17 m was calculated according to the given density, and the dry sand of the corresponding mass was filled into the model groove and compacted to the layered mark to realize the control of soil density.

The model piles are designed according to the similarity theory. Due to the limitation of materials, it is difficult to apply the external dimensions and bending stiffness of the prototype piles to the model piles in a similar ratio. Bending stiffness similarity requirements. To make the flexural stiffness as close to the theoretical value as possible, PVC material with a smaller elastic modulus is used. The model pile is finally determined as a PVC pipe with openings at both ends. The pile length is 0.8 m, the pile diameter is 7.5 cm, the wall thickness is 2.44 mm, and the burial depth is 5 times the pile diameter, i.e. 37.5 cm. The elastic modulus of PVC measured by the simply supported beam method is 3.24 GPa. According to the calculation, the bending stiffness of the model pile is 4 times the theoretical value, which meets the requirements of the test.

In this test, for sandy soil foundation conditions, a model soil with grading similar to Berlin soil is prepared by mixing and mixing Fujian quartz sand with various standard particle sizes. During the test, the sand density of each working condition is controlled to be 1.60 g/cm^3. According to the Standard for Soil Test Methods (GB/T 50123 - 1999) (《*Standards for Geotechnical Test Methods*》(*GB/T50123-2019*), 1999) indoor soil tests are carried out. The basic physical parameters of the model soil measured are shown in [Tab. 8-8].

[Tab. 8-8] Basic physical parameters of model soil, source: (LU Zhiyue et al., 2019)

Physical parameters	Values
Maximum dry density /g·cm^{-3}	1.75
Minimum dry density /g·cm^{-3}	1.43
Compactness D_r	58%
Average particle size d_{50}/mm	0.55
Nonuniform coefficient C_u	1.97
Curvature coefficient C_c	0.90

The loading equipment for this test is an electric servo actuator with a rated output of 1100 N and a maximum stroke of 200 mm. It can generate a sinusoidal cyclic load with a frequency of 1 Hz, so it can meet the loading requirements of this test. In the test, an LVDT high-precision displacement sensor with an accuracy of 0.03 mm was used. The measuring range was 0~30 mm, and the nonlinearity was less than 0.1%. The displacement sensor and servo actuator were connected to the data acquisition instrument to synchronously collect the displacement and load, and then the collected data was transferred to the data acquisition software for real-time recording.

(3) Loading and measuring point layout

The equipment connection of horizontal static and cyclic loading tests is the same, as shown in [Fig. 8-17]. Displacement sensors are arranged on both sides of the pile at different heights from the mud surface along the loading direction to ensure that the contact of the sensor is in close contact with the pile body. One side of the pile is connected to the electric cylinder with a steel wire rope. The level gauge is used to ensure that the loading point and the connection head of the electric cylinder are at the same height. The loading height of each working condition is 25.5 cm. This height is based on the similarity of the typical action height of the horizontal resultant force of the prototype pile (Cuéllar, 2011). In addition, based on the mass of the upper structure of a 5 MW fan in Song Yang (SONG Yang & WANG Ying, 2018) study, a 0.5 kg mass block is placed on the pile top according to the similarity ratio to make the vertical load of the whole structure more consistent with the similarity theory.

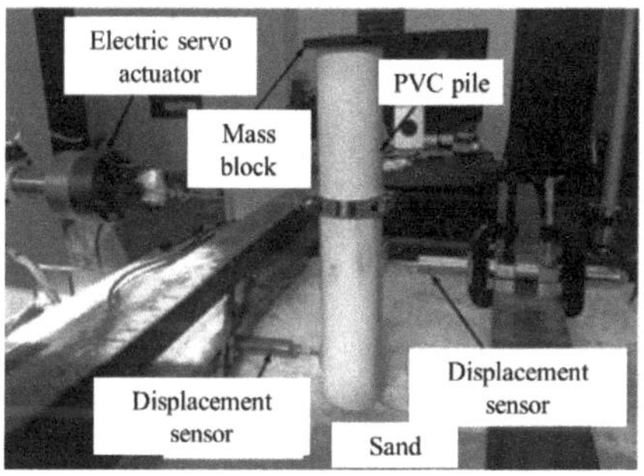

(a) Layout of test equipment

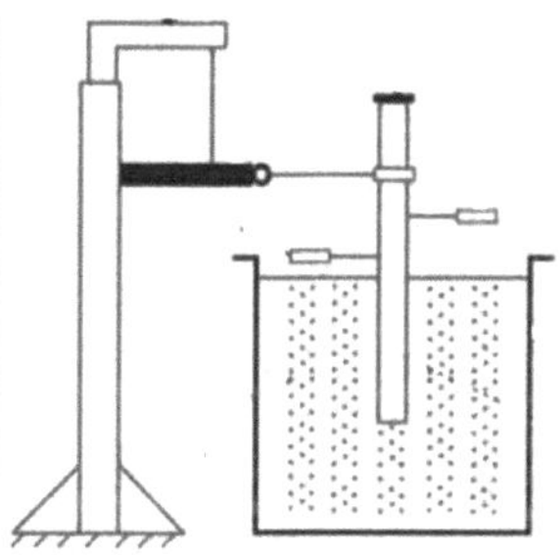

(b) Schematic diagram of equipment layout

[Fig. 8-17] Layout of test equipment and schematic diagram of equipment layout, source: (LU Zhiyue et al., 2019)

(4) Test loading scheme and working condition

The test load is divided into horizontal static load and cyclic load. The purpose of horizontal static load test is to obtain the ultimate bearing capacity of the pile. The load when the displacement of the pile at the mud surface is 0.1 D can be regarded as the ultimate bearing capacity of the pile, and D is the pile diameter. Use the servo actuator to load the pile step by step. Initially, 10 N is applied at each stage, with an interval of 3 minutes. When the cumulative load reaches 60 N, then each stage is lifted to 15 N, with an interval of 3 minutes. When the displacement at the mud surface exceeds 0.1 D, the loading is stopped.

The cyclic loading tests carried out are all one-way cyclic loads, namely $\xi_c \geq 0$.ξ_b of small amplitude cyclic load in the test is taken as 0.2, 0.3, 0.4 near 0.3, and ξ_b of the typhoon, the load is 0.7. The frequency of wind load under site conditions is lower than that of wave load, so the frequency of wave load is taken as the frequency of the overall cyclic load. The frequency of wave load is about 0.1 Hz (Arany et al., 2017), According to the similarity ratio, the frequency of all cyclic loads in this test is determined as 1 Hz. This paper focuses on the most dangerous instantaneous load, which originates from the super typhoon with the largest wind speed. In China, such typhoons mostly occur along the coast of Fujian. Luo Chao (LUO Chao & CAO Wensheng, 2013) has made statistical data on the 50 years of the typhoon from 1961 to 2012, which shows that there have been three times of super typhoons in Fujian, and the service time of offshore wind turbines is 20 to 25 years. Therefore, only one or two rounds of the typhoon action process are set in the working condition with the typhoon, At the same time, refer to Nadine et al. (Abadie, 2016) for the number of cycles of large amplitude load in their multi-amplitude cyclic loading test. The number of cycles of typhoon load in this test is 12. See the literature (LU Zhiyue et al., 2019).

8.1.3.2 Test results and analysis

(1) Horizontal static loading test

The load-displacement relationship curve of the ST1 and ST2 mud surface in the horizontal static load test is shown in [Fig. 8-18]. In this test, the pile diameter D is 75 mm, and 0.1 D is 7.5 mm. From the figure, it can be seen that the ultimate bearing capacity under ST1 working condition is about 177.5N, and that under ST2 working condition is about 182.5 N. The average value of the ultimate bearing capacity under the two working conditions is 180 N. Therefore, the ultimate bearing capacity of the pile in this test is taken as 180 N.

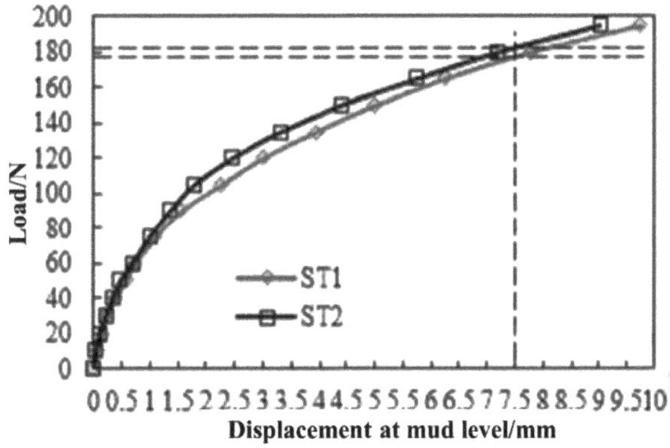

[Fig. 8-18] Load displacement curve at mud surface under ST1 and ST2 working conditions, source: (LU Zhiyue et al., 2019)

(2) Horizontal cyclic loading test of typhoon load

First, the fitting parameters of cumulative rotation angle under the action of typhoon load should be obtained. Because the number of typhoon load actions is small, the fitting parameters of typhoon load can be approximately obtained through the first 12 typhoon loads of CT2, CT5, and CT8. The fitting of the cumulative rotation angle of no typhoon working condition in the previous text started from the 6th cycle, so the curve of typhoon load is still fitted from this position. The fitting curves of the three working conditions are shown in [Fig. 8-19]. It can be seen that the fitting curves of CT2 and CT8 working conditions almost coincide, while the fitting curves of CT5 working conditions are staggered with them. Therefore, the average value of CT2 and CT8 working conditions is taken to approximate the typhoon load. The processing results are shown in [Tab. 8-9].

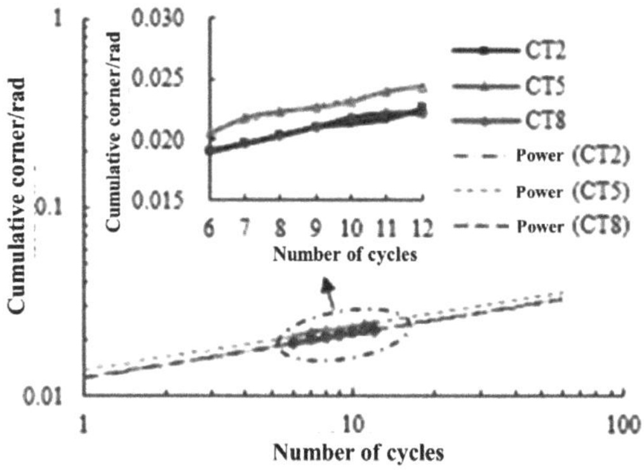

[Fig. 8-19] Fitting of the 6th~12th cycle of typhoon load under CT2, CT5 and CT8 working conditions, source: (LU Zhiyue et al., 2019)

[Tab. 8-9] Fitting parameters of typhoon load corresponding CT2 and CT8 working conditions, source: (LU Zhiyue et al, 2019)

working condition	β	α	R^2
CT2	0.012 4	0.239 1	0.987 8
CT8	0.012 6	0.233 1	0.977 8
Average value	0.012 5	0.236 1	-

There are 6 working conditions with the typhoon, namely CT2, CT3, CT5, CT6, CT8, and CT9. Among them, CT2, CT5, and CT8 set 12 typhoon loads at the beginning, while CT3, CT6, and CT9 set 12 typhoon loads at the positions of 5000 and 10000 small amplitude cyclic loads respectively. [Fig. 8-20] shows the relationship curve between the cumulative rotation angle and the number of cycles under CT2, CT5, and CT8 working conditions. It can be seen that the first 12 times of typhoon load make the rotation angle rapidly rise to a higher value, and then under the action of small amplitude cyclic load, there is a small fall back, and then continue to rise, and the ξ_b of small amplitude cyclic load. The smaller the b is, the slower the cumulative angle increases. It is shown in the figure that under the CT2 working condition, after the typhoon load, the small amplitude cyclic load can hardly increase the angle anymore. The cumulative angle curve presents a platform. Compared with the CT8 working condition, the cumulative angle still has a certain increase after the typhoon load.

[Fig. 8-21] shows the relationship curve between the cumulative rotation angle and the number of cycles under CT3, CT6, and CT9 working conditions. It can be seen that: (1) ξ_b of cyclic load with small amplitude increase, the steep rise of the corner caused by typhoon load is decreasing, which can be explained as the larger the small amplitude cyclic load is, the stronger the compaction effect on the soil mass is, the higher the lateral stiffness of the pile is, and the corner generated by the typhoon load of the same size is smaller; The steep rise of angle caused by the second round of typhoon load is smaller than that caused by the first round of typhoon load. In addition, through calculation, the proportion of the second steep rise of angle to the first steep rise of displacement under CT3, CT6, and CT9 working conditions is 27.9%, 36.4%, and 55.9% respectively. This proportion is rising constantly, indicating that with the small amplitude of the cyclic load ξ_b, the difference between the sharp rise of the corner caused by the front and rear typhoon loads is constantly decreasing, and the extreme case is that it is always affected by typhoon loads, so there is no sharp rise of displacement, and it becomes a smooth curve; (3) The growth rate of three small amplitude cyclic loads separated by two rounds of typhoon loads also increases with ξ_b The increase of b increases. After two cycles of typhoon load, the rotation angle of CT3 and CT6 working conditions has almost remained unchanged, showing a platform shape, but the CT9 working condition still has a large growth rate, indicating that when the small amplitude cyclic load is large, even after two cycles of typhoon load, it can still increase the cumulative rotation angle of the pile.

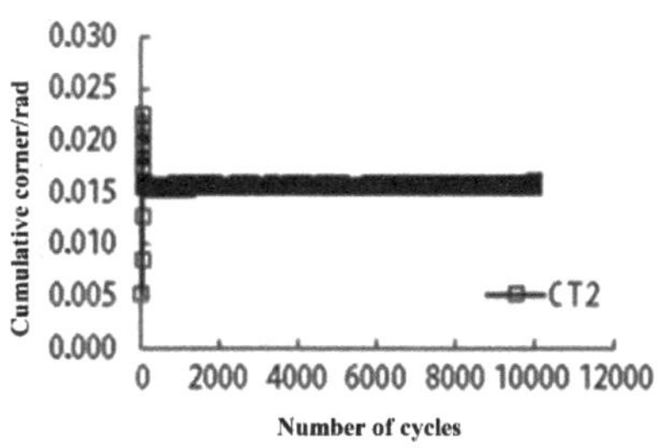

(a)CT2

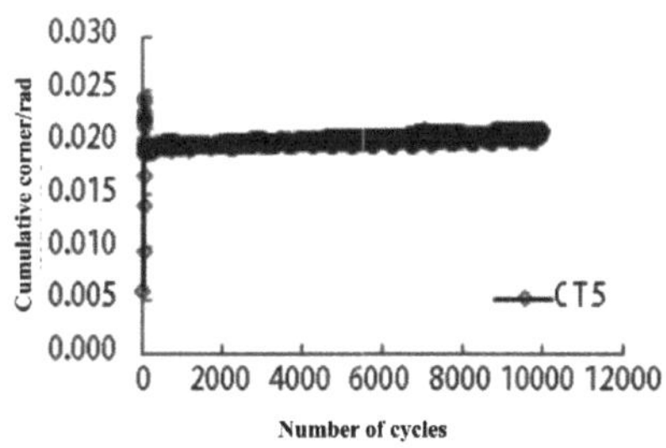

(b)CT5

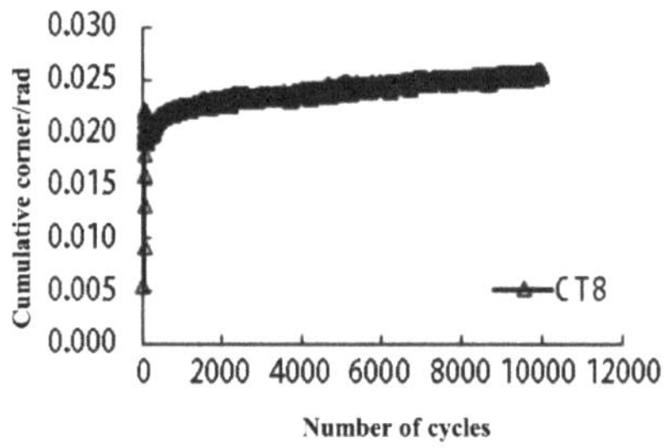

(c)CT8

[Fig. 8-20] Relation between cumulative rotation angle and number of cycles (CT2, CT5 , CT8), source: (LU Zhiyue et al., 2019)

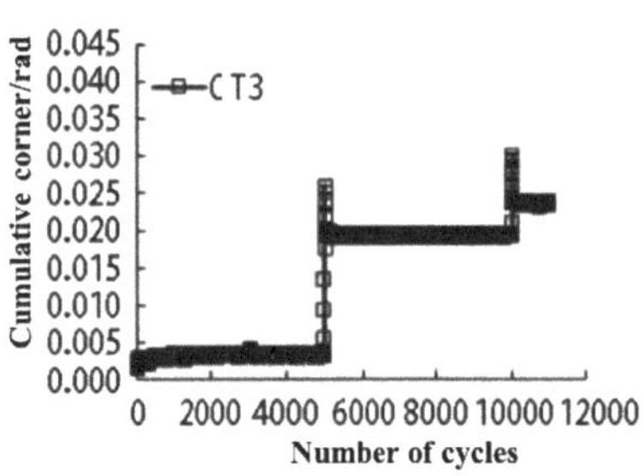

(a)CT3

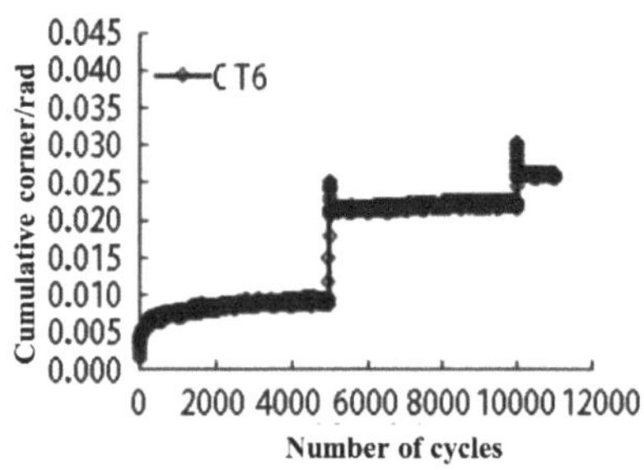

(b)CT6

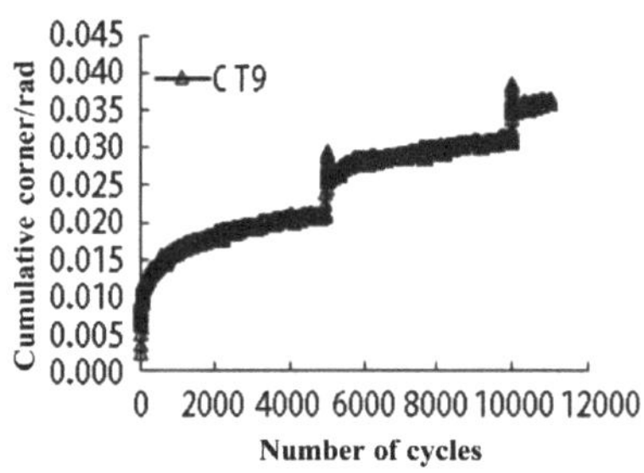

(c)CT9

[Fig. 8-21] Relation between cumulative rotation angle and number of cycles (CT3, CT6 and CT9), source: (LU Zhiyue et al., 2019)

8.2 Influence of wind load on bucket foundation

8.2.1 Influence of wind load on bucket foundation

A simplified fan tower foundation integral model is established to analyze the vibration characteristics of the tower steel-concrete composite structure tube foundation, and the flow-induced vibration response of the integral model under wind load is studied (LIU Xizhu & LIU Zhifeng, 2013).

8.2.1.1 Calculation model

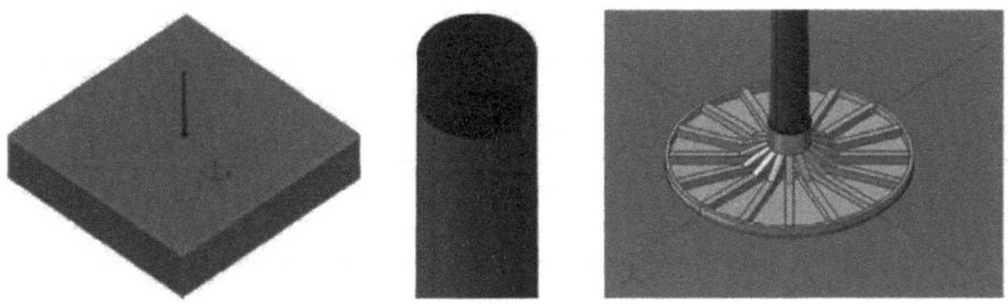

[Fig. 8-22] Calculation model, source: (LIU Xizhu & LIU Zhifeng, 2013)

8.2.1.2 Wind load simulation

The model of the target random process is approximated by a discrete spectrum. The random signal is effectively decomposed into sine wave with different frequencies and amplitudes through discrete Fourier analysis and transformation (HU Chunhong & ZHANG Zhihao, 2011).

Liu Xizhu selected Longyuan offshore wind field as a calculation example, which is a relatively large wind field in Rudong County, Jiangsu Province, China. During 1959-2000, the actual measured maximum wind speed of the wind field reached 20 m/s, and the maximum wind speed reached 31.5 m/s. The reference wind speed is determined as the wind speed of 38 m/s with a 50-year return period. Using the method described above and Matlab program for analysis, we can obtain the fluctuating wind speed time history, power spectrum, and other data in [Fig. 8-23] and [Fig. 8-24](LIU Xizhu & LIU Zhifeng, 2013).

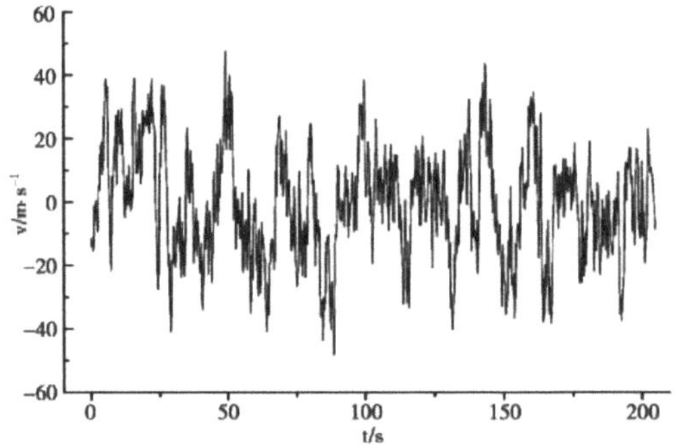

[Fig. 8-23] Time history of fluctuating wind speed, source:(LIU Xizhu & LIU Zhifeng, 2013)

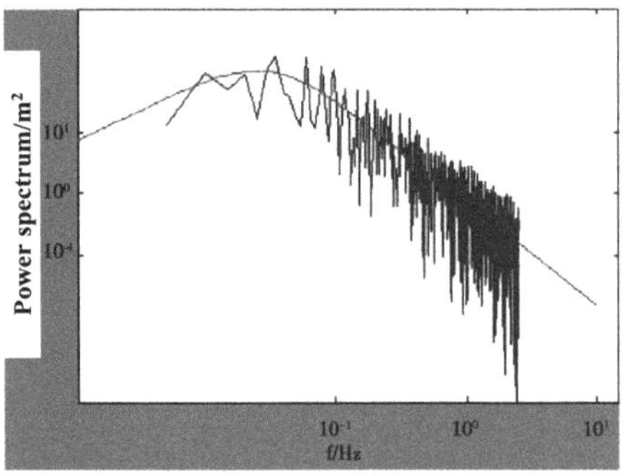

[Fig. 8-24] Target power spectrum and analog power spectrum, source: (LIU Xizhu & LIU Zhifeng, 2013)

8.2.1.3 Results of wind vibration

The calculation model is the same as the natural vibration characteristics of the structure. When selecting the time average load, the wind load borne by the crazy foundation and tower under the 50-year wind speed is mainly selected. It is divided into 15 sections from the top of the tower to the sea level. The positive X direction is determined as the downwind direction, the Y direction is determined as the vertical direction, and the Z direction is determined as the vertical direction.

(1) Displacement response analysis

Under the 50-year return period wind load condition, the structural vibration along the wind direction is severe. Since the foundation is no longer affected by the initial deformation, and drainage and consolidation measures have been taken, the vertical displacement of the whole structure under the wind load will not be large.

Maximum slope rate tan caused by differential settlement at the connection $\tan\theta = 0.001$, meeting the requirements of DNV and other offshore wind power infrastructure design specifications. The design gradient of the tube foundation of this example project under the 50-year return period wind load conforms to the safety of the engineering structure.

(2) Stress analysis of wind vibration response

It can be seen from [Fig. 8-25] that at a certain time, the overall structure of the tower base is analyzed according to the Mises stress distribution. The values with higher stress in the structure are usually concentrated on the tower, and most of them are not far from the middle of the tower, while the stress on the tower base is relatively small.

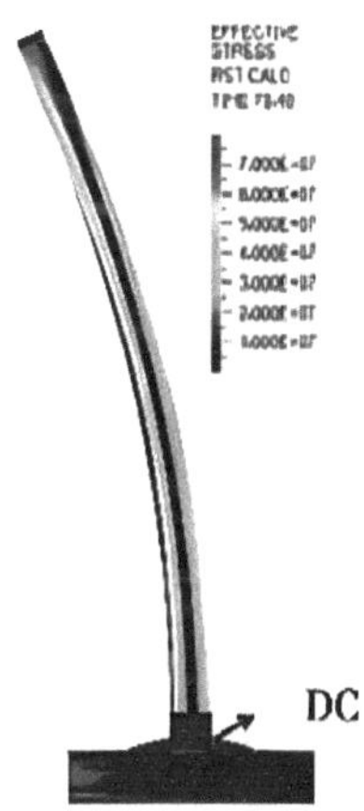

[Fig. 8-25] Structural stress, source: (LIU Xizhu & LIU Zhifeng, 2013)

From [Fig. 8-25] we can know that:

1) The stresses at the top of the tower are very small, so the damage to the steel at the top of the tower is not considered.
2) Because the force has a certain transfer effect and is constantly attenuated in the process of transfer, the stress of the outer tube skirt on the tube base is much smaller than that of the inner tube skirt, so you can choose a point on the inner tube skirt for stress effect analysis. Through calculation and analysis, it is found that each stress index at the maximum stress position on the skirt is within the allowable range of steel stress.
3) The inclined bracing of the tube foundation can effectively transmit the force because the stress at the position where the inclined bracing and the tube foundation tower are connected is usually relatively large, and it is easy to be damaged.

8.2.2 Response of large-scale offshore wind torch foundation under wind load in the offshore shoal area

Ding Hongyan, et al. (Ding Hongyan et al., 2013) converted the wind speed time history into wind load and studied the response of large-scale fan structures to wind load by using the finite element software ABAQUS: the treatment of load is to convert the measured wind speed time history into wind load time history and apply it to the fan model; For the nonlinear problem of soil, the density, strength, elastic modulus, cohesion and friction angle of soil are considered; The time history effect of soil in the dynamic process is studied by analyzing and comparing the responses of the tube and soil in different periods; The change of foundation in different soil properties is studied by changing the strength of the soil.

8.2.2.1 The finite element model

(1) Basic parameters

[Tab. 8-10] shows the basic parameters of materials in each part of the model. Taking sandy clay as an example, its cohesion $c = 15\,kPa$, internal friction angle $\varphi = 15°$, and expansion angle $\psi = 0°$, Considering the range of influence in the soil (DU Jie et al., 2005), the radial dimension of the boundary is 200 m (about 7 times the cylinder diameter), and the depth is 35 m (about 7 times the cylinder height). The Mohr-Coulomb model (DU Jie et al., 2005) .

[Tab. 8-10] Main parameters of the model, source: (Ding Hongyan et al., 2013)

Parts	Material	Destiny/ $kg \cdot m^{-3}$	Elastic modulus /MPa	Poisson's ratio
Tube foundation	Reinforced concrete	2400	36 500	0.17
Tower and engine room	Steel	7800	210 000	0.30
soil mass	Saturated soil	950(Floating density)	4.5	0.35

The tubular foundation is $h = 5.9\,m$ in height, $D = 30\,m$ in diameter, and the foundation cushion cap is 18 m in height. The tubular tower is divided into two parts: the lower part is a cylinder, 19.5 m in height, 4.34 m in outer diameter, and 50 mm in wall thickness; The upper part is a cone with a height of 58 m, an outer diameter of 4.34 m at the bottom, 2.68 m at the top and a wall thickness of 38 mm. Since only the reaction of the cylindrical foundation under the action of the horizontal load is considered, to simplify the calculation and save the calculation cost, the engine room and fan blades are replaced by mass blocks with a total mass of 127 t. Considering the eccentricity, the distance between the center of gravity and the central axis is 3.1 m.

[Fig. 8-26] shows the model size (including the load application mode) and the finite element model. The dimensions of the inner cylinder subdivision plate and the main and secondary beams are omitted from the sizing diagram.

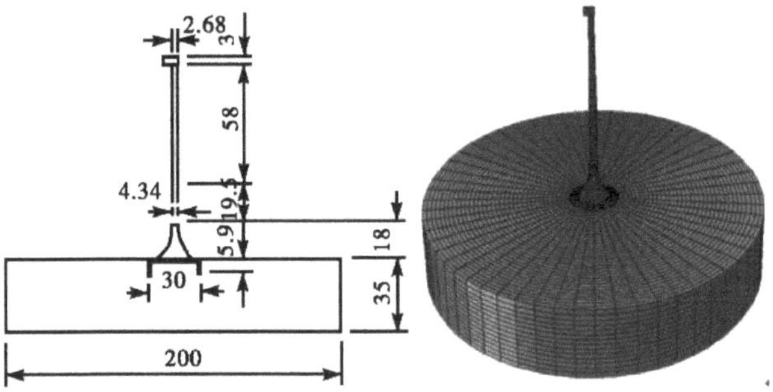

[Fig. 8-26] Size of model and finite element model, source: (Ding Hongyan et al., 2013)

8.2.2.2 Results analysis

Analysis of the change of the stress state in the cylinder and soil with time After the load is applied, analyze the stress state in the cylinder and soil in different periods. The change of the stress in the cylinder and soil when the load is applied to 1s, 6s, and 143s respectively is shown in [Fig. 8-27] ~ [Fig. 8-29].

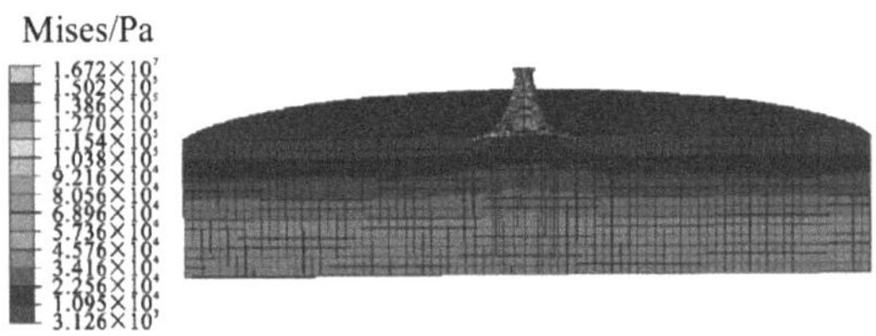

[Fig. 8-27] Stress of bucket and soil when loading to 1,s, source: (Ding Hongyan et al., 2013)

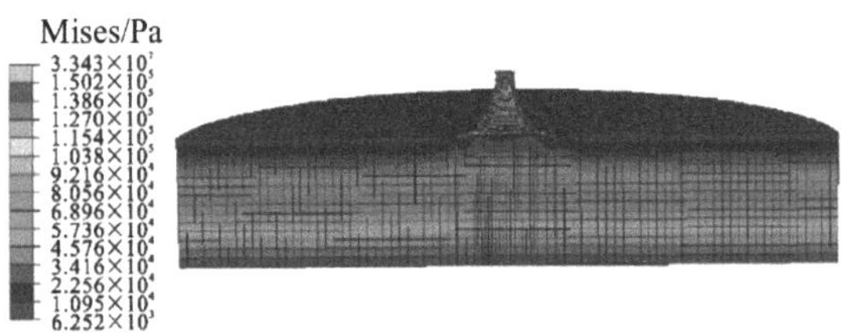

[Fig. 8-28] Stress of bucket and soil when loading to 6,s, source: (Ding Hongyan et al., 2013)

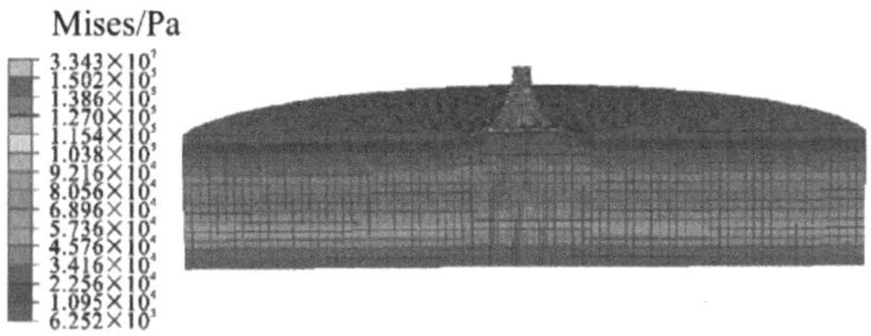

[Fig. 8-29] Stress of bucket and soil when loading to 143,s, source: (Ding Hongyan et al., 2013)

The results show that the stress at the bottom plate of the cylinder is the largest, but it has little change at any time. The stress in the soil is small at the initial stage of loading. With the continuous application of load, the stress value gradually increases, but after 6 seconds of loading, the stress and settlement in the soil and the cylinder gradually stabilize. On the one hand, the load amplitude increases slightly with time, and on the other hand, the wind load circulates, The pore water in the soil is more difficult to discharge, and the additional pore pressure gradually increases, so the stress in the soil gradually increases

(1) Reaction between cylinder and soil during loading

The results show that at the initial stage of loading, the foundation and soil have no obvious response due to the small load in the early stage and the insignificant cyclic cumulative effect; As the loading time continues, the soil layer around the cylinder gradually descends and forms a saddle-shaped pit (see Fig.8-30). The size of the pit in the loading direction is larger than the size perpendicular to the loading direction. Because the loading time is relatively short, the cylinder and soil have been gradually settling, but the settlement has been very slow and will tend to be stable after a long time, which will be discussed in depth in future studies.

In the loading direction, there is about a soil layer around the cylinder that is equivalent to the cylinder diameter, while in the direction perpendicular to the loading, there is about 80% of the cylinder diameter. The cylinder settlement is greater than the soil settlement, Therefore, the cylinder surface is lower than the soil layer surface at the end of the settlement. The shaping area of soil mass after the settlement is shown in [Fig. 8-31]. The equivalent plastic area shows that the soil mass begins to produce plasticity at about twice the height of the cylinder from the bottom of the cylinder at a deeper position under the foundation. However, due to its small plasticity and deep location, it has no impact on the stability of the foundation.

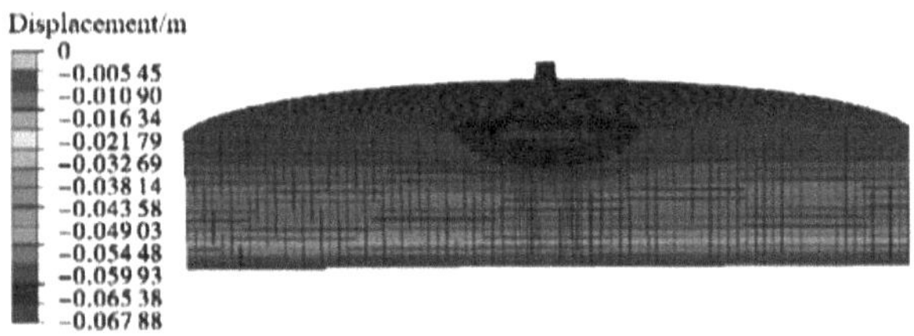

[Fig. 8-30] Settlement of bucket and soil and influence area after loading, source: (Ding Hongyan et al., 2013)

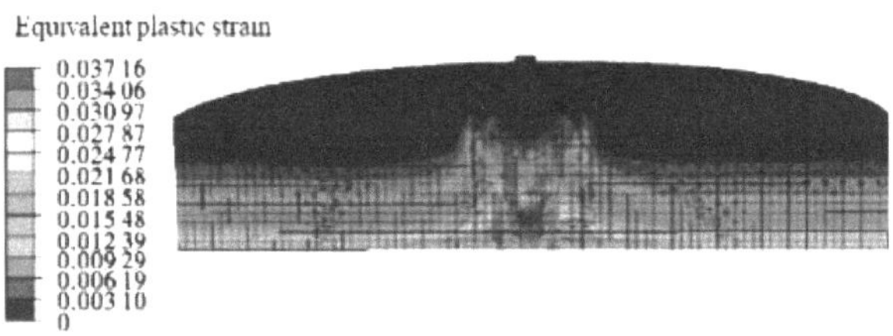

[Fig. 8-31] Equivalent plastic zone of soil after loading, source: (Ding Hongyan et al., 2013)

8.3 Influence of wind load on large-diameter steel pipe pile foundation

8.3.1 Pile soil interaction under horizontal static load and dynamic load in typhoon environment

Liu Junwei et al.(LIU Junwei et al., 2022) introduced the research status of pile-soil interaction under horizontal load and discussed the relationship between the natural frequency evolution of single pile support system and the cyclic weakening characteristics of soil mass in the marine environment.

The p-y curve method is the most commonly used theoretical analysis method of pile-soil interaction under horizontal load, which was adopted by API (Young et al., 1976) and DNV(Design of Offshore Wind Turbine Structures, 2014). This method is proposed by Matlock (Matlock, 1970) based on the test results of slender piles in soft clay. Therefore, when calculating large-diameter piles, stiffness is often overestimated and lateral deformation is underestimated (ZHU Bin et al., 2013)). For this, domestic and foreign scholars have modified the p-y curve based on centrifuge test and numerical simulation results of sand (Choo et al., 2014), (Choo & Kim, 2016), and soft clay (Pradhan, 2012), (C. Zhang et al., 2011) respectively. At present, the study of the p-y curve under multi-directional cyclic loading is very limited. Su, et al. (Su & Yan, 2013) based on the boundary surface theory, proposed a p-y curve that can consider any loading direction, but there are many model parameters and lack of physical significance, so it is difficult to calibrate.

The experimental study on the cumulative deformation of rigid single pile foundations is mainly based on the single degree of freedom loading model test. Leblanc et al. (Bakmar et al., 2010), and Klinkvort et al. (Klinkvort, 2013) successively carried out a 1g indoor model test and centrifuge model test for sand and put forward the calculation method of pile foundation deformation under arbitrary load combined with rain flow counting method. He Ben et al. (HE Ben et al., 2016)、Deng tao et al. (DENG Tao et al., 2020) conducted centrifugal model tests and field tests on cohesive soil. The research shows that the main factors affecting the residual deformation of pile foundations include the over-consolidation ratio of soil, the re-consolidation after circulation, the vertical load of the pile, and the stiffness of the pile-soil phase. Hu Anfeng et al. (HU Anfeng et al., 2020), and Jiang Zhenqiang et al. (Jiang Zhenqiang et al., 2021) respectively analyzed the influence of pile swelling deformation, soil properties, load direction, and other factors on the cumulative lateral displacement of single pile foundation by numerical simulation.

The actual measurement shows that although the typhoon has a short duration of action, its intensity is large and its spatial and temporal variation is complex.

For example, within 13 hours of the 2013 super typhoon "Tiantu" in China, the wind speed near the "Red Bay" wind farm rapidly increased from 33 m/s to 57 m/s, during which the wind direction changed 270 degrees (Chen et al., 2015),(Chen & Xu, 2016). Chen Chaohui et al. (Chen Zhaohui & Tang Haitao, 2008) carried out statistical analysis on the long-term data of the China Typhoon Yearbook, tested the fitting degree of the probability model of the key parameters of a typhoon, and established the probability model and key parameters of typhoon applicable to the southeast coast of China. The waves caused by typhoons have high intensity and strong nonlinearity, and the maximum effective wave height can reach 16.91 m (such as in typhoon Katrina). At the same time, it may also lead to the appearance of abnormal waves. The complexity of wave forces is far greater than that of ordinary waves. Conventional wave load models based on the Morison equation (such as Fan Huiyan). At present, semi-empirical methods are mainly used to predict the significant wave height and spectral peak frequency during typhoons (such as Ochi(Ochi, 2003)). In particular, the wave field and typhoon field are not coaxial, and the intersection angle changes with time and space, showing different characteristics in different regions (Hu & Chen, 2011) , as shown in [Fig. 8-32] and [Fig. 8-33].

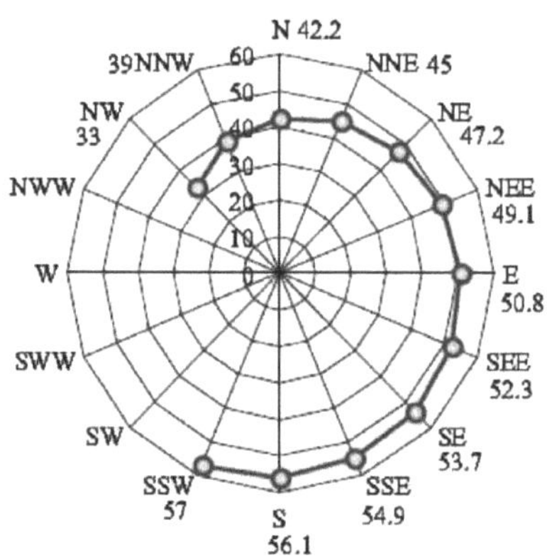

[Fig. 8-32] Variation of wind speed and direction of typhoon "Tiantu" gust, source: (LIU Junwei et al., 2022)

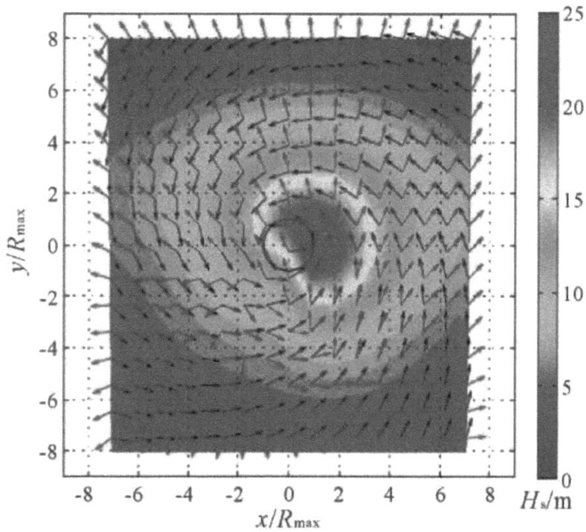

[Fig. 8-33] Distribution of wind direction and main wave direction, source: (WANG Jun et al., 2009)

Under the excitation of different wind speed spectrums, the wind-induced pulsation response of wind turbines is affected (CHU Chenhui & CHEN Shaolin, 2021). Under the combined action of typhoon and wave, the horizontal cyclic load on the pile foundation will change in time and space, which will inevitably lead to the change of the spatial deformation characteristics of the pile foundation, which has been verified in the limited regular multi-directional loading tests (such as figure 8, cross, etc.) (Su et al., 2014), (Rudolph et al., 2014) and it is found that the load change does increase the horizontal deformation of the pile foundation. However, the interaction mechanism between the open large-diameter pipe pile and the complex seabed under the superimposition of multi degrees of freedom and high-strength loads in the typhoon environment is still unclear, and targeted research needs to be carried out to establish a quantitative analysis method of "soil plug pipe pile soil" cyclic interaction in the complex soil under the typhoon environment.

8.4 Influence of wind load on high pile cap foundation

8.4.1 Analysis of horizontal bearing characteristics of fan pile foundation and research on measures to improve its bearing capacity

For three different foundation conditions, Han Jiwei studied the bearing characteristics of a single pile under horizontal load through a numerical method (HAN Jiwei et al., 2022).

8.4.1.1 Model establishment and working condition setting

ABAQUS software is used for 3D simulation. Set the elastic modulus of steel pipe pile $E = 210\ GPa$, Poisson's ratio $v = 0.25$, density $\rho = 7.85\ g/cm^3$, and wall thickness 8 cm. The density, strength index, elastic modulus, Poisson's ratio, soil layer thickness, pile length, and pile diameter of soil in the model are determined according to specific working conditions (Tab.8-11). The mohr-Coulomb model can simulate the Elastoplast city and the nonlinearity of soil mass. The model parameters are few and easy to obtain. It is the most commonly used soil constitutive model in geotechnical engineering. Therefore, the Mohr-Coulomb elastoplastic model is used for soil mass in this paper. The pile body adopts a linear elastic constitutive model.

To compare and analyze the influence of pile diameter, pile length, and foundation soil conditions on the horizontal bearing capacity of pile foundation, five pile diameters and pile lengths of different sizes are set respectively.

[Tab. 8-11] Test programs, source: (HAN Jiwei et al., 2022)

Soil layer distribution		The thickness of soil layer /m	density(/ kg/m^3)	Compressive modulus /MPa	Poisson's ratio	internal friction angle(/ °)	Cohesion /kPa
uniform soil		75	1900	12.0	0.35	20	20
Up bad, down good	Poor soil	15	1800	5.0	0.45	8	10
	Good soil	60	1950	12.0	0.30	20	20
Poor soil in the middle	Good soil	5	1950	12.0	0.30	20	20
	Poor soil	15	1800	5.0	0.45	8	10
	Good soil	55	1950	12.0	0.30	20	20

For foundation soil, the working conditions of homogeneous foundation, upper poor lower good, and middle poor soil are set. See [Tab. 8-11] for specific data.

The pile surface is the main surface, and the soil surface is the secondary surface. The normal direction of the contact property between the pile and the foundation soil is "hard contact", and the tangential type is "penalty". The model size is 100 m × 100 m × 75 m (Length × wide × High). The hexahedral grid is used to divide the pile and soil mass. The pile grid size is approximately 2 m, and the soil mass grid adopts a single precision grid method. The grid size of the soil mass close to the pile is approximately 0.75 m, and the grid size of the soil mass far from the pile is approximately 3 m. [Fig. 8-34] and [Fig. 8-35] show the grid division of the soil mass and the pile respectively.

[Fig. 8-34] Model sketch, source: (HAN Jiwei et al., 2022)

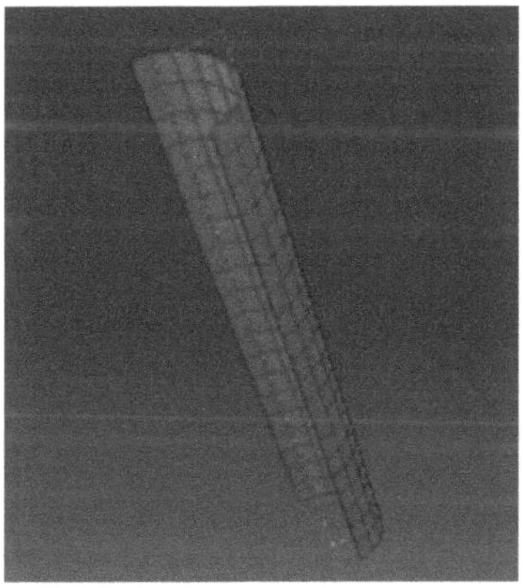

[Fig. 8-35] Finite element meshing of the pile, source: (HAN Jiwei et al., 2022)

8.4.1.2 Results and analysis

For the single pile foundation of offshore wind turbines, it is generally to control the displacement or angle at the mud surface to not exceed a certain limit. Since this paper mainly studies the regularity of the horizontal bearing characteristics of a single pile, and the limit value of the failure standard does not change the law, it can be seen in the following text that the load-displacement curve of the pile foundation under each working condition in this paper belongs to the slowly changing type, without obvious turning point, so this paper takes the horizontal displacement at the pile foundation mud surface to reach 5 cm as the failure standard to determine the horizontal bearing capacity of a single pile.

(1) Horizontal bearing characteristics of single pile foundation in homogeneous soil

1) Influence of pile diameter on horizontal bearing characteristics of single pile

[Fig. 8-36] shows the load-displacement curve of a single pile with a pile length of 40 m and different pile diameters. It can be seen from [Fig. 8-36] that with the increase of horizontal load, the displacement of the pile top increases gradually, but there is no obvious turning point; In addition, as the pile diameter increases, the horizontal bearing capacity of a single pile

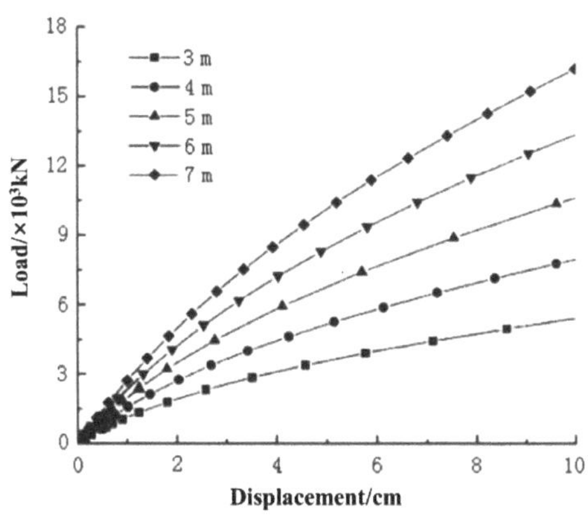

[Fig. 8-36] Horizontal load-displacement curves of the single pile with different pile diameters in the homogeneous soil, source: (HAN Jiwei et al., 2022)

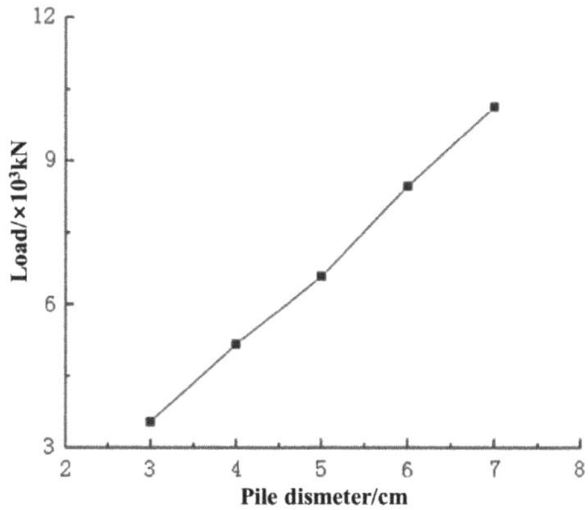

[Fig. 8-37] Relationship between pile diameter and horizontal bearing capacity of the single pile in the homogeneous soil, source: (HAN Jiwei et al., 2022)

increases. Under the failure criteria in this paper, when the pile diameter increases from 3 m to 7 m, the horizontal bearing capacity of a single pile increases linearly (Fig.8-37). The increase in pile diameter can significantly improve the horizontal bearing capacity of the pile, but in actual projects, the appropriate pile diameter should be determined in combination with production technology and construction technology.

2) Influence of pile length on the horizontal bearing capacity of a single pile

[Fig. 8-38] shows the load-displacement curve of a single pile foundation with a 5 m pile diameter and different pile lengths. It can be seen from [Fig. 8-38] that when the pile length increases from 20 m to 30 m, the horizontal bearing capacity of a single pile increases significantly, but when the pile length is increased later, the horizontal bearing capacity of a single pile increases little or even does not increase. For example, in [Fig. 8-38], three load-displacement curves with a pile length of 40 m, 50 m, and 60 m coincide, because the horizontal bearing capacity of a single pile mainly depends on the nature of the shallow soil around the pile, When the pile length increases to a certain extent, the pile end is in the embedded state.

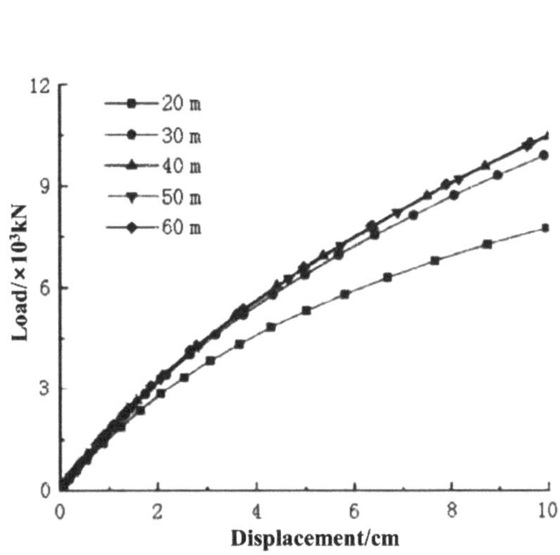

[Fig. 8-38] Horizontal load-displacement curves of the single pile with different pile lengths in the homogeneous soil, source: (HAN Jiwei et al., 2022)

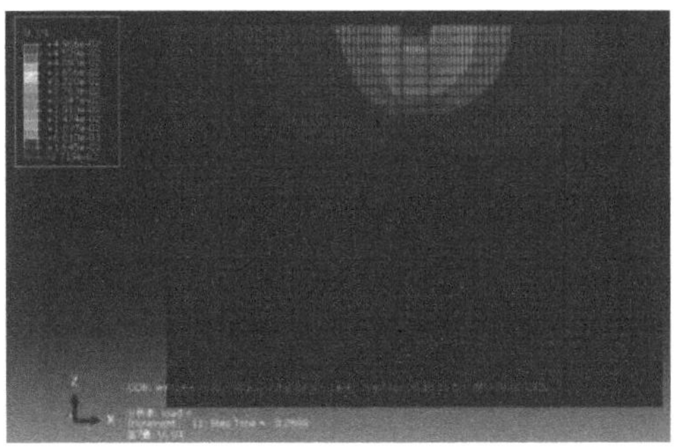

[Fig. 8-39] Displacement image of the pile foundation, source: (HAN Jiwei et al., 2022)

At this time, increasing the pile length will have little impact on the bearing capacity.[Fig. 8-39] shows the displacement program when the pile foundation is damaged under the condition of 60 m pile length. It can be seen that the displacement at the pile end is very small, and the extracted data shows that the displacement at the pile end is only 2.7 mm; It can also be seen from [Fig. 8-39] that the deformation of the soil around the pile mainly occurs within about 10 m below the mud surface (green and light blue areas in the figure), which again shows that the horizontal bearing capacity of a single pile mainly depends on the nature of the shallow soil around the pile. Therefore, it is unreasonable to increase the horizontal bearing capacity of a single pile simply by increasing the pile length.

(2) Horizontal bearing characteristics of single pile in layered soil

1) Horizontal bearing characteristics of single pile foundation in a good foundation

Due to the short sedimentation time, the poor upper and good lower foundation distribution is widespread. However, the existing research shows that the horizontal bearing capacity of pile foundation mainly depends on the nature of the soil within the range of 5 times the pile diameter depth under the mud surface. [Fig. 8-40] shows the horizontal load-displacement curves of a single pile with a pile diameter of 5 m and different pile lengths. It can be seen from the figure that when the pile length is short, increasing the pile length can significantly improve the horizontal bearing capacity of a single pile. This is because of the poor nature of the upper soil, increasing the pile length can obtain greater constraints in the lower soil, and then form a pile end embedded to improve the bearing capacity.

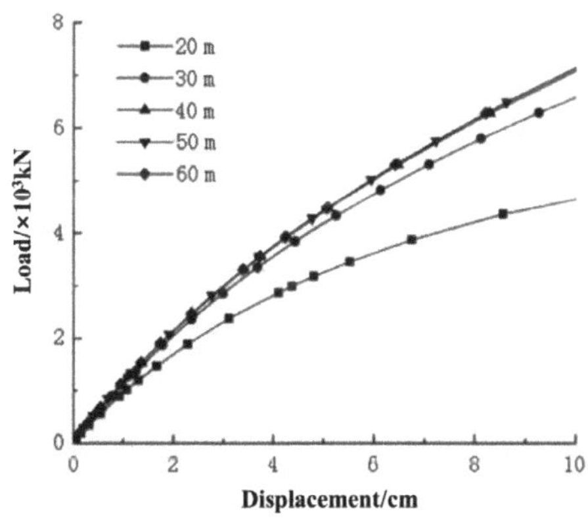

[Fig. 8-40] Horizontal load-displacement curves of the single pile with different pile lengths in the upper poor and lower good soil, source: (HAN Jiwei et al., 2022)

[Tab. 8-12] shows the data of the horizontal bearing capacity of a single pile with different pile diameters under two working conditions of homogeneous foundation and good foundation under upper and lower differences. It can be seen from [Tab. 8-12] that compared with a homogeneous foundations, the horizontal bearing capacity of a single piles is significantly reduced by about 35% due to the variation of upper soil.

[Tab. 8-13] shows the data of the horizontal bearing capacity of a single pile with different pile lengths under two working conditions: homogeneous foundation and good foundation under upper and lower differences. It can be seen from [Tab. 8-13] that compared with a homogeneous foundations, the horizontal bearing capacity of a single pile is significantly reduced by more than

[Tab. 8-12] Horizontal bearing capacity of the single pile with different diameters in the homogeneous, upper poor lower, and good soil, source: (HAN Jiwei et al., 2022)

Pile length /m	Pile diameter /m	Horizontal bearing capacity of single pile /kN		Decrease percentage /%
		Homogeneous foundation	The upper part is poor, and the lower part is good	
40	3	3548	2237	37.0
	4	5156	3296	36.1
	5	6585	4416	32.9
	6	8462	5536	34.6
	7	10134	6702	33.9

[Tab. 8-13] Horizontal bearing capacity of the single pile with different lengths in the homogeneous , upper poor, and lower good soil, source: (HAN Jiwei et al., 2022)

Pile length /m	Pile diameter /m	Horizontal bearing capacity of single pile /kN		Decrease percentage /%
		Homogeneous foundation	The upper part is poor, and the lower part is good	
	20	5282	3259	38.3
	30	6373	4166	34.6
	40	6585	4416	32.9
3	50	6648	4471	32.7
	60	6650	4477	32.7

30% due to the deterioration of the upper soil. However, it should be noted that when the pile diameter is fixed and the pile length increases to a certain value, the horizontal bearing capacity of a single pile is almost unchanged, Therefore, it can be seen from [Tab. 8-13] that the reduction percentage of three working conditions with pile diameters of 40 m, 50 m and 60 m is almost the same.

2) Horizontal bearing characteristics of a single pile in foundation with poor soil in the middle

[Fig. 8-41] shows the horizontal load-displacement curve of a single pile with different pile diameters when there is poor soil in the foundation. It can be seen from Figure 8 that for the foundation with differential soil in the middle, increasing the pile diameter can also significantly improve the horizontal bearing capacity of a single pile. [Fig. 8-42] shows the horizontal load-displacement curve of a single pile with different pile lengths in the foundation with poor soil in the middle, which is similar to the horizontal characteristics of a single pile in a homogeneous foundation. When the pile length increases to a certain value, the horizontal bearing capacity of a single pile cannot be improved by increasing the pile length.

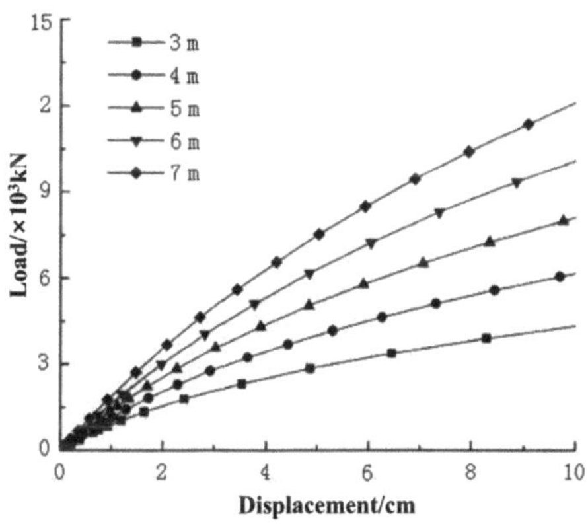

[Fig. 8-41] Horizontal load-displacement curves of the single pile with different pile diameters in the intermediate poor soil, source: (HAN Jiwei et al., 2022)

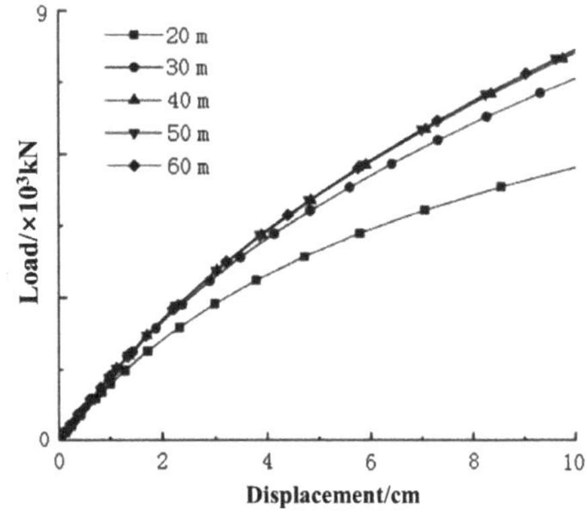

[Fig. 8-42] Horizontal load-displacement curves of the single pile with different pile lengths in the intermediate poor soil, source: (HAN Jiwei et al., 2022)

[Tab. 8-14] shows the data of the horizontal bearing capacity of a single pile with different pile diameters under the two working conditions of a homogeneous foundation and a foundation with poor soil inclusion in the middle. [Tab. 8-15] shows the data of the horizontal bearing capacity of a single pile with different pile lengths under the two working conditions of a homogeneous foundation and a foundation with poor soil in the middle. According to the analysis of the data in [Tab. 8-14] and [Tab. 8-15], when the foundation is mixed with poor soil, compared with the homogeneous stratum, the horizontal bearing capacity of a single pile is significantly reduced, no matter what the pile length and diameter are, with a reduction of more than 20%.

[Tab. 8-14] Horizontal bearing capacity of the single pile with different diameters in the homogeneous and intermediate poor soil, source: (HAN Jiwei et al., 2022)

Pile length /m	Pile diameter /m	Horizontal bearing capacity of single pile /kN		Decrease percentage /%
		Homogeneous foundation	The upper part is poor, and the lower part is good	
40	3	3548	2888	18.6
	4	5156	3981	22.8
	5	6585	5166	21.5
	6	8462	6283	25.8
	7	10134	7467	26.3

[Tab. 8-15] Horizontal bearing capacity of the single pile with different lengths in the homogeneous and intermediate poor soil, source: (HAN Jiwei et al., 2022)

Pile length /m	Pile diameter /m	Horizontal bearing capacity of single pile /kN		Decrease percentage /%
		Homogeneous foundation	The upper part is poor, and the lower part is good	
5	20	5282	4001	24.3
	30	6373	4929	22.6
	40	6585	5193	21.1
	50	6648	5193	22.0
	60	6650	5226	21.4

8.4.2 Numerical simulation of horizontal bearing capacity of offshore wind power high pile cap foundation in radial sandbar area

Based on the results of the horizontal static load test of a single pile in Jiangsu Province, the finite element numerical model of a high pile cap foundation is established at Jiaojin Perimeter, and the numerical simulation research on the horizontal bearing capacity of pile cap is carried out (TOU Jinzhou et al., 2022).

8.4.2.1 Stratum physical and mechanical distribution and numerical simulation parameters

[Tab. 8-16] Physical and mechanical distribution of formation and numerical simulation parameters, source:(Shen Xiaolei et al., 2021)

No.	Thickness / m	Elastic modulus / MPa a	Poisson's ratio	Cohesion /kPa	Friction angle /°	Penalty friction coefficient	Expansion angle /°	Effective gravity/ $kN \cdot m^3$
② Silty sand	4.5	145	0.3	4.83	33.9	0.49	3.9	8.6
③ Silt mixed with silt	7.5	138	0.3	2	32.0	0.56	2.0	9.3
④-1 Silty clay	4.5	72	0.3	31	30.0	0.49	0.1	9.1

No.	Thickness / m	Elastic modulus/ MPa a	Poisson's ratio	Cohesion /kPa	Friction angle /°	Penalty friction coefficient	Expansion angle /°	Effective gravity/ $kN \cdot m^3$
④-2 Silty clay mixed with silt	9.5	69	0.3	31	30.0	0.44	0.1	8.4
⑤ Silt mixed with silty clay	4.6	69	0.3	8.8	29.8	0.52	0.1	9.1
⑥-1 Silty sand	10.0	133	0.3	4.82	33.4	-	3.4	9.6
⑥-3 Medium sand	17.0	123	0.3	5.8	33.1	-	3.1	9.9
⑦ Medium sand	2.4	157	0.3	3.2	34.5	-	4.5	10.1

8.4.2.2 Horizontal bearing characteristics of single pile foundation

There is a lack of trial installation data of stratum bearing capacity characteristics in the offshore radial sandbar area of Jiangsu Province, and the horizontal static load test of pile group foundation of high pile cap is expensive and difficult. Radial sandbar 1 Field test on the horizontal bearing capacity of an 8 m steel pipe pile is shown in [Fig. 8-43]. The on-site test pile S1 has a pile length of 51 m and a pile diameter of 1 8 m, a wall thickness of 25 mm, and a pile buried depth of 29 m. The test pile takes the soil layer ⑤ - silty soil mixed with silty clay as the bearing stratum, and the horizontal force acting point of the test pile is located at 0.5% below the pile top At 5 m, horizontal static load test pile top horizontal load displacement (Q-s) curve, as shown in [Fig. 8-45].

ABAQUS analysis software is used to establish the numerical model of a single pile foundation, as shown in [Fig. 8-44]. The model is selected as a semi-cylinder, with a radius of 30 m and a depth of 60 m. The bottom and the four sides are all constrained. The normal displacement of the symmetry plane is constrained, and the top surface is free. Hexahedral element-type grid is used for division. For the contact between the steel pipe pile and the surrounding soil, the "penalty" function is used to assign the corresponding pile-soil friction relationship. According to the stratum distribution and soil physical and mechanical data measured during the on-site pile test, the pile-soil penalty friction coefficient is obtained by optimizing and adjusting the numerical simulation parameters through the numerical simulation calculation of the horizontal load displacement of the numerical model. It can be seen from [Fig. 8-45] that the horizontal load-displacement curves obtained from the field test pile and ABAQUS numerical simulation are consistent. At present, the diameter of a single pile in the field horizontal static load test conducted in this sea area is 1 8 m, steel pipe pile of high pile cap is 1 5 m and 2 Therefore, the soil physical and mechanical parameters obtained from the static load test and the pile-soil contact parameters obtained from the numerical simulation test can be used to analyze the bearing characteristics of the high pile cap foundation in this sea area.

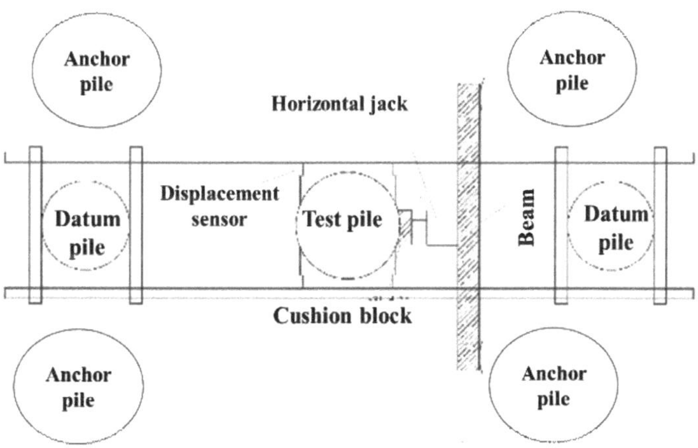

[Fig. 8-43] Horizontal static load test, source: (TOU Jinzhou et al., 2022)

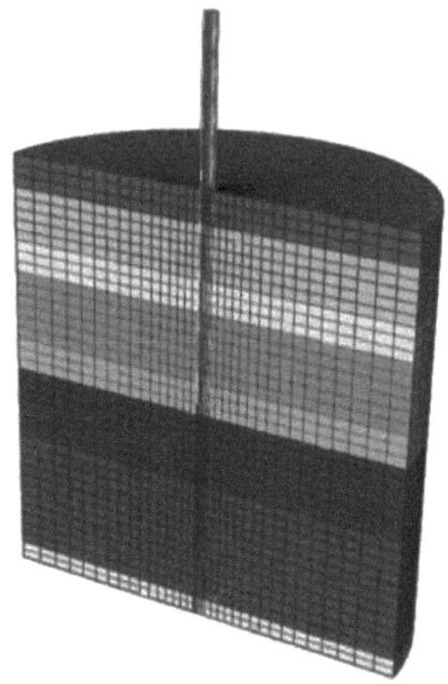

[Fig. 8-44] Numerical model of single pile foundation in ABAQUS, source: (TOU Jinzhou et al., 2022)

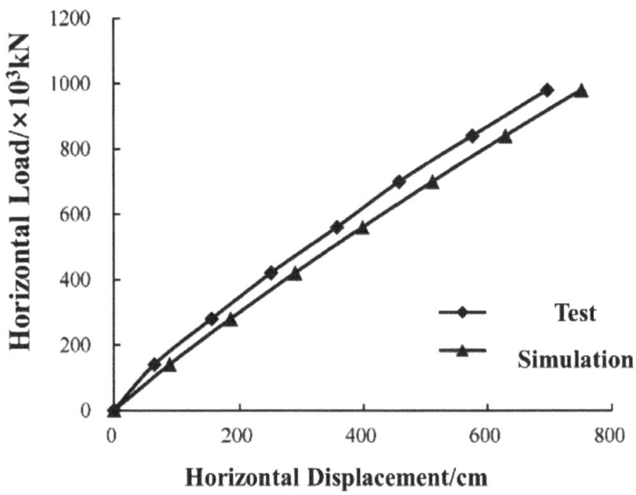

[Fig. 8-45] Pile top horizontal Q-s curve, source: (TOU Jinzhou et al., 2022)

8.4.2.3 Horizontal bearing characteristics of high pile caps

(1) ABAQUS numerical model

The numerical model of the ABAQUS high pile cap foundation is established, as shown in [Fig. 8-46. The cushion cap material is C45 concrete, and the C3D8R unit is used for modeling; The steel pipe pile group is an elastic pile made of Q335C; the Mohr - Coulomb model is usedfor soil mass; The pile and platform are considered as a whole, without considering the friction between the pile and the platform; The general contact is adopted for the pile-soil contact, the contact property is normal linear contact, and the tangent is given the corresponding friction coefficient through the "penalty" function.

[Fig. 8-46] ABAQUS numerical model of pile foundation of high pile platform, source: (TOU Jinzhou et al., 2022)

(2) Numerical analysis results

The field static load test of high pile cap foundation is not only expensive but also extremely difficult to carry out offshore pile tests. Therefore, the numerical simulation study on the horizontal bearing capacity of high pile cap foundations can provide technical support for the subsequent construction of wind farms in the sea area. As the stress loading method can only obtain the displacement of the platform under a specific load and obtain discrete Q-s points, to obtain a continuous Q-s curve, this numerical simulation test uses the displacement loading method to conduct a numerical simulation analysis of the cushion cap under horizontal loading. Displacement loading refers to the method of applying displacement boundary conditions to obtain the corresponding bearing capacity; Compared with stress loading, this method can avoid numerical calculation nonconvergence caused by excessive applied stress. The displacement loading numerical simulation analysis is carried out for the upper platform of the pile cap in the long axis direction (AC direction), and the horizontal load displacement (Q-s) curve of the high pile cap foundation is obtained, as shown in [Fig. 8-47]. There are three inflection points on the curve: point A, point B and point C. The PEEQVAVG diagram of the equivalent plastic strain of soil mass (average volume in a single Euler element) and PEEQ diagram of the equivalent plastic strain of pile shaft (Fig. 8-48~Fig. 8-50) at three points.

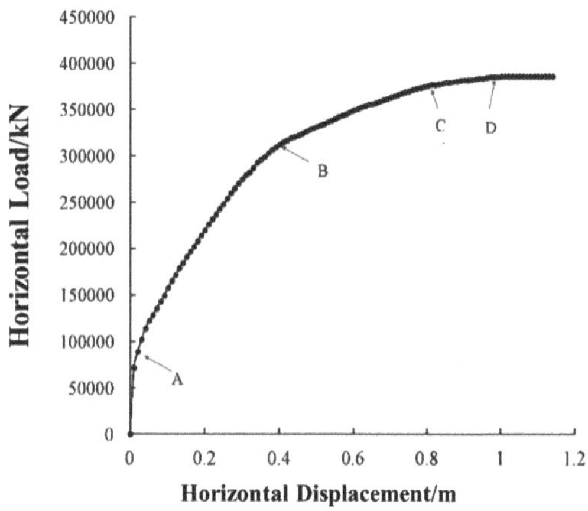

[Fig. 8-47] The relationship between horizontal displacement and bearing capacity of the cap, source: (TOU Jinzhou et al., 2022)

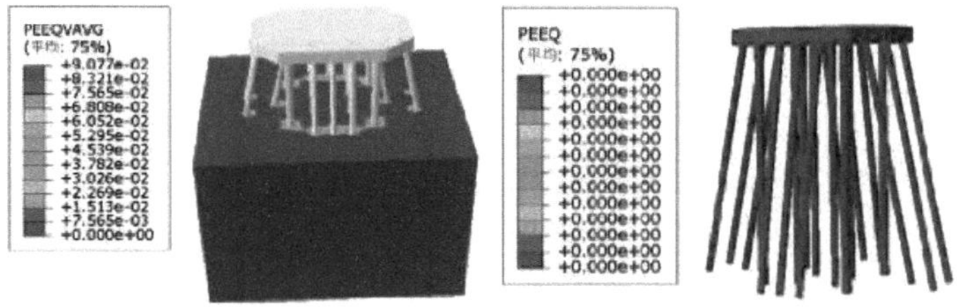

(a) Soil (b) Steel pipe pile

source: (TOU Jinzhou et al., 2022)

[Fig. 8-48] The equivalent plastic strain of the pile and the soil (point A),

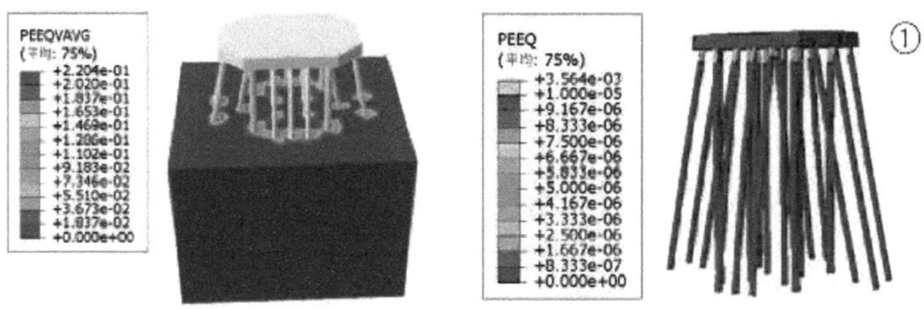

(a) Soil (b) Steel pipe pile

[Fig. 8-49] The equivalent plastic strain of the pile and the soul (point B), source: (TOU Jinzhou et al., 2022)

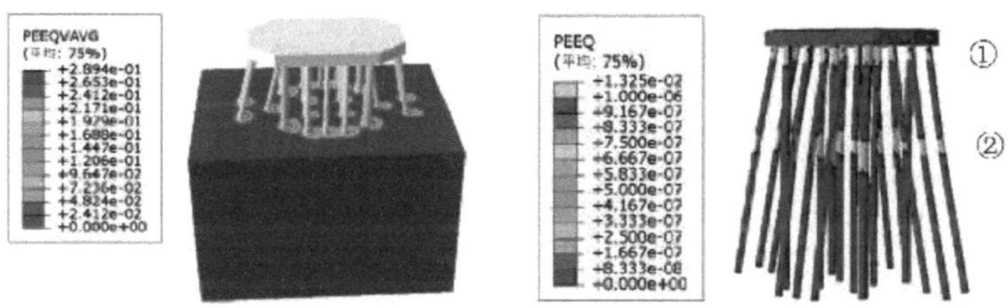

(a) Soil (b) Steel pipe pile

[Fig. 8-50] The equivalent plastic strain of the pile and the soil (point C), source: (TOU Jinzhou et al., 2022)

Influence of wave load on wind power structure foundation

9

9.1 Influence of wave load on horizontal deformation of wind power single pile foundation

Large diameter monopile foundation is a foundation form widely used in offshore wind power. Strictly controlling the displacement of the monopile is the key factor to ensure the stability of the foundation and the safe operation of the wind turbine. A three-dimensional monopile-seabed numerical model is established. The elastic-plastic constitutive model, which can describe the over-consolidation and structure of marine sand, is embedded into the finite element software ABAQUS by the UMAT subroutine. The wave load on the monopile foundation is calculated by the Morison equation. Three cases, including no wave load, wave load acting only on the seabed, and wave load acting on both monopile foundation and seabed, are adopted to analyze the dynamic response of seabed soil and horizontal displacement of monopile foundation(WANG Tao et al., 2022).

9.1.1 Pile foundation seabed soil numerical model

The elastoplastic constitutive model of seabed soil is developed from the upper and lower load surfaces, which can take into account the over-consolidation and structure of soil (the fabric of marine sand is represented by the size, porosity, and arrangement of sand particles). Five parameters of the constitutive model are the same as those of the Cambridge model, namely, the slope M of the critical state line and the initial void ratio e_0. Poisson's ratio v, compression index λ, expansion index x, and three other parameters are respectively the control parameter m of over consolidation m_R (controlling the dissipation rate of over-consolidation), structural control parameter m_R * (controlling the rate of structural decay), the anisotropic control parameter b_r (controls the rate of anisotropy development). The initial state parameters include the initial over-consolidation ratio R_0 and initial structural R_0^*. [Fig. 9-1] shows the upper and lower load surfaces and the normal yield surface of the constitutive model.

The ratio of the normal yield surface to the upper load surface is defined as R^*, and the ratio of the lower load surface to the upper load surface is defined as R. There are:

$$R^* = \frac{\tilde{p}'}{\bar{p}'} = \frac{\tilde{q}}{\bar{q}},\ 0 < R^* \leqslant 1,\ \frac{\tilde{q}}{\tilde{p}'} = \frac{\bar{q}}{\bar{p}'}$$

$$R = \frac{p'}{\bar{p}'} = \frac{q}{\bar{q}},\ 0 < R \leqslant 1,\ \frac{q}{p'} = \frac{\bar{q}}{\bar{p}'}$$

where: $(p',q),(\tilde{p},\tilde{q}),(\bar{p}',\bar{q})$ respectively represents the current stress state, the corresponding normal consolidation stress state, and the stress state of natural soil on the p-q plane.

The current stress state is always on the load surface, and its expression is:

$$f = \ln\frac{\tilde{p}}{\tilde{p}_0} + \ln\frac{M^2 - \xi^2 + \tilde{\eta}^{*2}}{M^2 - \zeta^2} + \ln R^* - \ln R - \frac{\epsilon_v^p}{C_P} = 0$$

where: $\tilde{p}$ is average effective stress; $\tilde{p}_0 = 98kPa$ = is reference stress; ϵ_v^p is plastic volumetric strain; ζ is a quantity describing the magnitude of stress-induced anisotropy, $\zeta = \sqrt{\frac{3}{2}\beta_{ij}\beta_{ij}}$, β_{ij} is anisotropic stress tensor; $\tilde{\eta}^* = \eta^* = \sqrt{\frac{3}{2}\hat{\eta}_{ij}\hat{\eta}_{ij}}$

Its expression equation is:

$$C_p = \frac{\lambda - x}{1 + e_0}$$

The associated flow rule is adopted in the model:

$$d\epsilon_{ij}^p = \Lambda\frac{\partial f}{\partial \sigma_\eta}$$

The coordination equation of the lower load surface is:

$$df = 0 \Rightarrow \frac{\partial f}{\partial \sigma_i} d\sigma_{ij} + \frac{\partial f}{\partial \beta_{ij}} d\beta_{ij} + \frac{1}{R^*} dR - \frac{1}{R} dR - \frac{1}{C_p} ds_v^p = 0$$

In addition, the development formula of the three parameters in the constitutive model:

$$dR = U\|d\epsilon_{\psi j}^v\| + R\frac{\eta \partial f}{M \partial \beta_j} d\beta_q$$

where: $\|ds_j^p\| = \sqrt{ds_{jj}^p s_j^p} = \Lambda \sqrt{\frac{\partial f}{\partial \sigma_{jj} \partial \sigma_q}} = \Lambda \frac{\sqrt{6\eta^{*2} + \frac{1}{3}(M^2 - \eta^2)^2}}{(M^2 - \zeta^2 + \eta^{*2})\sigma_n}$;

$$U = \frac{mM}{C_{\bar{p}}}\left[\frac{(\bar{p}/\bar{p}_0)^2}{(\bar{p}/\bar{p}_0)^2 + 1}\right]\ln R_0$$

Structural state parameter development formula:

$$R^* = U^* \ d\epsilon_d^p, U^* = \frac{m_R \cdot M}{C_P} R^*(1 - R^*)$$

Development of stress-induced anisotropy tensor:

$$d\beta = \frac{M}{C_p} b_r (M - \zeta) d\epsilon_d^p \frac{\eta_{ij}}{\| \eta_{ij} \|} = \sqrt{\frac{3}{2}\frac{M}{C_p}} b_p (M - \zeta) ds_d^p \frac{\hat{\eta}_q}{\eta^*}$$

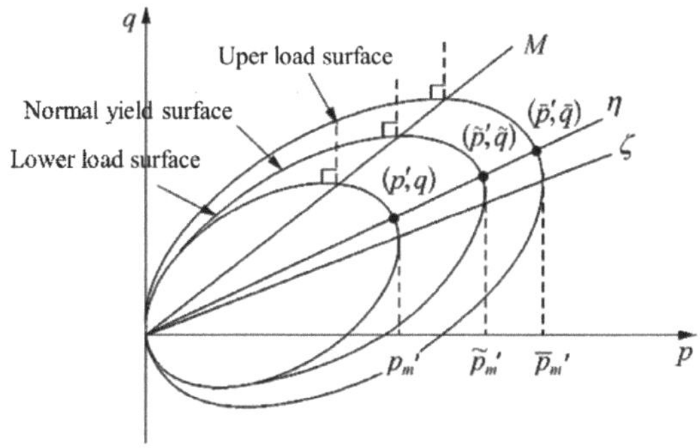

[Fig. 9-1] The relationship between the sub-loading surface, the normal field surface, and the super loading surface, source: (WANG Tao et al., 2022)

9.1.2 Boundary condition

The numerical model calculation diagram is shown in [Fig. 9-2]. The four surfaces of the seabed soil and the bottom of the seabed are set as impermeable boundaries, the lateral constraints on horizontal displacement, and the bottom of the seabed is a fixed boundary. The seabed surface is set as the permeable boundary, and vertical cyclic wave load is applied, namely:

$$p(x, y, t) = P_b(x, t) = P_0 \sin(kx - \omega t), \tau_{xt} = \tau_x = 0$$

where , $P_0 = \dfrac{\gamma_\omega H}{\cosh(kd)}$, k is wave number , ω is angle frequency , t is time , γ_ω is the specific gravity of seawater , H is wave height , d is wave depth. The pore water pressure on the seabed surface is the pressure generated by waves. At the same time, the shear force exerted by waves on the seabed surface is ignored. The wave load is exerted on the seabed surface through the pore water pressure boundary and the pressure boundary at the same time to ensure that the effective stress of the soil mass on the seabed surface is zero.

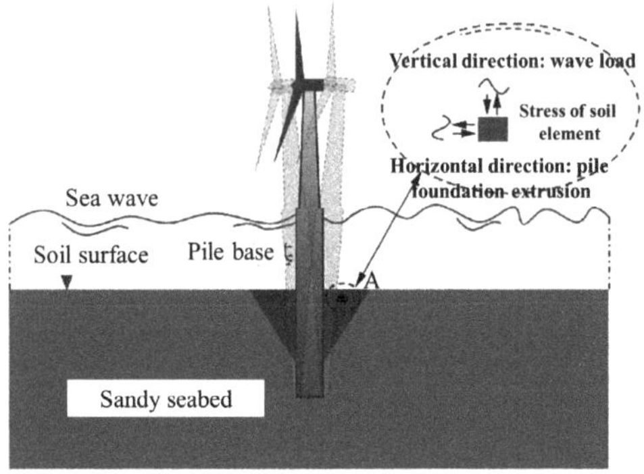

[Fig. 9-2] Numerical model calculation diagram, source: (WANG Tao et al., 2022)

The wave load acting on the pile foundation is calculated by the Morison formula. The wave forces on pile foundations include the drag force caused by the horizontal velocity of wave water particles and the inertial force caused by the horizontal acceleration of wave water particles. The wave load on the pile foundation is distributed on the side surface of the pile foundation above the mud surface and below the water surface. The horizontal load on the dz length can be calculated according to the following formula, and the schematic diagram of the horizontal load is shown in [Fig. 9-3].

$$dF_H = f_H\, dz = \frac{1}{2} C_D \rho D u_x |u_x| dz + C_D \rho \frac{\pi D^2}{4} \frac{\partial u_x}{\partial t}\, dz$$

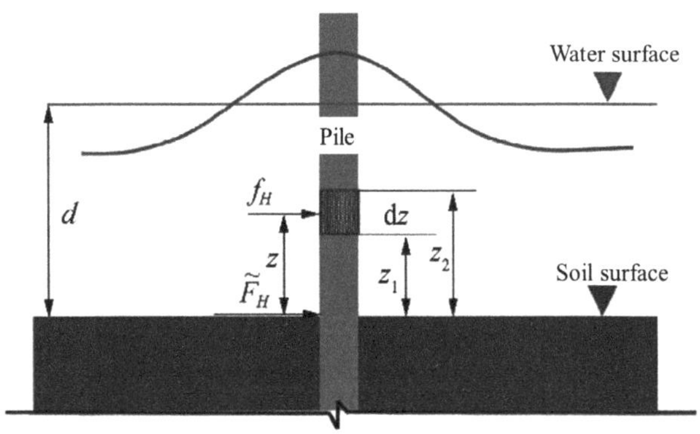

[Fig. 9-3] Diagram of horizontal wave loads acting on piles

The total force of wave load on the pile foundation can be obtained by integration:

$$F_H = C_D \frac{\gamma D H^2}{2} K_1' \cos\theta |\cos\theta| + C_M \frac{\gamma \pi D^2 H}{8} K_2' \sin\theta$$

$$\theta = kx - \omega t,\ k = \frac{2\pi}{L},\ \omega = \frac{2\pi}{T}$$

$$K_1' = \frac{2k\left(d + \frac{H}{2}\right) + sh\, 2k\left(d + \frac{H}{2}\right)}{8 sh\, 2kd}$$

$$K_2' = \frac{sh\, kd}{ch\, kd}$$

where: F_H is the resultant force of wave force on the pile in the length range above the mud surface; γ is the weight of water; D is the diameter of the pile foundation; L is the Wavelength; T is the period of the wave; H is wave height; d is water depth above mud level; Drag force coefficient and inertia force coefficient related to section shape respectively for C_D and C_M, Refer to the DNV specification C_D = 0.7, C_M = . To simplify the modeling and calculation, only the pile foundation and seabed models below the mud surface are established. Since the bending moment of the wave force at 10 mathrm {} m water depth at the mud surface contributes little to the horizontal displacement of the pile foundation, the following only considers that the cyclic wave force obtained by integral calculation acts on the pile top at the mud surface in the form of concentrated force, and does not consider the influence of the bending moment at the pile top at the mud surface.

The contact between the pile foundation and the seabed adopts the master-slave contact algorithm. The normal direction is hard to contact, the tangential direction is the Mohr-Coulomb friction penalty function, and the sliding friction coefficient of the interface is calculated as $u = \tan(0.75\varphi)$, where φ is the internal friction angle.

9.1.3 Establishment of numerical model

The elastic-plastic constitutive model of seabed soil is embedded into the finite element software ABAQUS through the UMAT subroutine, and the numerical model of pile foundation seabed under wave action is established. Considering the symmetry of the model, a half model is established. The C3D8P element that can consider the pore pressure is used for soil, and the linear elastic model is used for pile foundation simulation. According to the marine hydrological data of the East China Sea offshore project, the recurrence period of wave data used in this paper is 50 years. The specific parameters of the numerical model are shown in [Tab. 9-1].

[Tab. 9-1] Numerical model parameters, source: (WANG Tao et al., 2022)

physical quantity	value	physical quantity	value
Seawater density / $kg \cdot m^{-3}$	1000	Soil compression index	0.05
Wave period /s	7.76	Soil rebound index	0.0064
Wavelength /m	74.1	Soil anisotropy	1
Water depth /m	10	Overconsolidation attenuation coefficient	0.1
Permeability coefficient / $m \cdot s^{-1}$	0.0001	Structural attenuation coefficient	2.2
Soil density / $kg \cdot m^{-3}$	1800	Buried depth of pile foundation /m	20
Initial void ratio of the soil mass	1.0	Pile diameter /m	6
Poisson's ratio of the soil mass	0.3	Pile density / $kg \cdot m^{-3}$	2500
Wave height /m	1.5	Elastic modulus of pile /GPa	210
Critical state slope	1.2	Poisson's ratio of pile	0.17

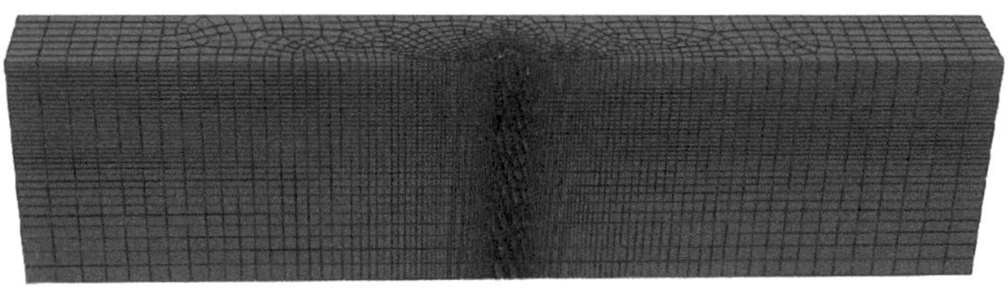

[Fig. 9-4] Meshing of the numerical model, source: (WANG Tao et al., 2022)

9.1.4 Response of seabed soil and horizontal displacement of pile foundation

To analyze the development law of seabed liquefaction caused by waves, the cloud chart display of the excess pore pressure ratio of seabed soil is realized through the UVARM subroutine, and the user-defined variable b is b_{UVARM3} represents excess pore pressure ratio:

$$b_{UVARM3} = \frac{P_e}{\sigma'_{z0}}$$

Where , P_e is excess pore pressure of seabed soil , σ'_{z0} is Initial vertical effective stress. So , when $b_{UVARM3} > 1$, it is considered that the soil mass has been liquefied(Jeng, 1997). [Fig. 9-5] shows the development trend of seabed liquefaction. The soil and pile foundation parameters are shown in [Table 9-1], and the over-consolidation parameters and structural parameters are both 1. The liquefaction depth and range of seabed soil mass gradually increase with time, and the liquefaction depth does not change after the 300s. [Fig. 9-6] shows the development law of the excess pore water pressure ratio of the joints on the pile side along the depth direction. The part beyond the liquefaction boundary in the figure indicates that the soil has been liquefied. It can be found that the excess pore water pressure ratio increases rapidly in the first 200 seconds, then gradually slows down, and the excess pore water pressure ratio of seabed soil tends to be stable within 15 m.

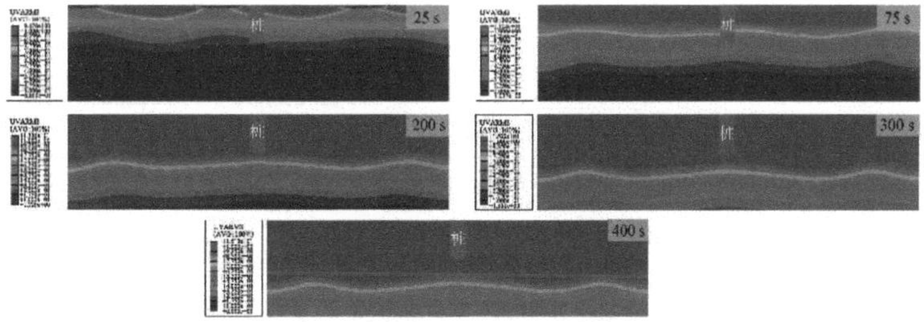

[Fig. 9-5] Development process of liquefaction, source: (WANG Tao et al., 2022)

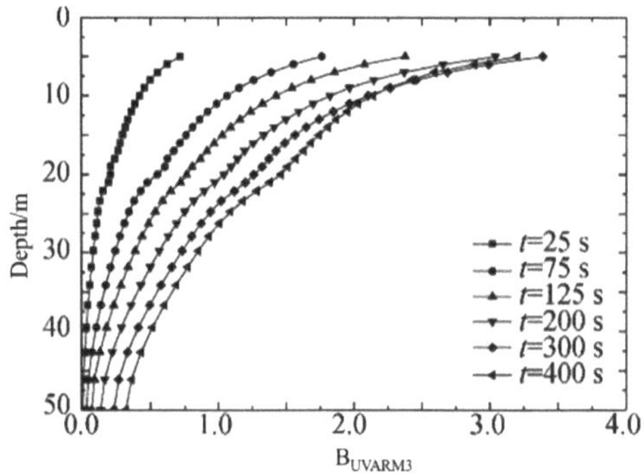

[Fig. 9-6] Development of EPWPR along the depth direction, source: (WANG Tao et al., 2022)

The wave load acting on the seabed will generate a vertical cyclic load on the seabed surface in the process of continuous propagation, while the load acting on the pile foundation will be transferred to the surrounding soil through the pile foundation, generating a horizontal cyclic load. Previous studies often ignored the horizontal cyclic load transferred to the seabed soil through the pile foundation, which would underestimate the effect of wave load on the pile foundation and seabed. Therefore, the comparison of three working conditions is considered respectively, and the specific working conditions are as follows:

1) Case 1 - without considering wave load, apply unidirectional horizontal load directly on the pile top ;
2) Case 2-Only the vertical wave load acting on the seabed surface is considered. When the soil liquefaction depth no longer changes, a one-way horizontal load is applied to the pile top ;
3) Case 3-At the same time, the vertical and horizontal wave loads acting on the pile foundation and seabed shall be considered. When the soil liquefaction depth no longer changes, the one-way horizontal load shall be applied on the pile top.

The schematic diagram of the three working conditions is shown in [Fig. 9-7] . Compare the horizontal displacement of the pile top under three working conditions and analyze the difference between the dynamic response of seabed soil and the horizontal displacement of the pile foundation.

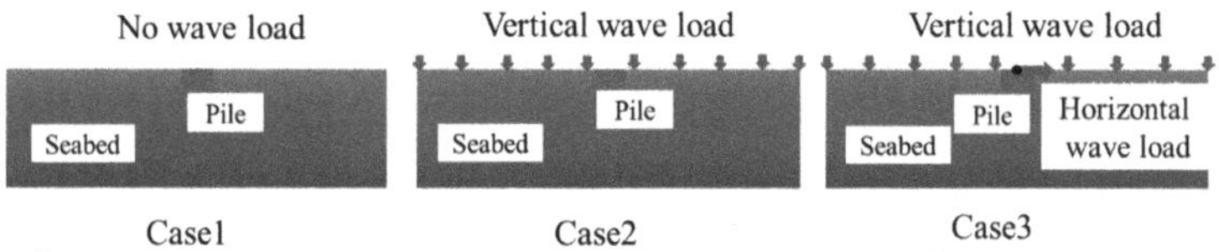

[Fig. 9-7] Wave load influence analysis of three cases, source: (WANG Tao et al., 2022)

[Fig. 9-8] shows the results of the development law of the pore pressure around the pile with time under Case 2 and Case 3. The development trend of excess pore water pressure of soil under the two working conditions is similar, and the pore water pressure results after 400s are similar. However, when considering the effect of horizontal wave load, the growth rate of excess pore pressure of seabed soil around the pile is faster than that of condition 2, which indicates that the horizontal cyclic load will accelerate the development of liquefaction of seabed soil.

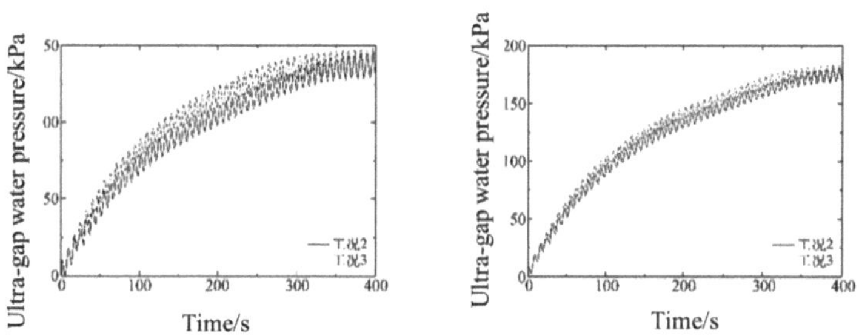

[Fig. 9-8] Development of EPWP of the soil around the pile in case 2 and case 3, source: (WANG Tao et al., 2022)

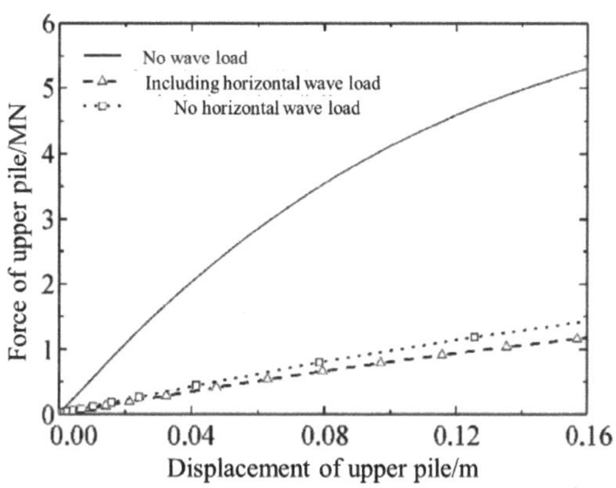

[Fig. 9-9] Comparison of horizontal displacement of pile top, source: (WANG Tao et al., 2022)

[Fig. 9-9] shows the results of pile top horizontal displacement under three working conditions. As the seabed liquefaction is caused by wave load, the horizontal displacement of the pile top in combination with cases 2 and 3 under wave load will reach 0.16 m under the horizontal load of about 1000 kN, while the horizontal load will exceed 5000 kN when the horizontal displacement of pile top under no wave load (combination mode 1) reaches 0.16m, which indicates that the seabed liquefaction caused by wave load has a significant impact on the horizontal displacement of pile foundation. By comparing the results of case 2 and case 3, when considering the horizontal cyclic load, the impact of the horizontal load on the pile foundation will be transferred to the surrounding seabed soil through the pile shaft, which will further reduce the horizontal bearing capacity of the seabed soil, and the horizontal displacement of case 3 will be greater under the same horizontal load.

9.2 Experimental influence of wave load on offshore wind power composite bucket foundation

9.2.1 Test Overview

With the development of computer technology, the numerical calculation does greatly enhance the ability to solve complex engineering problems, but its calculation accuracy remains to be discussed. On the one hand, the offshore foundation soil is mainly composed of silty clay, silty sand, and other soils. It is difficult to select a structural parameter that can not only quantitatively describe the soil structure type, but also be reflected in the model constitutive equation in the absence of model tests. Only when the model test research and numerical calculation work are carried out at the same time, can the results of the two be compared with each other, Only then can we determine the rationality of soil parameter selection and the accuracy of calculation results in numerical calculation; On the other hand, because the soil is not a linear elastic material, the stress level of the soil is very low in the small scale model test. Many soils show strong dilatancy and strain-softening characteristics under low confining pressure, and the strength index of the soil has been significantly improved. The friction strength of cohesive soil is generally linearly related to the confining pressure, while its cohesive strength is not affected by the confining pressure. Therefore, the cohesive strength of small-scale models cannot be reduced in proportion correctly. Therefore, small-scale tests are generally of little significance. Therefore, only large-scale model tests or full-scale tests can have high reliability(LIU Yonggang et al., 2016). Therefore, Wang Xuyue conducted an experimental study on the 1:10 prototype composite bucket foundation model in silty clay (WANG Xuyue, 2017).

9.2.1.1 Test soil mass

The silty clay is transported from Xiangshui Haishang Wind Farm to the test site and is gradually filled into the pre-dug pool with an excavator layer by layer, and leveled layer by layer manually with a shovel. The silty clay is about 100 m^2 in total, and then it is cured for more than 6 months. During this period, water is supplied continuously to keep the soil saturated. The soil around and at the bottom of the soil pool is sandy soil with good permeability, which is beneficial to the maintenance of silty clay in the pool(LIU Yonggang et al., 2016) The length and width of the test soil pool are about 10 m × 10 m, as shown in [Fig. 9-10]. [Fig. 9-11] shows the static penetration test.

[Fig. 9-10] Silty clay . test site, source: (WANG Xuyue, 2017)

[Fig. 9-11] Double bridge static penetration test, source: (WANG Xuyue, 2017)

Before the test, a double bridge static penetration test is carried out in the soil pool. As shown in [Fig. 9-11], the cone head of the penetrometer is separated from the friction cylinder, and the cone head resistance and side wall friction can be measured simultaneously. The cone head area is 10 cm^2, and the surface area of the friction cylinder is 200 cm^2. Static penetration test results are shown in [Fig. 9-12].

According to the results of a static penetration test, the soil mass can be preliminarily divided into two layers. The end resistance qc curve is smooth at the depth of 0~1.2 m, nearly vertical, and the qc value is low, averaging about

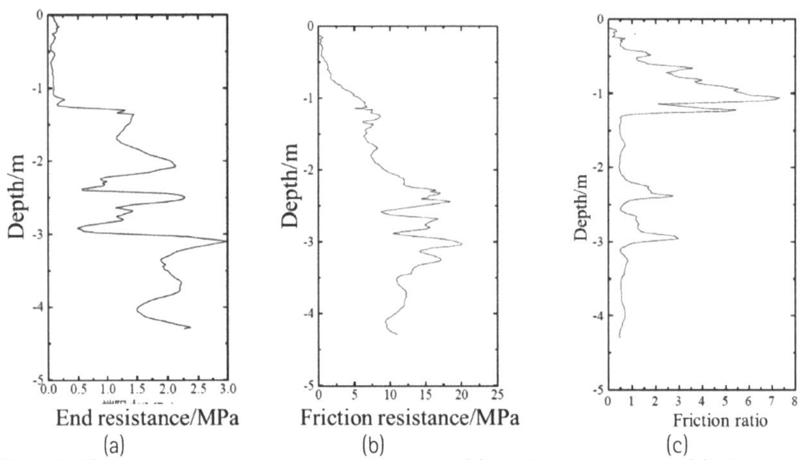

[Fig. 9-12] Results of a static penetration test: (a) End resistance qc, (b) Side wall friction fs, (c) Friction resistance ratio FR., source: (WANG Xuyue, 2017)

0.1 MPa. The friction ratio FR=fs/qc value is greater than 2% after the probe completely enters the mud surface, which conforms to the characteristics of silty clay. The thickness of silty clay is about 1.2 m, which is consistent with the height of actual soil preparation. The qc value of the soil layer below 1.2 m is larger than that of the upper layer, and the change frequency and amplitude are both large. The friction resistance curve is zigzag, the peak and valley are sharp, and the friction ratio FR value is less than 1, which conforms to the characteristics of sand and is consistent with the actual situation.

To further define various physical and mechanical property indexes of soil mass, indoor soil tests were carried out, mainly including a water content test, soil density test, limit water content test, rapid consolidation test, and direct shear test. Based on the results of the double bridge static cone penetration test and indoor geotechnical test, it is obtained that the density of silty clay is 1780 kg/m^3, the moisture content is 36.8%, the compression modulus is 3.688 MPa, the cohesion is 3.84 kPa, the internal friction angle is 7.14°, the plasticity index is 11.75, the liquidity index is 1.19, and the soil is in the flow plastic state. The e - p curve and f relationship of soil mass are shown in [Fig. 9-13] and [Fig. 9-14]. The main parameters of soil mass are listed in [Tab. 9-2].

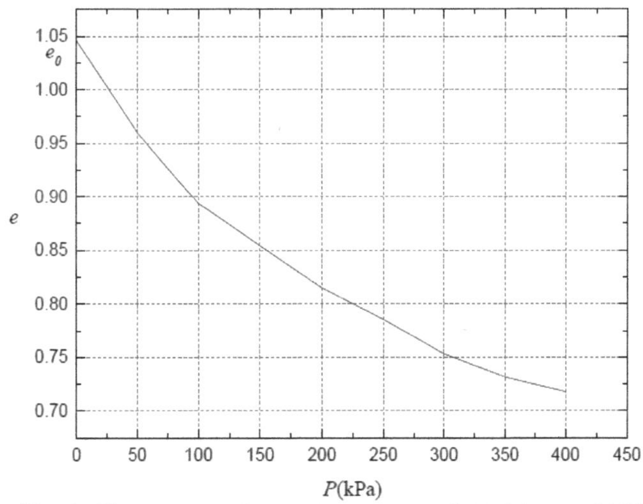

[Fig. 9-13] E-p curve of silty clay, source: (WANG Xuyue, 2017)

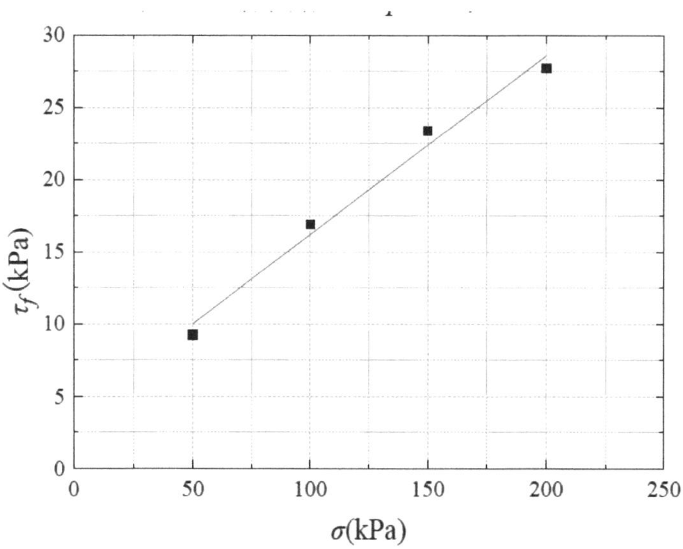

[Fig. 9-14] Relationship between vertical stress and shear stress in direct shear test, source: (WANG Xuyue, 2017)

[Tab. 9-2] Main parameters of test soil mass, source: (WANG Xuyue, 2017)

Density kg/m^3	Moisture content	Compressive modulus (Es)	Cohesion (c)	internal friction angle (υ)	Liquidity index	Plasticity index	Soil state
1780	36.8%	3.688 MPa	3.84 kPa	7.14°	1.19	11.75	Flow plastic state

9.2.1.2 test model

The tubular foundation used in the test is a full steel structure, and the dimensions of the tubular foundation model are listed in [Tab. 9-3]. The foundation is 3.5 m in diameter and 0.9 m in height; The tower has a diameter of 0.66 m and a height of 5 m. The barrel foundation is divided into 7 tanks, as shown in [Fig. 9-13]. The thickness of the cylinder wall, subdivision plate, cylinder top cover, and tower wall is 8 mm; Lifting lugs shall be set symmetrically at the height of 2 m, 3 m, 4 m, and 5 m of the tower barrel to facilitate load application; The connection between the tower barrel and the foundation is strengthened by ribbed plates. The total weight of the bucket foundation is about 2.8t. The test site and test model are shown in [Fig. 9-15] and [Fig. 9-16].

[Tab. 9-3] Dimensions of bucket foundation model, source: (WANG Xuyue, 2017)

Diameter of cylinder	Wall height	Cylinder wall thickness	Diameter of the tower drum	Tower height	Model quality
3.5 m	0.9 m	8 mm	0.66 m	5 m	2.8 t

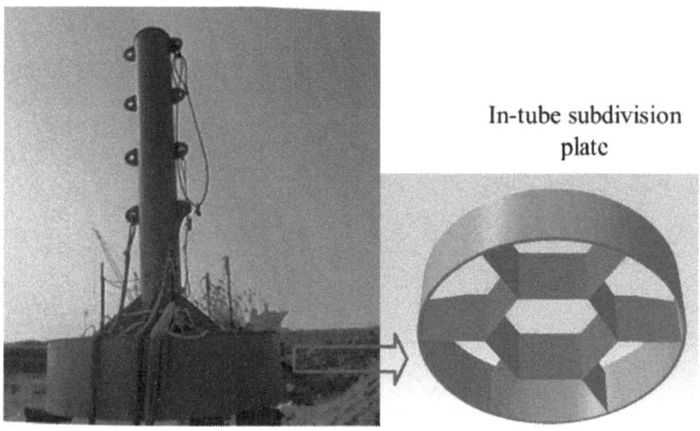

[Fig. 9-15] Subdivision plate structure of bucket foundation model, source: (WANG Xuyue, 2017)

[Fig. 9-16] Test site and test model, source: (WANG Xuyue, 2017)

The top cover of the bucket foundation model is provided with two removable holes with a diameter of 30 cm, as shown in [Fig. 9-17]. When the bucket foundation is sinking and loaded, it is sealed by bolts, which can be opened at any time to observe the conditions in the bucket, and a static penetration test can be conducted to measure the soil parameters in the bucket. A negative pressure hole is opened in the middle of each hatch top cover, which can be connected to the vacuum negative pressure equipment through a steel wire rubber hose, as shown in [Fig. 9-18].

[Fig. 9-17] Removable opening of the top cover, source: (WANG Xuyue, 2017)

[Fig. 9-18] Vacuum pump and air compressor, source: (WANG Xuyue, 2017)

9.2.2 Test content

9.2.2.1 Test device and layout

The method for calculating the force on the entire offshore fan and the load P and height y. This method has been used in previous studies (Bhattacharya & Adhikari, 2011), in which the cycle loading device used needs to be reconnected when carrying out the next group of cycle loading, which not only brings some inconvenience to the test but also causes inevitable interference to the soil around the foundation (Yu et al., 2015). Moreover, the amplitude of cyclic load applied on large-scale models is very large. In actual engineering, wind load and wave load are changing, so the following cyclic loading mode is designed in this test to make the amplitude of cyclic load change.

Two cyclic loading tests (constant amplitude cyclic loading test and variable amplitude cyclic loading test) are conducted under the same soil conditions. The bucket foundation model is installed in the soil by vacuum negative pressure equipment (Zhu et al., 2011) , (Erbrich & Tjelta, 1999). Before each cyclic loading test, place the bucket foundation for enough time after installation to make it reach a stable state. The cyclic load is applied to the lifting lug at a height of five meters on both sides of the tower.

There are 8 earth pressure sensors on the top cover of the bucket foundation, 16 on the side wall, and 2 on the bottom, totaling 26. It is mainly arranged along the load loading direction, in which 24 earth pressure sensors are in one plane to measure the earth pressure distribution in the passive zone, active zone, and top cover of the bucket foundation during load loading. The top cover earth pressure sensor is type A; Type B is adopted for the side wall; The bottom end is type C, as shown in [Fig. 9-19]. The thickness of the earth pressure box is 8 mm, which is consistent with the thickness of the cylinder wall. During installation, holes shall be opened at the corresponding positions on the cylinder foundation, with a diameter slightly larger than the diameter of the earth pressure box. Then the earth pressure box shall be put into the hole and fixed with gypsum and silica gel. The force-bearing surface of the earth pressure box shall be flush with the cylinder wall. The layout of the tank foundation subdivision and earth pressure sensor is shown in [Fig. 9-20].

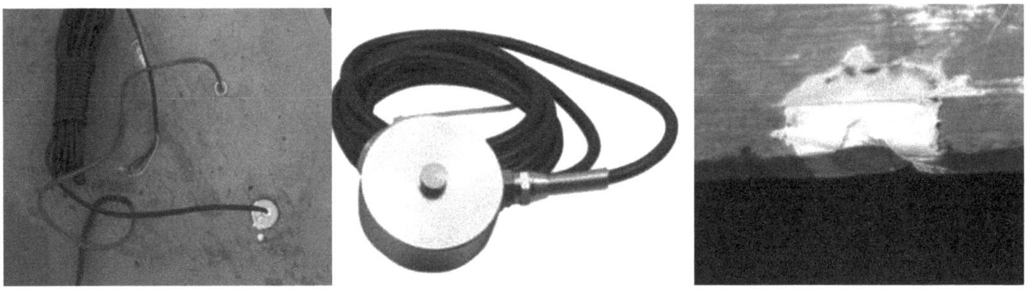

Type A Type B Type C

[Fig. 9-19] Soil pressure sensor, source: (WANG Xuyue, 2017)

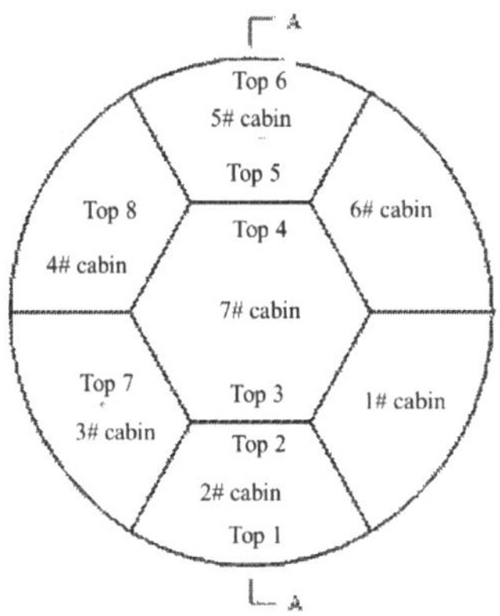

(a) Layout of the soil pressure sensor at the top of the cylinder viewed from top to bottom, source: (WANG Xuyue, 2017)

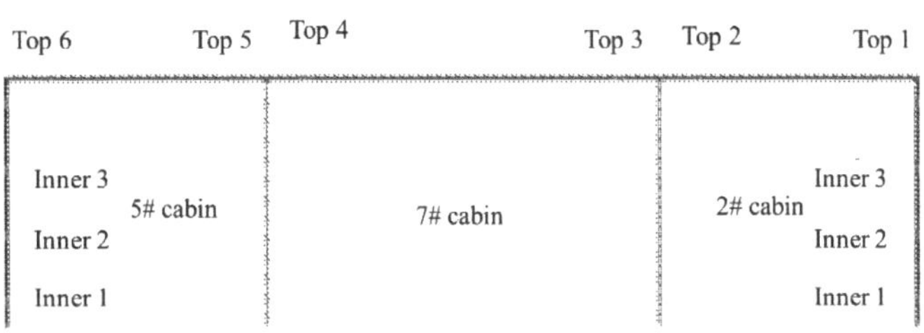

(b) A-A Layout of the profile pressure sensor

[Fig. 9-20] Layout of tank foundation subdivision and earth pressure sensor, source: (WANG Xuyue, 2017)

9.2.2.2 Test procedure and design

The first test is a cyclic load test with a variable amplitude value, and the second test is a cyclic load test with a constant amplitude value. The two-cycle loading process is shown in [Fig. 9-21]. The average cycle of the first test is 900 seconds, and the second cycle is about 200 seconds. The maximum frequency of the typical wave is 0.12 Hz, which is larger than the maximum frequency of wind load (Yu et al., 2015). The loading frequency of both tests is smaller than 0.12 Hz. The first cycle loading frequency is 0.001 Hz, and the second cycle loading frequency is 0.005 Hz. Based on the static test results before the cyclic loading test, the horizontal limit load of the bucket foundation model is 3.0 for the horizontal one-way load test applied to the 5m lifting 3.0×10^4 kN(Ding et al., 2015). This horizontal limit load corresponds to the overturning load, that is, under this load, the bucket foundation will overturn, which is very difficult to encounter in practical projects. Under the limit load of 50 year return period, the estimated limit load is 1.2×10^4 kN ~ 2.0×10^4 kN, so 2.0×10^4 kN is selected for convenience and conservatism is the ultimate load. During the second cyclic loading test, the cyclic load amplitude remains at 1.6×10^4 kN.

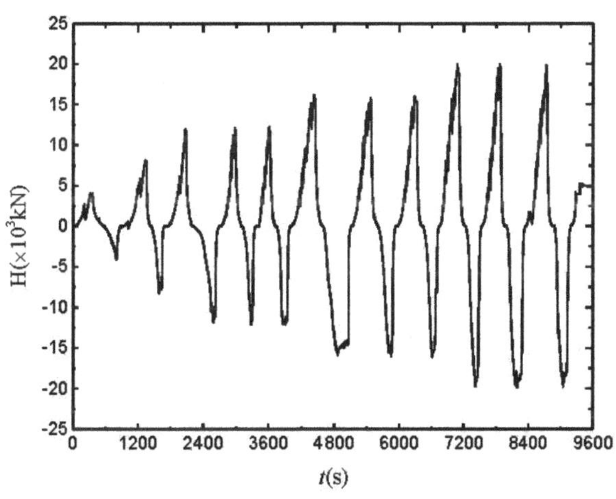

(a) The first experiment, source: (WANG Xuyue, 2017)

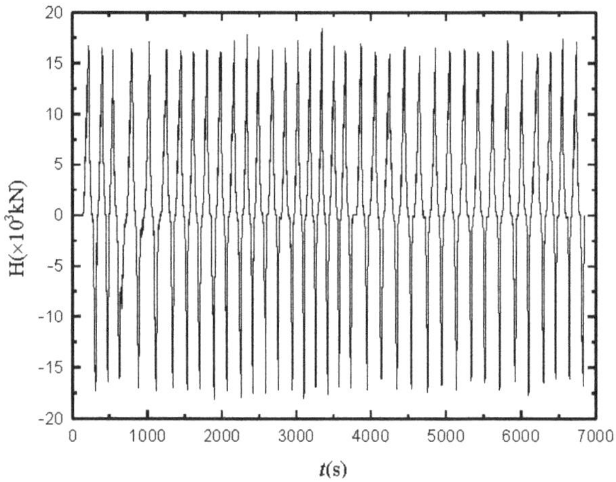

(b) The second experiment
[Fig. 9-21] Cyclic loading process, source: (WANG Xuyue, 2017)

9.2.3 Test results and analysis

9.2.3.1 Hysteresis curve

During the test, for the hysteretic curve based on the load-displacement relationship, see Formula (2-3) and Formula (2-4) in the literature(WANG Xuyue, 2017) to calculate the hysteretic curve of the moment rotation relationship, as shown in [Fig. 9-22]. In the first cyclic loading test with variable amplitude, the area of the hysteresis curve increases with the increase of the bending moment. The hysteresis loop in the figure increases with the increase of the load amplitude, while the stiffness decreases with the load cycle. These cyclic hysteresis loops are very valuable for the prototype foundation. Under such a large displacement, the prototype foundation will not "fail" in the whole process. It is worth noting that from the shape characteristics of the hysteresis loop, a serious hardening plastic reaction was observed after gradual and thorough unloading. This behavior is a typical "popping" reaction. In this reaction, hardening occurs as the gap between the cylinder and the soil (generated in the last cycle) is closed, and the gap gradually becomes larger. In conclusion, in the first test, the system stiffness decreases with the increase of cyclic load, and the hysteresis loop increases with the increase of cyclic load.

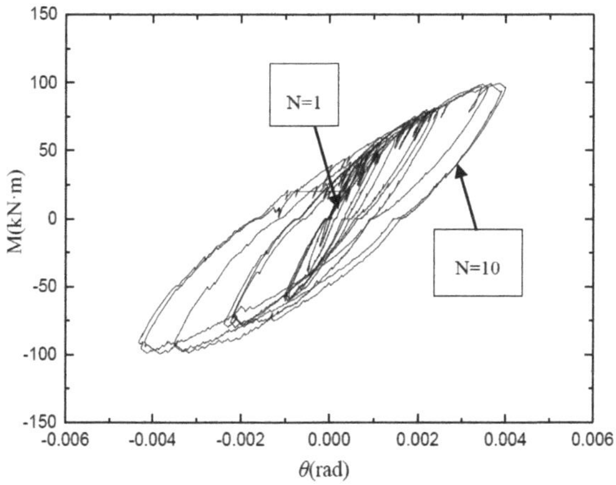

(a) Hysteresis curve of the first test

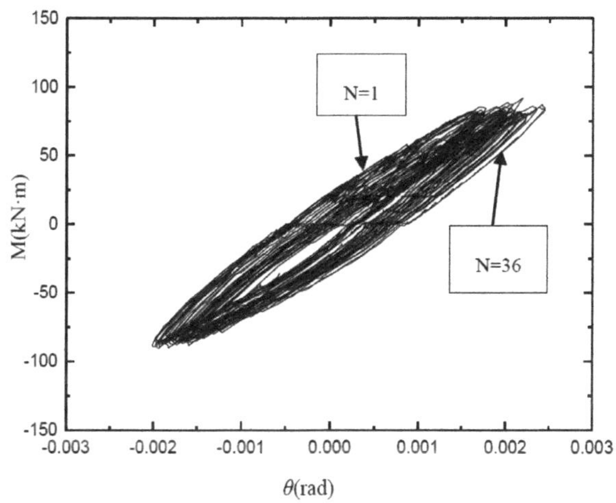

(b) Hysteresis curve of the second test

[Fig. 9-22] Moment rotation hysteresis curve, source: (WANG Xuyue, 2017)

9.2.3.2 Stiffness degradation

The stiffness calculation model of the composite barrel foundation is shown in [Fig. 9-23], considering the horizontal lateral stiffness and rotational stiffness. According to the load, displacement, bending moment, and rotation angle during

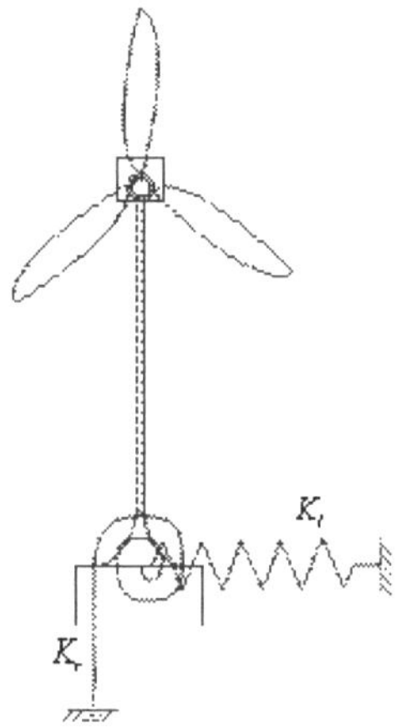

[Fig. 9-23] Stiffness Calculation Model of Composite Barrel Foundation, source: (WANG Xuyue, 2017)

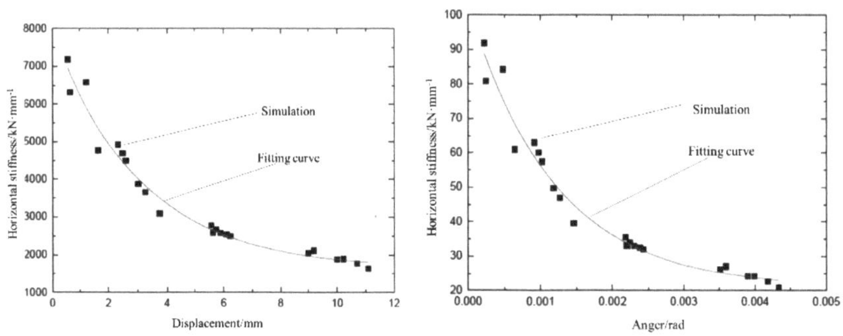

(a) Horizontal lateral stiffness K_1 (b) Rotational stiffness K_r

[Fig. 9-24] Stiffness degradation curve, source: (WANG Xuyue, 2017)

loading, the degradation curves of horizontal lateral stiffness K_1 and rotational stiffness K_r of the first test are calculated, as shown in [Fig. 9-24]. It is obvious from the figure that the stiffness degradation curve decreases slowly at an exponential decay rate, which is caused by cyclic loading. Stiffness can be expressed as a function of the angle. The stiffness change curve is simply fitted based on the exponential relationship between the bending moment and rotation angle in [Fig. 9-24].

The parameters of the soil will change during loading. The Poisson's ratio of clay ranges from 0.4 to 0.5, which has little effect on the stiffness. The secant shear modulus of clay calculated from the test ranges from 0 to 16 MPa (G. T. Houlsby et al., 2005), which has a great impact on the rotational stiffness. When stiffness degradation is predicted, the method of continuously extracting negative pressure can be adopted in engineering to increase the stiffness of the whole structure.

9.2.3.3 Damping coefficient

The motion matrix equation of a structural system subjected to dynamic excitation is :

$$[M] \cdot \{\ddot{u}(t)\} + [C] \cdot \{\dot{u}(t)\} + [K] \cdot \{u(t)\} = -[M] \cdot \{I\}\ddot{u}_g(t)$$

where , [M]——mass matrix ; [K]——Stiffness matrix ; [C] ——damping matrix ; $\ddot{u}(t)$——Acceleration vector relative to system foundation ; [I]——Unit vector ; $\ddot{u}_g(t)$——Absolute acceleration of structure foundation. For the equation of motion of a system, left [M] is invariant, [K] As shown in [Fig. 9-23].

Structural stiffness degradation means that the energy of the whole system is changing, mainly due to the existence of damping. The energy change of the dynamic response system composed of bucket foundation and soil is related to structural damping. The hysteresis loop in [Fig. 9-22] is turned back due to the existence of damping. There are three possible reasons: viscous material damping; Plastic dissipation of soil energy; Radiation damping (Bhattacharya et al., 2011). The existing research method of single-cylinder non-compartmentalized suction cylinder foundation is to compare the theoretical calculation and experimental data of the performance of the simplified circular foundation in elastic materials.

The curve of the equivalent viscous damping coefficient is shown in [Fig. 9-25]. It can be seen from the figure that in the initial stage of loading, the equivalent viscous damping coefficient first decreases, and then increases with the increase of displacement. It can also be seen that with the increasing area of the hysteresis loop, the equivalent viscous damping coefficient increases gradually,

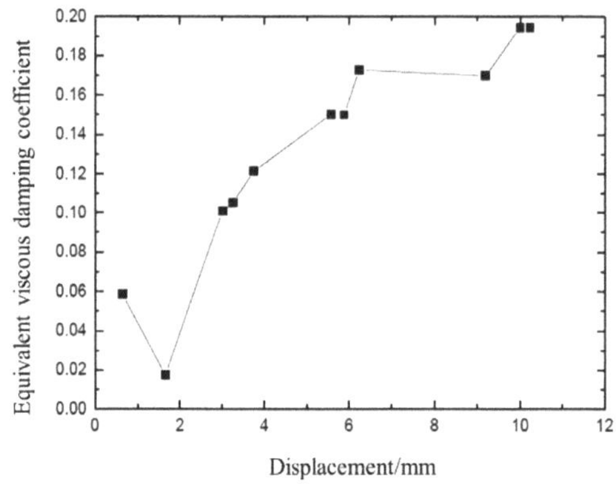

[Fig. 9-25] Curve of equivalent viscous damping coefficient, source: (WANG Xuyue, 2017)

and the energy dissipation capacity becomes larger and larger. The equivalent viscous damping coefficient is below 0.2 in the whole loading process. The basic reason for the change of hysteresis loop area is the existence of damping, and the reason for the change of damping is the damping of viscous materials, the dissipation of plastic energy of soil, radiation damping, etc. The equivalent viscous damping coefficient decreases at the initial loading stage, mainly due to the plastic deformation of the soil mass, which leads to the change of damping.

9.2.3.4 Cumulative deformation

The offshore wind turbine structure is subject to the action of wind, wave, current, and other cyclic loads for a long time, and its foundation soil will produce cyclic cumulative deformation (WANG Jianhua et al., 1999). As the wind, wave, and current loads on the composite bucket foundation have cyclic characteristics, it is very important to evaluate the cyclic characteristics of the composite bucket foundation. Cyclic loading leads to irreversible plastic deformation of soil volume around the foundation, resulting in the accumulation of foundation displacement and rotation angle. The vertical cumulative settlement may affect the safety of the whole structure. The cumulative deformation values of the two tests are obtained by calculating the displacement values of 3 # and 4 # LVDT. The central displacement of the cylinder top is taken as the cumulative

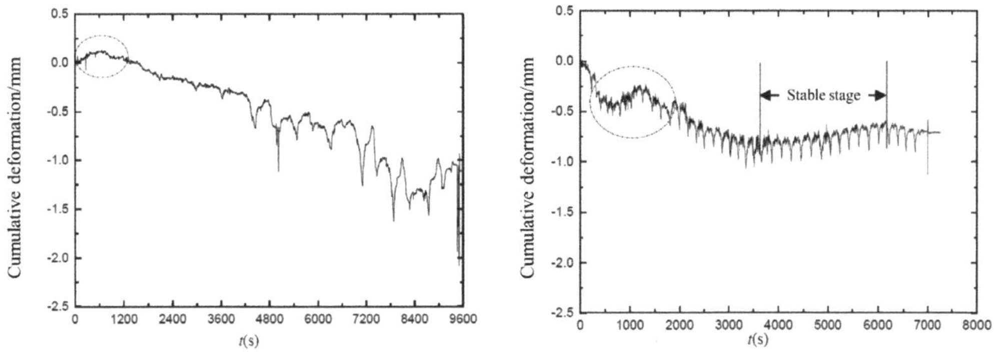

(a)The first test (b)The second test

[Fig. 9-26] Cumulative deformation curve, source: (WANG Xuyue, 2017)

deformation value. As shown in [Fig. 9-26] (a), in the first cyclic loading test with variable amplitude, the final value of cumulative deformation is 2.08 mm, which is very small and has little impact on the bearing capacity of the bucket foundation. The cumulative settlement increases with the increase of cyclic load amplitude. At the end of each cyclic loading, the soil will produce irreversible plastic residual deformation. With the increase in the number of cyclic loads, the growth rate of the cumulative deformation value gradually increases, which indicates that there is a certain relationship between the cumulative deformation and the amplitude of the horizontal cyclic load. It can also be seen from the figure that the displacement of the cylinder top changes from a positive value to a negative value at the initial stage of loading, indicating that when the cylinder foundation is subjected to external load at the initial stage, the soil in the cylinder is squeezed and arched, which results in the initial value of the cumulative deformation being positive, and there is a gap between the cylinder and the soil. This can be explained by the fact that the initial stiffness of the whole system composed of tube and soil is very large, so the whole structure is not easy to deform at the initial stage of loading. With the increase of the cycle period, the soil in the tube is gradually compacted to produce consolidation settlement, which eventually leads to the accumulation of plastic deformation of the soil. At the same time, the stiffness of the whole system, which is composed of a tubular foundation and soil, becomes larger and larger. This also proves that the tube soil joint bearing mode is beneficial to the stiffness growth of the whole structure. On the contrary, if the soil becomes looser and looser under external load, the stiffness of the whole system will become smaller and smaller.

As shown in [Fig. 9-26] (b), in the second constant amplitude cyclic loading test, the final value of cumulative deformation is 1.13 mm, which is as small as the cumulative deformation value in the first cyclic loading test and has little impact on the bearing capacity of the bucket foundation. The number of cycles in the second test is more than that in the first test, but the cumulative cyclic deformation value in the second test is small, which indicates that the cumulative cyclic deformation is not related to the number of cycles but related to the amplitude of the horizontal cyclic load. When the horizontal cyclic load is very small, the variation range of displacement value of each cycle is almost the same. With the increase in loading times, the cyclic cumulative deformation value tends to be stable at 3500 seconds. Under the horizontal cyclic load, the cumulative deformation of the foundation is caused by the deformation of the soil, mainly due to the creep characteristics of the soil. It can also be seen from the figure that in the second test, the value of vertical cumulative deformation did not always increase with the number of cyclic loads at the relative instantaneous deformation stage of soil creep (before the cumulative deformation reached the stable stage). Between 600 seconds and 1200 seconds, the accumulated deformation value rebounds. Excluding the error of the test equipment, it is similar to the initial stage of the first test, when the soil in the cylinder is extruded and arched, which leads to the increase of the cumulative deformation value in the initial stage. As the amplitude of the cyclic load is smaller than that of the first test, the rebound phenomenon occurs later.

To explore the impact of cumulative deformation on the safety of composite barrel foundations, the displacement values of 3 # and 4 # LVDTs are shown in [Fig. 9-27. In the Code for Foundation Design (DNV-OS-J101), it is mentioned that "the risk of uneven settlement related to the allowable slope rate of offshore wind turbine supporting structures should be considered. (The risk of uneven

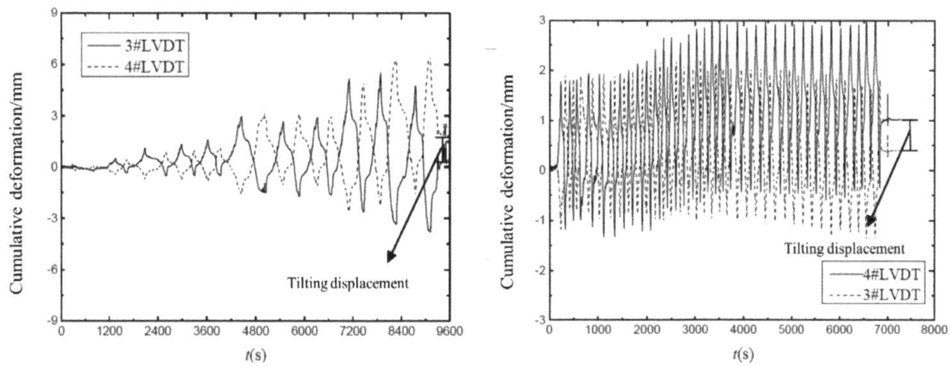

(a)The first test (b)The second test

[Fig. 9-27] Displacement curves, source: (WANG Xuyue, 2017)

settlements should be considered to the tolerable tilt of the wind turbine support structure.) The slope rate of offshore wind turbine composite bucket foundation can be calculated by two displacement values, and the slope rate is defined as

$$k_t = \frac{|z_3 - z_4|}{D}$$

where , k_t—Tilt rate , z_3, z_4—the displacement of 3#LVDT and 4#LVDT , D—Diameter of bucket foundation. As shown in [Fig. 9-27], the inclination rate of the bucket foundation in the first cyclic loading test varies from 0 to 0.291%, and that in the second cyclic loading test varies from 0 to 0.13%. These slope rates are acceptable for offshore wind turbine support structures and have little impact on the stability of the bucket foundation.

9.2.3.5 Frequency domain impedance

The dynamic response data of bending moment and rotation angle are first processed in the frequency domain through FFT, and then the complex frequency-related impedance is obtained by comparing the two groups of FFT data. The real part of impedance represents the stiffness and inertia effect, and the imaginary part represents damping. The real part and imaginary part data of the bending moment load in the first test after a fast Fourier transformation are shown in [Fig. 9-28] (a), and the real part and imaginary part data of the corner after a fast Fourier transformation are shown in [Fig. 9-28](b). The real and imaginary parts of the bending moment are divided by the real and imaginary parts of the rotation angle to obtain the real and imaginary parts of the impedance, as shown in [Fig. 9-28] (c). Use the same method to calculate the data of the second test, and get the real part and imaginary part of the impedance, as shown in [Fig. 9-29].

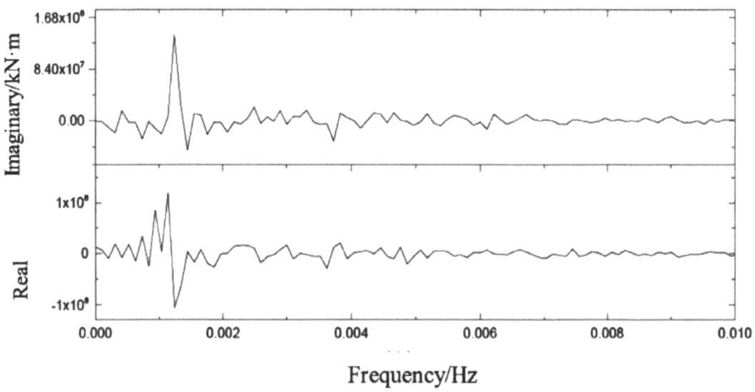

(a) Real and imaginary parts of bending moment (unit : $kN \cdot m$)

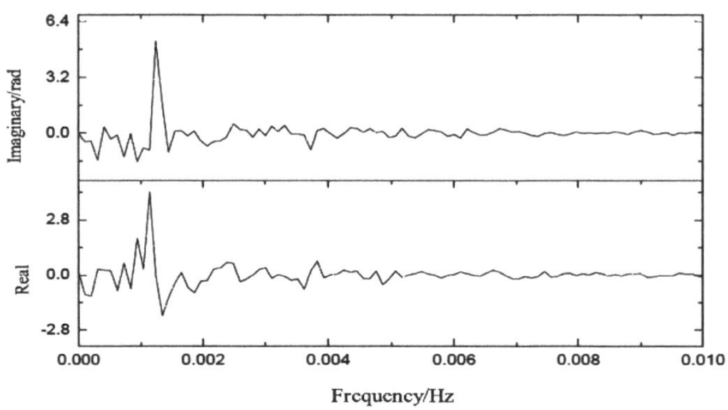

(b) Real and imaginary parts of corners (unit : rad)

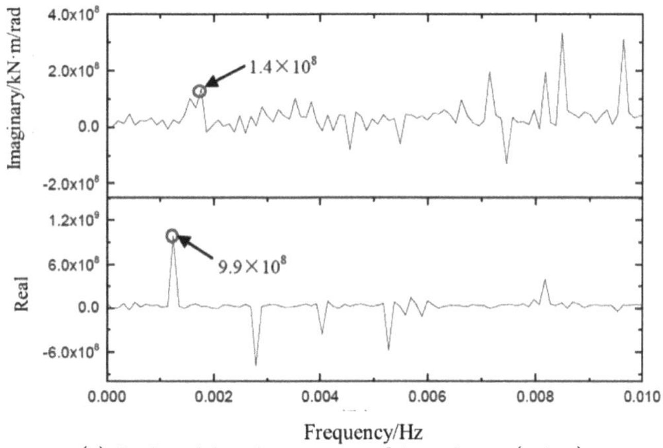

(c) Real and imaginary parts of impedance (unit :)

[Fig. 9-28] The real part and imaginary part of the impedance of the first test, source: (WANG Xuyue, 2017)

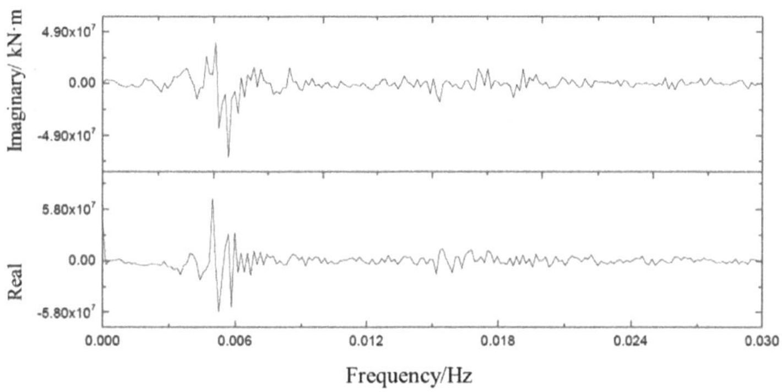

(a) Real and imaginary parts of bending moment (unit : kN·m)

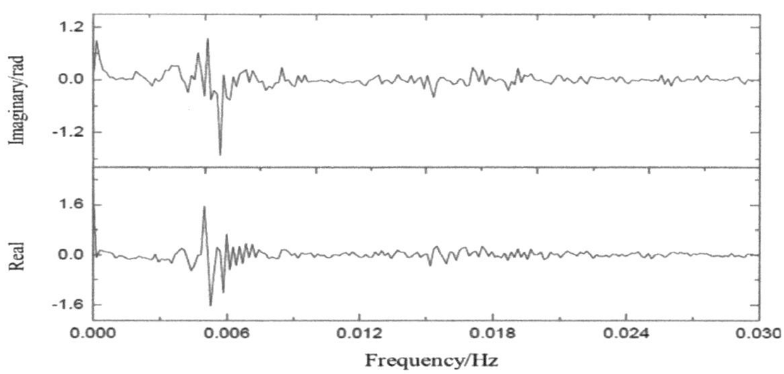

(b) Real and imaginary parts of corners (unit : rad)

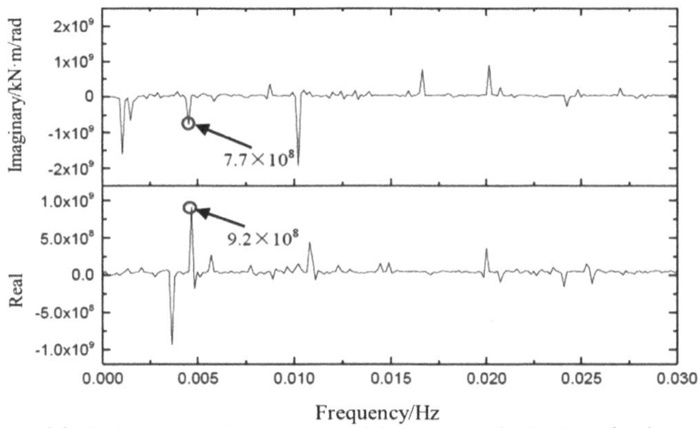

(c) Real and imaginary parts of impedance (unit : kN·m/rad)

[Fig. 9-29] The real part and imaginary part of the impedance of the second test, source: (WANG Xuyue, 2017)

9.2.3.6 Pressure change of soil

There are many earth pressure sensors. Only the results of typical earth pressure sensors inside, outside, and on the top of the cabin are listed, as shown in [Fig. 9-30], [Fig. 9-31], and [Fig. 9-32]. It can be seen from the figure that the changing trend of the earth pressure sensors 1, 2, and 3 in the 2 # cabin is the same. The closer to the cylinder top, the smaller the earth pressure is. The greater the soil extrusion deformation under the cylinder foundation, the greater the earth pressure is. The changing trend of 2 # external earth pressure sensors 1, 2, and 3 is the same, and the closer to the cylinder top, the smaller the earth pressure is; The soil pressure sensor 4 is close to the top of the cylinder, and it is a tensile force in the process of cyclic loading, which indicates that the adsorption force on the top of the cylinder appears in the process of cyclic loading, which is beneficial to the bearing capacity of the cylinder foundation. The measured values of 2 # cabin top and 5 # cabin top soil pressure sensors are opposite. When the 2 # cabin top is tensioned during the test, the pressure of the 5 # cabin top conforms to the actual observation. The measured value of the soil pressure on the top of the 7 # tank shows that the top of the 7 # tank is tensioned, indicating that there is obvious adsorption force in the middle tank during the cyclic loading process, which is beneficial to the combined cylinder soil bearing mode.

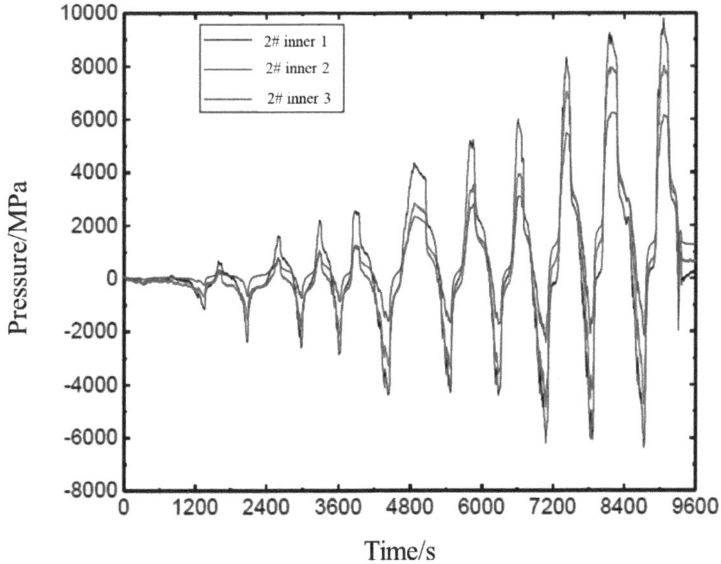

[Fig. 9-30] Measured value of 2 # tank earth pressure, source: (WANG Xuyue, 2017)

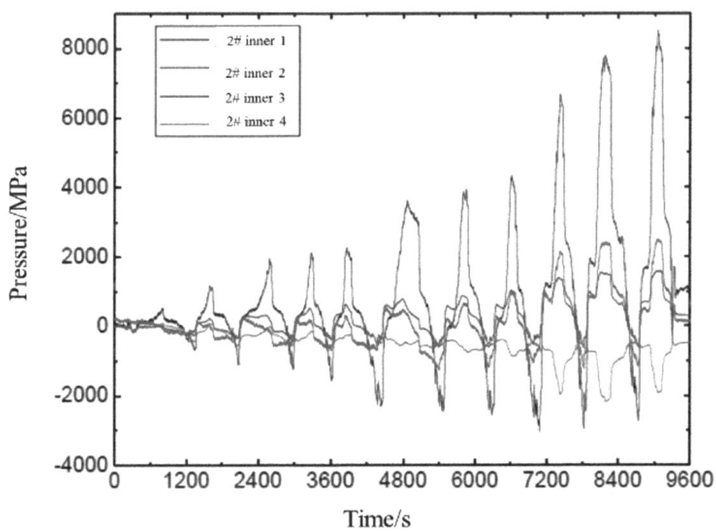

[Fig. 9-31] Measured value of earth pressure outside 2 # cabin, source: (WANG Xuyue, 2017)

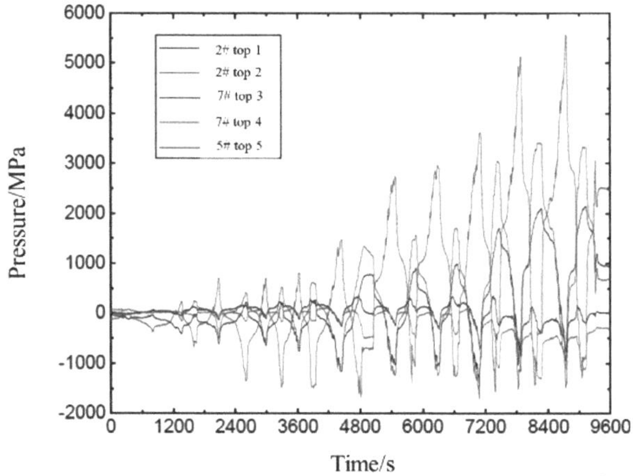

[Fig. 9-32] Measured earth pressure of top cover in loading direction, source: (WANG Xuyue, 2017)

9.3 Numerical Simulation of Hydrodynamic Loads on Composite Pile Foundation with High Pile Caps for Offshore Wind Power

Li Zhenglin and Hu Jinpeng built a three-dimensional wave numerical flume in Fluent based on the VOF method for wave action on wind power composite pile foundation. First, they simulated the process of regular waves impacting the lower pile separately to verify the effectiveness of the model in calculating wave loads on small-scale rods; Then the incident wave force, diffracted wave force, and total force acting on the composite pile foundation are obtained through numerical calculation (LI Zhenglin & HU Jinpeng, 2020).

9.3.1 Establishment of numerical model

9.3.1.1 geometric model

The research in this paper mainly focuses on the influence of large-scale structures on the wave forces of small-scale structures. Considering the computational complexity of the three-dimensional model, for the sake of simplification, the structure model in this paper only considers a single pile, and a more complete structure model will be carried out in the next step of the research.

Regarding the design conditions of the offshore wind pile foundation in the deep-water wind farm in East Guangdong (water depth is 50 m, which is the offshore extreme wave-prone area) currently designed by Guangdong Electric Power Design and Research Institute, the geometric schematic diagram of the three-dimensional numerical wave tank model in this paper is shown in [Fig. 9-33]. The design dimensions of the sink are: length 20 m, height 4 m, width 4 m; The water depth of the flume is 2 m; The rest is air, and the wave attenuation area is located at the end of the flume at 16-20m; The radius of the upper large-scale cylinder of the composite structure is a=0.4 m, the cylinder draft T=1 m, the pile radius b=0.1 m, the pile length L=1 m, and the pile is on the left side of the upper cylinder near the incident wave. The final grid is shown in [Fig. 9-34]. The structural grid divided in ICEM is imported into Fluent, the top is the pressure inlet boundary, and the upper part of the simulation tank relates to the outside atmosphere.

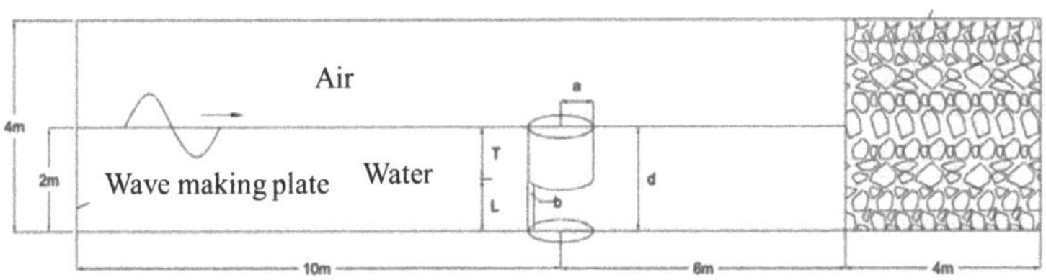

[Fig. 9-33] Three-dimensional numerical wave flume, source: (LI Zhenglin & HU Jinpeng, 2020)

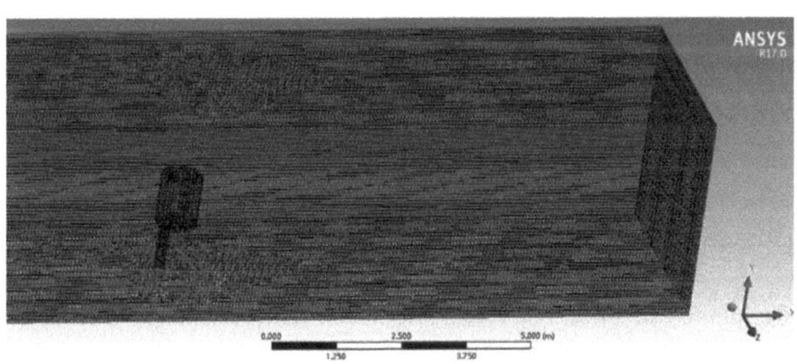

[Fig. 9-34] Grid division diagram of the numerical flume, source: (LI Zhenglin & HU Jinpeng, 2020)

9.3.1.2 Parameter setting

UDEFINE via UDF_ CG_ MOTION compiles the motion law of moving boundary and DEFINE_ SOURCE Control the regional momentum to realize the source term wave attenuation, and import the three-dimensional numerical wave tank. The left wall of the tank is equivalent to the wave-making plate, and the wall movement is realized through the dynamic grid model. The VOF method is used to obtain the free water surface, the pressure velocity term is calculated by the Piso algorithm, and the dynamic layer method (Layering) is used in combination with the grid reconstruction to realize the grid movement, with a time interval of 0.01s and a simulation duration of 20s.

9.3.1.3 Model validation

The reliability of the model was confirmed by simulating the process of regular waves impacting the lower pile separately, and the numerical simulation stress was compared with the results of the Morrison formula. The initial incident wave is the linear regular wave, the incident wave height is set as 0.15 m, and the period is 1.6 s. It can be seen from [Fig. 9-35] that the wave force obtained by numerical simulation is roughly consistent with the result of the Morrison formula. Therefore, it can be preliminarily determined that FLUENT has considerable accuracy in the numerical simulation of wave interaction with small-scale vertical column structures.

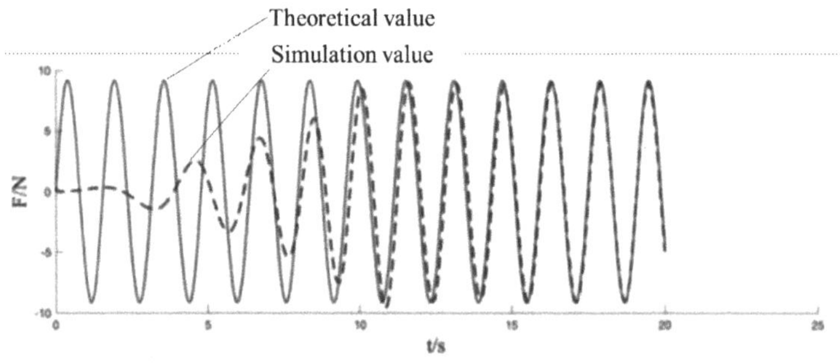

[Fig. 9-35] Comparison Chart of Wave Force Curves under Regular Waves, source: (LI Zhenglin & HU Jinpeng, 2020)

9.3.2 Analysis of numerical simulation results

9.3.2.1 Influence of upper cap size on wave force of pile foundation

When the radius of the upper cylinder takes different multiples of a, assume that the radius of different cylinders is $R = na$, and n is $n = 1, 2, 3$, and 4, respectively, to investigate the impact of the diffraction field generated by the cushion cap on the wave load of the lower pile foundation under different cylinder radii. The characteristics of incident waves are regular waves with amplitude $A/a = 0.75$ and wave number $ka = 0.8$. [Fig. 9-36] shows the comparison of the incident wave force, diffracted wave force, and resultant force amplitude of the pile when the radius R of the upper cylinder is different.

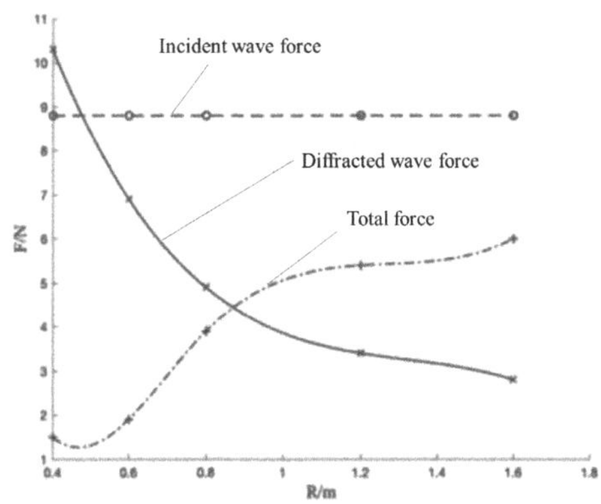

[Fig. 9-36] Total force, incident wave force and diffracted wave force under different cushion cap sizes, source: (LI Zhenglin & HU Jinpeng, 2020)

It can be seen from [Fig. 9-36] that the increase of the radius of the upper cushion cap structure leads to the more obvious diffraction effect of the flow field, and the wave load of the pile foundation under the influence of the incident and diffracted wave fields also gradually increases, but the increase of the wave force gradually slows down with the increase of the cylinder radius; At the same time, due to the difference between the phase of the diffracted force and the incident force, the amplitude of the resultant force is less than the incident wave force in most cases.

9.3.2.2 Influence of extreme wave on wave force of pile foundation

An abnormal wave is a special nonlinear extreme wave, which occurs in both good and bad sea conditions. Ocean abnormal waves have the characteristics of highly concentrated energy and randomness, which pose a great threat to offshore structures, so it is necessary to carry out relevant research.

Based on the Longuet Higgins random wave model, Liu Zanqiang et al. (LIU Zan-qiang et al., 2010) proposed a method to adjust the initial phase according to the interval, to generate abnormal waves at a specific time and point; Lin Jand hu et al. (LIN Jiongzeng & HU Jinpeng, 2019) and others changed the initial phase of the wave surface equation after the superposition of different regular waves so that the superimposed waves superposed at the predetermined focal point position and focal time to form a large wave. The spectrum used in this paper is the JONWAP spectrum, with an effective wave height of 0.15 m and a peak period of 1.6 s. The spectrum is divided into 100 parts according to the equal frequency method, and the wave frequency range is [0.5~3.5]. The time history curve of the deformed wave surface generated by this model is shown in [Fig. 9-37].

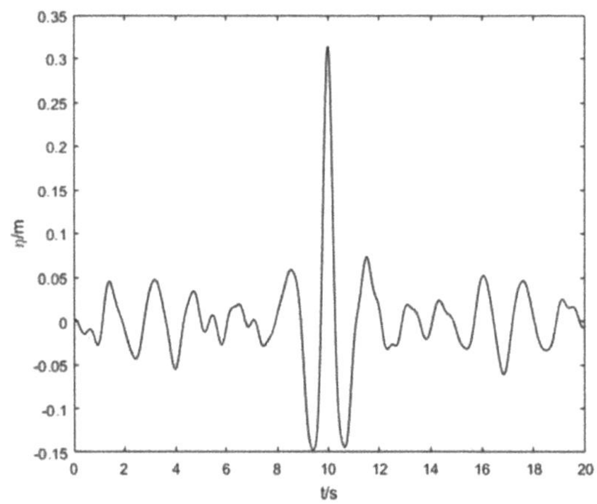

[Fig. 9-37] Time history curve of abnormal wave surface, source: (LI Zhenglin & HU Jinpeng, 2020)

[Fig. 9-38] shows the comparison of the total force, incident wave force, and diffracted wave force on the composite pile foundation of the pile cap when the abnormal wave is incident. The maximum stress of a pile foundation under the influence of an incident wave field is 26 N; The maximum stress of the pile foundation under the influence of diffracted wave field is 12 N, which is 46.2% of the incident wave force; The maximum stress of the pile foundation under the combined influence of incident and diffracted wave fields is 32 N, which is 1.23 times of that under the influence of incident wave fields. It can be seen from this that the size of the diffracted wave force almost reaches the size of the wave force under normal regular wave conditions, indicating the necessity of considering the diffracted wave force under extreme sea conditions. Although there is a certain difference between the phase of the wave field generated by diffraction and the phase of the incident wave field, the contribution of the diffraction force to the resultant force is positive, and the total force is three times the wave force under normal regular wave conditions. The phase difference between the incident wave force and the diffraction wave force is related to the pile position under the pile cap and the direction of the incident wave.

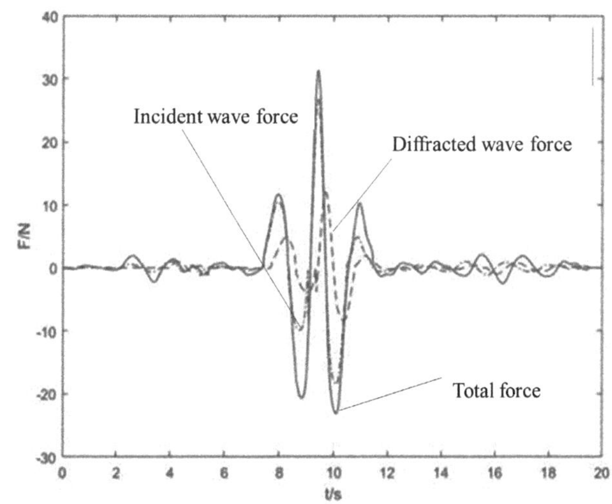

[Fig. 9-38] Comparison of the total force, incident wave force, and diffracted wave force under the abnormal wave, source: (LI Zhenglin & HU Jinpeng, 2020)

9.4 Wave dynamic response of offshore wind turbine with a jacket under random wave action

Xu Honglu and Peng Ya carried out a wave dynamic response analysis of offshore wind turbines with a jacket based on random wave theory using the mode superposition method. The results show that the dynamic load is greater than the static load; The dynamic amplification coefficient calculated from the theoretical formula of dynamic amplification coefficient is greater than that obtained from wave dynamic analysis (XU Honglu & PENG Ya, 2021).

9.4.1 Finite element model

Taking a domestic wind farm as an example, the wave dynamic response of the wind turbine foundation is calculated and analyzed. The wind turbine unit capacity is 6 MW, and the foundation form is four pile jacket foundation. The tower of the fan is 85 m high, the diameter of the tower bottom is 6.5 m, and the elevation of the foundation working platform is 18 m. SACS software is used for finite element analysis, and pole element is used for finite element modeling of jacket and fan tower structures. The engine room, wheel hub, and impeller are applied at the top of the tower as a concentrated mass. Simulation of wave loads using random wave theory (XU Honglu & PENG Ya, 2021).

9.4.2 modal analysis

Before wave dynamic analysis, a modal analysis of the wind turbine shall be carried out. Firstly, the pile-soil system is linearized, that is, the pile-soil interaction is equivalent to a linear spring system.

For structural systems, meaningful dynamic responses generally occur in low modes. Therefore, the appropriate mode order can be selected in the wave dynamic response analysis, which can not only ensure the calculation accuracy but also save the calculation cost. In this paper, the first 30 modes are used for dynamic response calculation. The first 10 mode parameters and the first 3 mode shapes are shown in [Tab. 9-4] and [Fig. 9-39] respectively. It can be seen from the modal analysis results that the mass participation coefficient of the first 10 modes is large enough to fully describe the vibration of the structure.

[Tab. 9-4] Wind turbine modal parameters, source: (XU Honglu & PENG Ya, 2021)

Mode	Frequency/Hz	Mass	characteristic value	Cumulative coefficients	Cumulative coefficients
1	0.306615	5.70E+02	2.69E-01	0.000392	0.260049
2	0.307441	5.63E+02	2.68 E-01	0.268827	0.260418
3	0.751868	3.12 E+03	4.48 E-02	0.510796	0.730980
4	0.754638	3.49 E+03	4.45E-01	0.978756	0.977405
5	1.1404772	1.41 E+03	2.08 E-02	0.980599	0.980285
6	1.842419	4.29 E+02	7.46 E-03	0.980600	0.984045
7	2.009303	3.65 E+02	6.27 E-03	0.983994	0.984046
8	3.184529	1.28 E+03	2.50 E-03	0.984014	0.984049
9	3.523778	7.65 E+02	2.04 E-03	0.984040	0.985740
10	3.552951	9.15 E+02	2.01 E-03	0.985750	0.985761

First mode Second mode Third mode

[Fig. 9-39] First three modal shapes of a wind turbine, source: (XU Honglu & PENG Ya, 2021)

9.4.3 Dynamic response analysis

The calculation example of this project uses 30 groups of random waves to analyze the wave dynamic response of offshore wind power foundations based on the P-M spectrum. The wave simulation duration is 3600s, and the step length is 0.5s. The random wave surface profile generated by the wave spectrum and the comparison between the fitting spectrum of linear wave superposition and the theoretical spectrum is shown in [Fig. 9-40] and [Fig. 9-41] respectively. It can be seen from the figure that the theoretical wave spectrum fits well with the calculated wave spectrum, the simulation time is sufficient, and the wave input conditions meet the requirements.

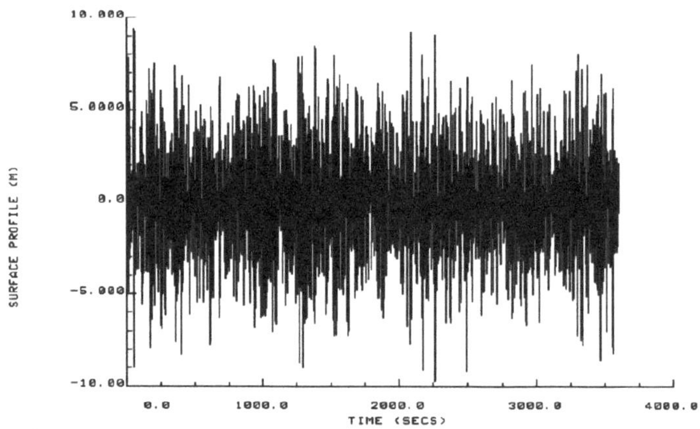

[Fig. 9-40] Wave surface time history curve, source: (XU Honglu & PENG Ya, 2021)

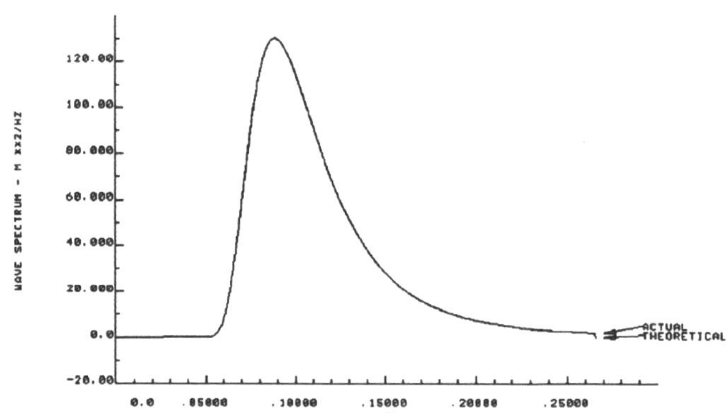

[Fig. 9-41] Actual wave spectrum and fitted wave spectrum, source: (XU Honglu & PENG Ya, 2021)

The foundation shear force and the bending moment of the wind turbine generator are obtained through wave dynamic analysis, as shown in [Fig. 9-42] ~ [Fig. 9-43]. It can be seen from the figure that the shear force and bending moment of the moving base are greater than those of the static base, so it is necessary to analyze the wave dynamic response of the wind turbine.

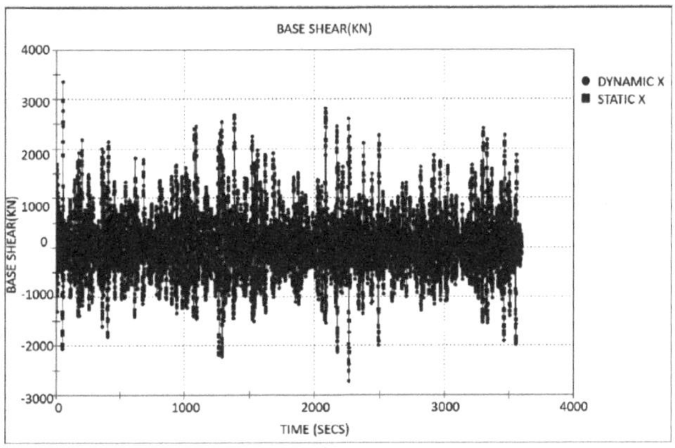

[Fig. 9-42] Base shear, source: (XU Honglu & PENG Ya, 2021)

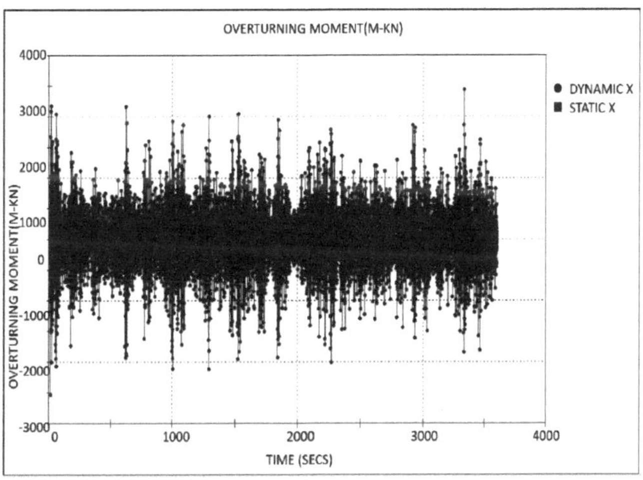

[Fig. 9-43] Base bending moment, source: (XU Honglu & PENG Ya, 2021)

9.4.4 Dynamic amplification factor

From the analysis in the previous section, the natural vibration period of the wind turbine is 3.26s, and the wave dynamic response should be considered when designing the wind turbine structure according to relevant specifications. The inertial load caused by waves can be directly obtained from the wave dynamic response analysis and can also be derived from the wave load and dynamic amplification factor DAF of static calculation.

Generally, before detailed dynamic analysis, the dynamic amplification factor of the wind turbines can be estimated through the single degree of freedom system formula. The dynamic amplification coefficient of single degree of freedom system is calculated as follows:

$$DAF=\frac{1}{\sqrt{(1-\Omega^2)^2+(2\cdot\zeta\cdot\Omega)^2}}$$

where ζ is the damping ratio, Generally it is 2%~3%, in this paper it is3%. Ω is the ratio of structural natural vibration period to wave period.

According to the calculation results in the previous section and the above formula, the dynamic amplification factor DAF of the wind turbine is 1.0961.

A more accurate calculation method for the dynamic amplification factor DAF is to analyze the dynamic response of the structure, and then compare the maximum dynamic base shear with the maximum static base shear. This example generates 30 groups of random waves for dynamic analysis of wind turbines. The dynamic amplification coefficient of each group of waves is calculated, and then the average dynamic amplification coefficient of the wind turbine is obtained. The calculation result of the dynamic amplification coefficient is 1.0422.

It can be seen from the calculation results that the dynamic amplification coefficient obtained from the wave dynamic response analysis is smaller than that calculated from the formula. Therefore, it will be more conservative to design the dynamic amplification coefficient obtained through formula calculation. This is because only the first mode is considered in the formula calculation, and the first mode period is the largest, which is closer to the main wave period.

10 Influence of wind, wave, current, ice cone, and other coupling or combined actions on offshore wind power foundation

10.1 Influence of wind and wave on the horizontal bearing capacity of offshore wind power pile foundation

Hongjun Liu (LIU Hongjun et al., 2019) designed four groups of wind load and wave load combinations, as shown in [Tab. 10-1].

[Tab. 10-1] Load combinations, source: (LIU Hongjun et al., 2019)

	Wind load	Wave load
Load combination 1	Normal wind load	Normal wave load
Load combination 2	Normal wind load	Extreme wave load
Load combination 3	Extreme wind load	Normal wave load
Load combination 4	Extreme wind load	Extreme wave load

Among them, wave loads are listed in [Tab. 10-2].

[Tab. 10-2] Wave design parameters, source: (LIU Hongjun et al., 2019)

	Depth of water d/m	Wave height H/m	Cycle T/s	Wavelength L/m
Normal sea condition	10	0.9	4.5	31
Extreme sea condition	10	5	7.85	70

Wherein, the design wind speed of 14 m/s under normal sea conditions and 33 m/s under extreme sea conditions are selected for wind load in combination with the wind conditions in the Yellow Sea and the Bohai Sea (Hanjia-Xin, 2015).

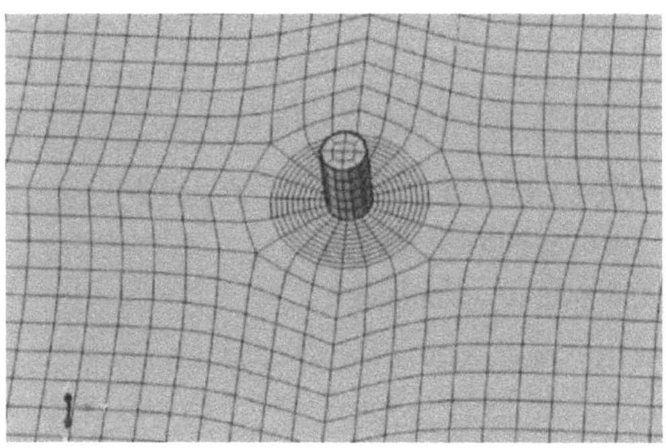

[Fig. 10-1] Numerical calculation model, source: (LIU Hongjun et al., 2019)

As shown in [Tab. 10-1], wind load and wave load are combined. The wind load is considered a stable continuous load, and the wind speed is converted into wind pressure to calculate the wind load acting on the swept area. The bending moment caused by wind load is applied to the pile foundation, and the calculated wave force is applied to the pile foundation to realize the combined action of wind and wave. The environmental load is taken as load combination 1, and the mud surface displacement of the pile foundation is obtained as shown in [Fig. 10-2].

It can be seen from the analysis of [Fig. 10-2] that the continuous wind load under the combined load causes relatively large and stable displacement of the pile foundation, and the cyclic wave load causes the pile foundation to swing circularly in a small range.

Martínez-Chaluisant, Schneider, et al. (Martínez-Chaluisant et al., 2012) also took a 3.6 MW wind turbine as an example in the dynamic response analysis of wind turbine under the combined action of wind and waves. The diameter of the wind turbine is 116m, the height from the hub center to the tower bottom is 80m, and the height from the hub center to the sea level is 90 m. The total weight of the unit is 284 t, the diameter of the steel pipe pile is 6 m, the pile length is 65 m, the wall thickness is 75 mm, and the buried depth is about 50 m.

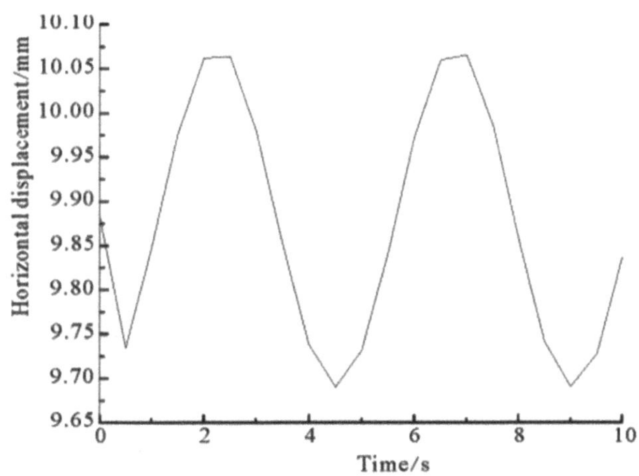

[Fig. 10-2] Pile foundation displacement, source: (LIU Hongjun et al., 2019)

Take the seabed of a sea area in the East China Sea as an example. The basic wind speed is 31.2 m/s, and the random wind speed is used for loading. The variation range of wave force borne by the pile at sea level of the model is approximately -100~100 kN/m. Considering the bottom damping characteristics of the wind turbine generator, the peak horizontal displacement of the pile foundation at mud level is 32.5 mm. Using the numerical model in this paper, combined with the environmental loads under conventional and extreme working conditions, it is analyzed that the approximate variation range of the pile foundation mud surface displacement is 10~50 mm, and the results of the two models are different on countless orders of magnitude.

10.2 Combined action of wind, wave, and current

10.2.1 Dynamic response and bearing capacity analysis of offshore wind power foundation under wind wave current interaction

10.2.1.1 Load cases

In this section, the dynamic response of the wind turbine system of Xiangshui Wind Farm under the combined action of wind, wave, and current loads is numerically simulated based on the project of Xiangshui Wind Farm in Yancheng, Jiangsu Province. Jiangsu Province is located on the east coast of China, with a coastline of nearly 1000 km long, and the coastal beach area reaches 6500 km^2. Xiangshui is in Yancheng City, Jiangsu Province, with a total area of 1378 km^2, a coastline of 43 km, and a coastal beach area of 1600 km^2. The offshore sea area is open, the seabed terrain is flat, and there is no large fluctuation. The 5 m isobath is about 5~10 km from the coastline, and the 10 m isobath is about 7~15 km from the coastline. The submarine geological conditions in the region are good, the regional geology is relatively stable, the waves and ocean currents are small, there are few aquacultures, ports, navigation channels, military, and other facilities nearby, and wind energy resources are rich, so there are good conditions for the construction of large offshore wind farms. In this area, the measured wind speed at 10 m above the sea is

13.4 m/s, the water depth is 5 m, the wave height is 2.68 m, the wavelength is 74.1 m, and the current velocity is 0.88 m/s.

10.2.1.2 boundary condition

For the seabed around offshore wind power foundation, the boundary conditions of the rigid seabed, seabed surface, and side are considered respectively.

Seabed surface: when the friction force of the seabed surface is ignored, the vertical effective stress and shear stress of the seabed surface may not be considered, namely

$$\sigma_z' = \tau_{xz} = 0, \quad z = 0$$

Bottom of seabed: the bottom of the seabed is an impermeable rigid foundation bed, the displacement of soil is 0, and the normal flow is 0, that is

$$u = v = w = 0 \quad \frac{\partial p}{\partial z} = 0, \quad z = -h$$

Seabed side: considering that the influence of the lateral boundary conditions is only limited to the vicinity of the boundary, if the calculation conditions permit, a large horizontal calculation domain can be used to reduce the influence of the lateral boundary, so the impermeable fixed boundary is also used for the seabed side, namely

$$u = v = w = 0 \quad \frac{\partial p}{\partial z} = 0, x = \pm l$$
$$u = v = w = 0 \quad \frac{\partial p}{\partial z} = 0, y = \pm l$$

10.2.1.3 computational model

This section relies on Jiangsu Yancheng Xiangshui Wind Farm Project and uses the large-scale commercial finite element software COMSOL Multiphysics to establish a calculation model for the dynamic response of offshore wind power foundation and surrounding seabed under the combined action of wind, wave, and current, in which the wind turbine parameters and soil layer parameters are taken from the actual data in the project. The single pile foundation adopts a steel pipe concrete pile, which is 70 m long, 42.4 m buried in the sand layer, and the fan tower is 66.68 m high; The detailed parameters of the foundation wall and tower barrel materials are shown in [Tab. 10-3]. The seabed is mainly composed of muddy soil and silt. The main physical property parameters of soil are shown in [Tab. 10-4]. The isotropic linear elastic constitutive model is adopted for the concrete-filled steel tubular pile and the fuselage, and the porous elastic constitutive model is adopted for the seabed. In the calculation model, the contact between the single pile foundation and the surrounding seabed is constrained by the specified displacement, and the pile is bound to the seabed to make the displacement and deformation of the seabed consistent with the pile foundation. In this way, the interaction between the foundation and the seabed is considered. The finite element model has 80068 nodes and 43749 elements.

[Fig. 10-3] shows the offshore wind power foundation model and load diagram, in which the position and direction of action of wind load, wave current load, and wave load and the positions and materials of components of seabed soil and

[Tab. 10-3] Main component parameters of the wind turbine, source: (LIU Jianxiu, 2018)

Serial No	equipment	Size, length×wide×high	Material Science
1	engine room	8.96×3.62×3.68	compound material
2	Tower drum	$\varnothing 4.3 \times 0.045 \times 62.3$	Q345 Steel
3	pile	$\varnothing 4.3 \times 0.045 \times 70$	—

[Tab. 10-4] Soil physical property parameters, source: (LIU Jianxiu, 2018)

Serial No	Local name	thickness /m	density / g/cm^3	Young's modulus /kPa	Poisson's ratio
1	mucky soil	17.6	1.724	2180	0.38
2	Silty sand	〉12.4	2.13	80000	0.33

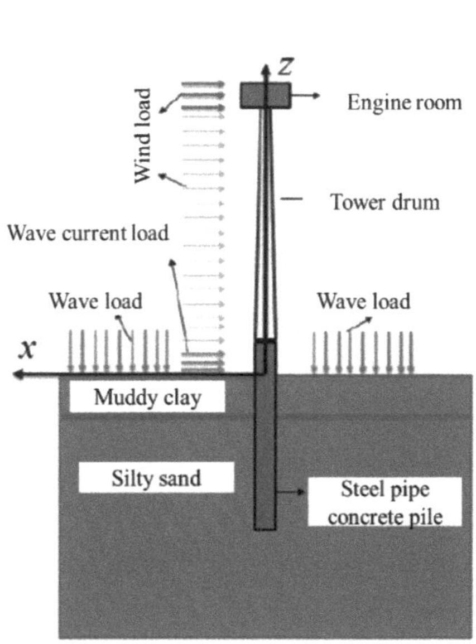

[Fig. 10-3] The diagram of model and loadings, source: (LIU Jianxiu, 2018)

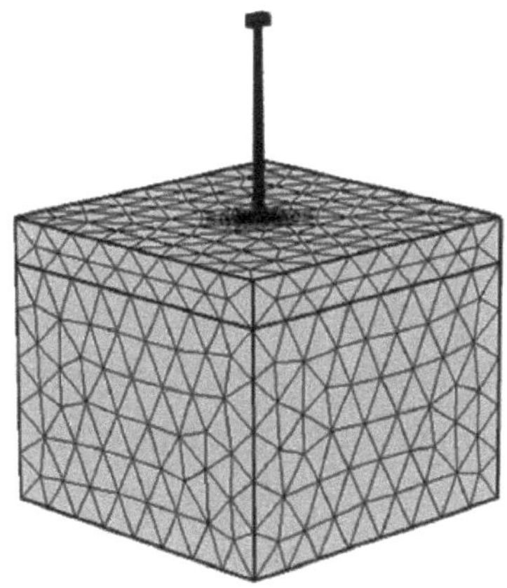

[Fig. 10-4] The view of the three-dimensional model, source: (LIU Jianxiu, 2018)

wind turbine structure are simply shown. [Fig. 10-4] shows the three-dimensional model of offshore wind power foundation established in the finite element software COMSOL Multiphysics.

10.2.1.4 Wind wave current load combination

There is no unified theoretical formula for the load calculation of offshore wind power foundations due to the lack of codes and specifications for offshore wind power foundation design in China. At present, for the dynamic response calculation of wind turbine structure and seabed under the combined action of multiple loads in engineering, the simple linear superposition method, and peak superposition method are often used to combine loads. Such methods ignore the most unfavorable load combinations that may exist and are mostly used for the static calculation of extreme strength conditions. Therefore, given the dynamic problem of offshore wind power found under the wind, wave, and current loads, (LIU Jianxiu, 2018) adopts Turkstra criterion of engineering load combination(JIN

Weiliang, 2006) to combine the three loads, and on this basis, the comparative analysis of the three working conditions and the basic working conditions is added.

10.2.1.4 Wind wave current load combination

There is no unified theoretical formula for the load calculation of offshore wind power foundations due to the lack of codes and specifications for offshore wind power foundation design in China. At present, for the dynamic response calculation of wind turbine structure and seabed under the combined action of multiple loads in engineering, the simple linear superposition method, and peak superposition method are often used to combine loads. Such methods ignore the most unfavorable load combinations that may exist and are mostly used for the static calculation of extreme strength conditions. Therefore, given the dynamic problem of offshore wind power found under the wind, wave, and current loads, (LIU Jianxiu, 2018) adopts Turkstra criterion of engineering load combination(JIN Weiliang, 2006) to combine the three loads, and on this basis, the comparative analysis of the three working conditions and the basic working conditions is added.

When there are three variable loads involved in the combination, the Turkstra combination rule is adopted to combine the extreme value of one load effect at [0, T] with the instantaneous value of the other two load effects. Since the instantaneous value of the two load effects does not reach the maximum at the same time, it is possible to miss the extreme case that three loads reach the maximum at the same time, Given this, the combination of the extreme value of two load effects at [0, T] and the instantaneous value of the other load effects are added in this paper, and compared with the three cases obtained by using Turkstra combination rules, the following six combination methods are obtained(LIU Jianxiu, 2018) :

Combination condition 1:

$$F_1(t) = \max_{0 \le t \le T}\{F_{wind}(t)\} + F_{current}(t) + F_{wave}(t)$$

Combination condition 2:

$$F_2(t) = F_{wind}(t) + \max_{0 \leq t \leq T}\{F_{current}(t)\} + F_{wave}(t)$$

Combination condition 3:

$$F_3(t) = F_{wind}(t) + F_{current}(t) + \max_{0 \leq t \leq T}\{F_{wave}(t)\}$$

Combination condition 4:

$$F_4(t) = \max_{0 \leq t \leq T}\{F_{wind}(t)\} + \max_{0 \leq t \leq T}\{F_{current}(t)\} + \max_{0 \leq t \leq T}\{F_{wave}(t)\}$$

Combination condition 5:

$$F_5(t) = \max_{0 \leq t \leq T}\{F_{wind}(t)\} + F_{current}(t) + \max_{0 \leq t \leq T}\{F_{wave}(t)\}$$

Combination condition 6:

$$F_6(t) = F_{wind}(t) + \max_{0 \leq t \leq T}\{F_{current}(t)\} + \max_{0 \leq t \leq T}\{F_{wave}(t)\}$$

10.2.1.5 Dynamic response results

Liu Jianxiu calculated the dynamic response of offshore wind power foundation and seabed under the action of wind, wave, and current load. In the calculation model, the intersection point of the seabed plane and the central axis of the pile foundation is taken as the coordinate origin, the direction opposite to the direction of wind load is taken as the positive direction of the x-axis, and the direction along the tower height is the positive direction of the z-axis. Since the wind, wave, and current load can reach the maximum within 300s, the calculation time is taken as 300s (LIU Jianxiu, 2018).

(1) Horizontal Displacement Response Analysis of the Top of the Fan Tower and the Top of the Foundation

[Fig. 10-5] shows the horizontal displacement time history curve of the upper tower top of the offshore wind power found under the combined action of wind, wave, and current loads. As shown in the figure, when one load maximum value is combined with the other two load time histories, in turn, i.e. combinations I, II, and III are used for calculation. The results show that combination I, i.e. combination of maximum wind load plus wave current load time histories and wave load time histories, produces the maximum horizontal displacement at the top of the tower, which is about 0.560 m; When two maximum loads are combined with the remaining one load time history, in turn, that is, combination IV, V and VI are calculated. It is found that when combination IV is used, that is, the maximum wind load superimposes the maximum wave current load and wave load time history, the horizontal displacement at the top of the tower is the largest, and the maximum value is 0.568 m.

It can be seen from the results in [Fig. 10-5] that when the wind load is combined with load time history, the dynamic change of the horizontal displacement at the top of the fan tower is like the dynamic change of the wind load. The horizontal displacement value fluctuates greatly with time, and its value changes in the range of -0.25 m~-0.55 m along the negative x-axis (such as combination II, III, and VI); When the wind load is combined with the maximum value, the horizontal displacement at the top of the tower drum fluctuates less with time. When the wave current load is combined with the load time history, the dynamic change of the horizontal

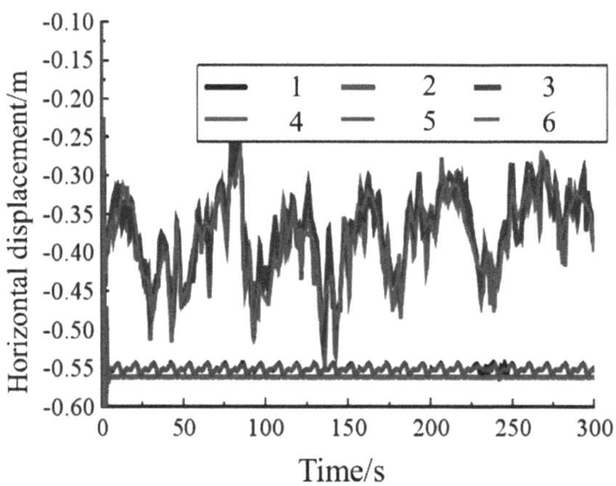

[Fig. 10-5] Time history curve of horizontal displacement at top of the tower, source:(LIU Jianxiu, 2018)

displacement at the top of the tower drum is close to the change law of the wave-current load, and its value changes within the range of 0.545 m~-0.560 m in the negative direction of the x-axis (such as combination I and V). When the wave current load is combined with the maximum value, The horizontal displacement at the top of the cylinder tends to be gentle with time (such as combination IV). Wind load and wave current load has a greater impact on the dynamic response of the upper tower of the wind power foundation.

It can also be seen from the horizontal displacement time history curve of the top of the tower on the offshore wind power foundation under various combinations in [Fig. 10-5] that when the wind load is applied to the wind turbine in load time history, the fluctuation amplitude of the horizontal displacement of the top of the tower is much larger than that of wave current load and wave load applied to the wind turbine in load time history, as shown in combination IV, V and VI, Therefore, the dynamic characteristics of wind load have a much greater impact on the upper structure of the fan than wave current load and wave load.

[Fig. 10-6] shows the time history curve of horizontal displacement at the top of offshore wind power found under the combined action of wind, wave, and current. It can be seen from the figure that when combinations

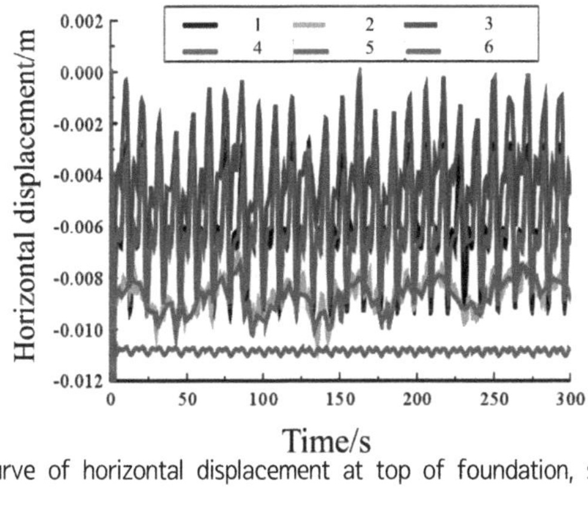

[Fig. 10-6] Time history curve of horizontal displacement at top of foundation, source: (LIU Jianxiu, 2018)

I, II and III are used for calculation, combination II is the combination of the maximum wave current load superimposed wind load time history and wave load time history, the top of the wind power foundation has the

largest horizontal displacement, which is in the negative direction along the x-axis, and the maximum value is about - 0.0105 m; This is due to the large horizontal force generated by wave current load at the top of the foundation, while the small horizontal force generated by wind load and wave load at the top of the foundation. When combinations IV, V, and VI are used for dynamic calculation of foundation, combination IV, i.e. the maximum wind load plus the maximum wave current load and wave load time history, has the maximum displacement, the direction is negative along the x-axis, and the maximum value is -0.0110 m.

It can be seen from the calculation results in [Fig. 10-6] that when wave current load is applied to the foundation in load time history, the fluctuation amplitude of the horizontal displacement at the top of the foundation is greater than that when wave current load and wind load is applied to the foundation in load time history respectively (as shown in the results of Combination IV, V and VI in the figure), so the dynamic characteristics of wave current load have a greater impact on the foundation than wind load and wave load.

Under the action of wind, wave, and current load, the horizontal displacement change curve of the fan tower and pile shaft along the height direction is shown in [Fig. 10-7], where the height z=10 m is the dividing height of the tower and pile shaft. It can be seen from the figure that the horizontal displacement of the pile shaft and tower barrel increases gradually along the height direction. The horizontal displacement of the pile shaft along the height direction is very small. The horizontal displacement at the top of the foundation is about -0.01 m, and the direction is negative along the x-axis; For the tower, the displacement increment of horizontal displacement along the height direction increases significantly, and the maximum horizontal displacement at the top of the tower is about -0.55 m

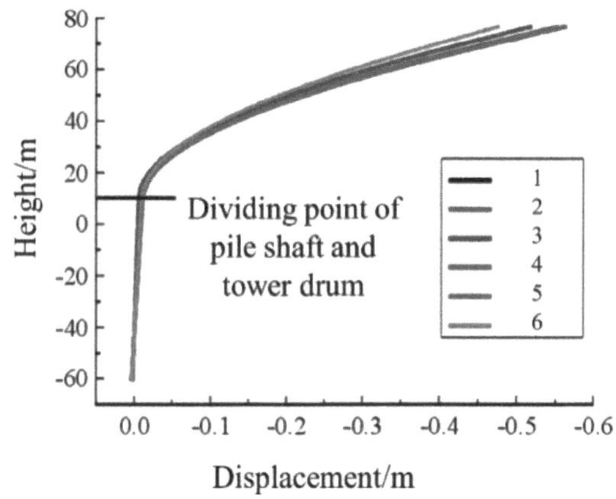

[Fig. 10-7] Horizontal displacement of the pile and tower along the height, source: (LIU Jianxiu, 2018)

along the negative x-axis. This is because the restraint effect of the seabed around the foundation on the foundation limits the displacement of the foundation, and the large wind load directly acts on the top of the tower, resulting in a large displacement of the top of the tower. Therefore, the horizontal displacement of the top of the foundation is far less than the horizontal displacement of the top of the tower.

(2) Vertical stress response analysis of tower bottom

[Fig. 10-8] shows the vertical stress time history curve of the bottom of the upper tower of the wind power found under the combined action of wind, wave, and current. It can be seen from the calculation results that when the maximum wind load plus the maximum wave current load and wave load time history of combination IV is adopted, the vertical stress at the tower bottom is the maximum, and the maximum value is 134 MPa. It can be seen from [Fig. 10-8] that when combinations I, II, and III are used for calculation, the vertical stress at the bottom of the combination I tower is the maximum, and the maximum value is 133 MPa. This is because the main load acting on the tower is the wind load, and the wind load has the greatest impact on the vertical stress response at the bottom of the tower.

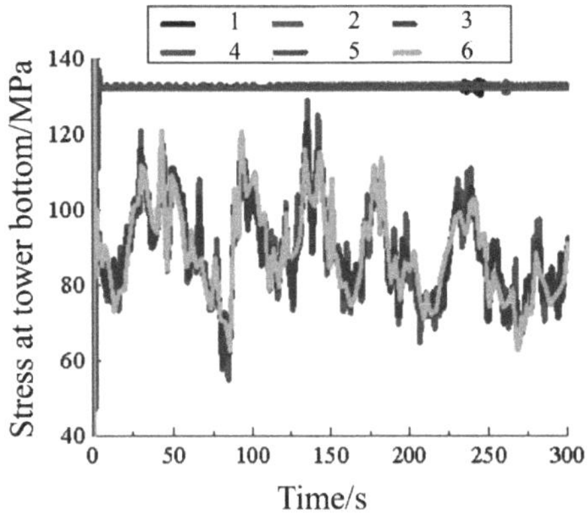

[Fig. 10-8] Time history curve of vertical stress at bottom of the tower, source: (LIU Jianxiu, 2018)

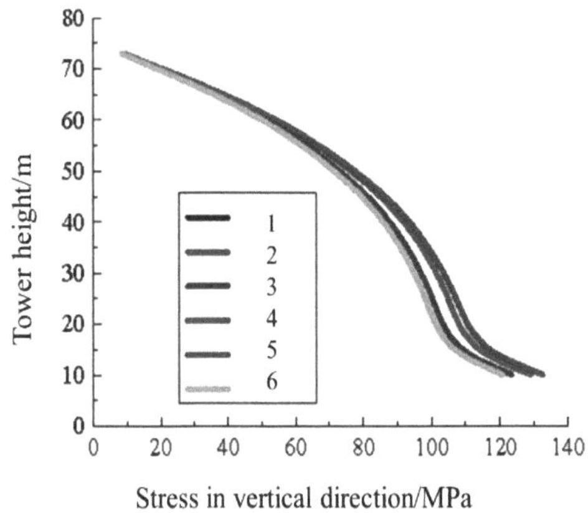

[Fig. 10-9] Vertical stress of the tower along the height, source: (LIU Jianxiu, 2018)

[Fig. 10-9] shows the vertical stress change curve of the windward face of the fan tower along the tower height direction under six combined loads. It can be seen from the figure that, no matter what load combination mode is adopted, the vertical stress on the windward side of the tower presents a decreasing trend along the height direction of the tower. The vertical stress on the top of the tower is the minimum, and its maximum value is about 9.13 MPa; The stress at the intersection between the bottom of the tower and the foundation is the maximum (z=10 m), about 134 MPa, which is 15 times of the vertical stress at the top of the tower; In addition, stress concentration is easy to occur here. Therefore, during design, attention should be paid to the stress change at the intersection of the tower and foundation, and the strength of materials and structures at this location should be strengthened.

(3) Analys of pile bending moment and pile shear response

[Fig. 10-10] shows the change of bending moment of wind power foundation pile shaft with height under different load combinations. It can be seen from the figure that the bending moment of the pile shaft of the wind power found under the six combinations changes with the height following the law. First, the pile bending moment increases continuously in the depth of the coastal bed, the maximum bending moment occurs below the mud surface and then decreases with the increase of the depth of the seabed. When it is near the bottom of the pile shaft, the pile bending moment increases slightly in the reverse direction, then decreases, and finally decreases to zero at the bottom of the pile shaft. This is mainly due to the restraint of the surrounding seabed on the pile foundation and the combined effect of wave-current load acting on the pile and wave load acting on the seabed.

[Fig. 10-11] shows the variation of pile shear force of wind power foundation with height under different load combinations. It can be seen that the variation of pile shear force and pile bending moment with height is the same, and the maximum shear force occurs below the mud surface.

It can be seen from the results in [Fig. 10-10] and [Fig. 10-11] that the pile bending moment and pile shear force are the largest when the combination four, namely, the maximum wind load and the maximum wave current load plus the wave load time history, is used in the six combination methods. That is when the wind load and wave current load acting on the fan system take the maximum value, and the pile bending moment and pile shear force are the largest. When the maximum value of a single load and the load time

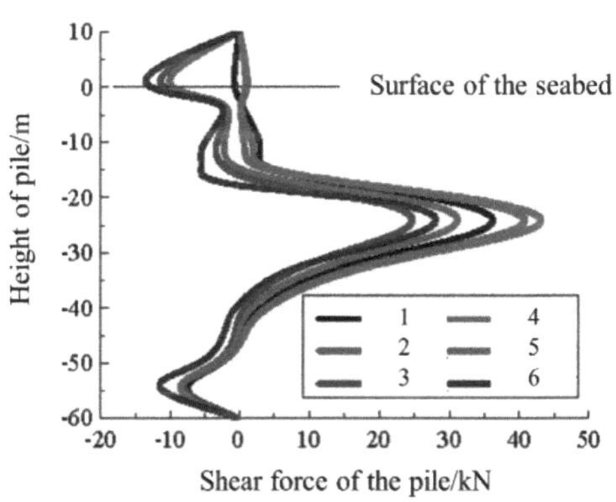

[Fig. 10-11] The curve of shear force of the pile along the height, source: (LIU Jianxiu, 2018)

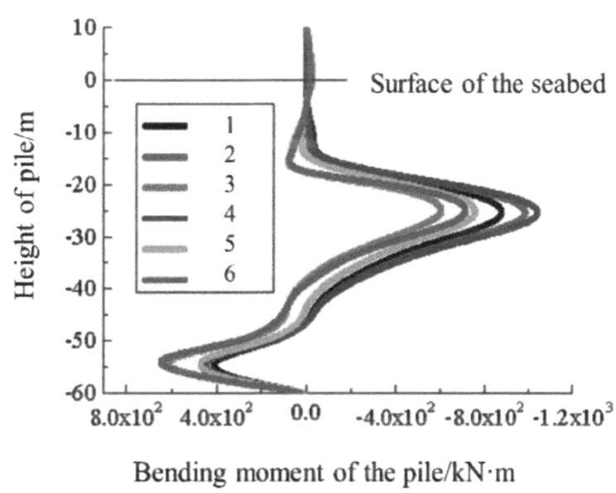

[Fig. 10-10] The curve of the bending moment of the pile along the height, source: (LIU Jianxiu, 2018)

history combination of other loads (i.e. combination I, II, and III) are used for calculation, it is found that the pile bending moment and pile shear force are maximum when the combination II is used, followed by a combination I and combination III is minimum, that is, when the wave current load takes the maximum value, the pile bending moment and pile shear force are maximum, the wind load takes the second place, and the wave load takes the maximum value, the result is minimum. Wave current load has the greatest impact on the bending moment and shear force of the pile shaft of the wind power foundation, followed by wind load and wave load.

10.2.2 Analysis of jacket foundation structure characteristics under wind wave current load

10.2.2.1 Project Overview

For an offshore wind power project in Yangjiang, the water depth in the sea area of the project site is 25 m~28 m, and the surface layer of the seabed is silt with uneven thickness. The four-pile jacket foundation scheme with 22 m spacing is adopted. The diameter of the rock socketed pile is 2.4 m, the elevation of the pile top is - 23 m, the elevation of the pile bottom is - 53 m, the original mud surface is - 26 m, and the depth of the pile foundation is 27 m. The upper fan adopts the MySE5.5-155 model, with a rated capacity of 5.5 MW, and the total weight of the engine room, gearbox, generator, blade, and hub is 553 t. The height of the pre-installed hub is 105.7 m, and the length of the jacket foundation tower is 78.03 m. According to the data of Yangjiang Meteorological Station, the design maximum wave is a 50-year return period of H1%, wave height is 13.95 m, wave period T is 14.2 s, and wavelength L=228.7 m. The maximum 10 min average wind speed is 34.6 m/s. The surface current velocity is 1.5 m/s, the middle current velocity is 1.0 m/s, and the bottom current velocity is 0.65 m/s.

10.2.2.2 Load introduction

The offshore wind power pipe rack foundation is located in the marine environment and mainly bears the wind, wave, and current loads, as well as the aerodynamic loads caused by the rotation of the fan rotor, which may be subjected to seismic loads, impact loads of floating objects such as ships or sea ice. The project is located in the Yangjiang sea area, where there is no freezing condition and the seismic intensity is below Grade 6. Therefore, ice load and seismic load need not be considered, and they are mainly subjected to wind turbine aerodynamic load, tower wind load, wave load, current load, and gravity load on the foundation part below the water surface. For a brief introduction, see the conference (GUO Mengyuan & FAN Lichan, 2019)

10.2.2.3 P—Y curve introduction and calculation results

At present, the P-Y curve method is a widely used pile foundation design method in offshore engineering. It is the relationship curve between the soil reaction P at the depth Z below the mud surface and the deflection Y of the pile at this point under the action of the horizontal force H, which comprehensively reflects the nonlinear characteristics of the soil around the pile, the stiffness of the pile and the nature of the external load (Shi et al., 2015).

See Table 2 for geological data of the Yangzijiang sea area and parameters required for P-Y curve calculation(Shi et al., 2015). The soil layer is divided into a series of discrete nonlinear springs every 1 m, and their loading properties are described by the P-Y curve method. See [Fig. 10-12] and [Fig. 10-13] for examples of P-Y curves of layer ② clay and layer ④ sand.

It can be seen from [Fig. 10-12] and [Fig. 10-13] that when the same displacement occurs, the soil reaction of the pile in the sand is much greater than that of the pile in clay, and when the soil reaction of the pile in sand reaches a stable level, the displacement occurs less.

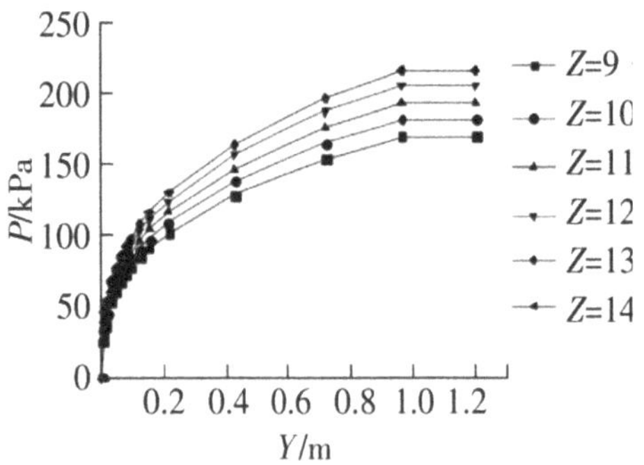

[Fig. 10-12] The P-Y curves of the 2th soil

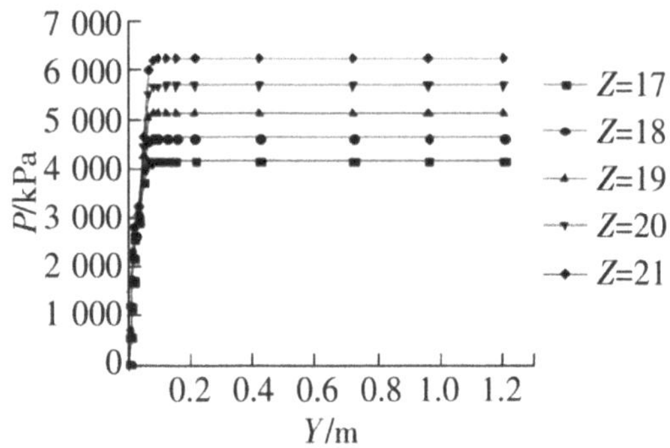

[Fig. 10-13] The P-Y curves of the 4th soil

The integral model of the jacket foundation is established by using ANSYS. The four piles are distributed in a square shape, with a root opening of 11 m at the top and 22 m at the bottom of the jacket. The diagonal bracing is arranged in an X-shape. One layer of horizontal cross brace is arranged at the top layer and the bottom layer respectively, and the jacket substructure relates to the tower through the transition section. The transition section and the diagonal brace of the transition section are tapered cylinder structures with a large diameter in the upper part and a small diameter in the lower part. (Shi et al., 2015).

Pipe59 unit is used for jacket foundation and pile foundation on mud surface, Pipe16 unit is used for pile foundation under mud surface, Beam188 unit is used for transition section and tower, hub, engine room, and rotor are simplified as concentrated mass points, which are applied on the top of the tower and simulated by Mass21 unit. The pile-soil interaction is simulated by the Combine39 element, and the horizontal spring stiffness is determined by the P-Y curve method.

10.2.2.4 Static Calculation of Jacket Foundation

We apply the aerodynamic load of the fan to the flange surface of the fan in the form of static force, and the wind load of the tower is uniformly applied to the height direction of the tower in the form of concentrated force. Stokes's fifth-order wave is adopted for wave theory. When using this theory, the phase

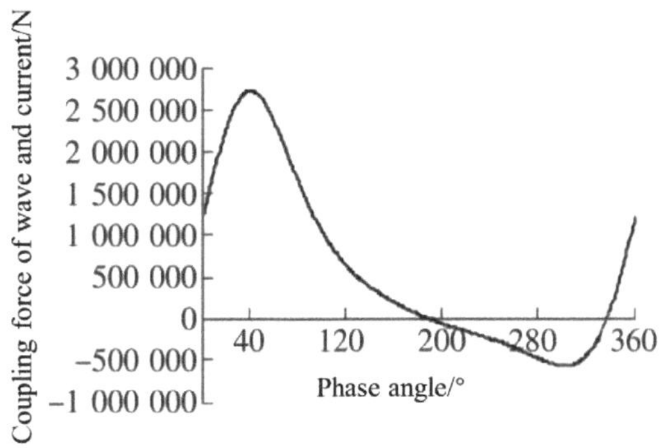

[Fig. 10-14] The relationship between Coupling force of wave and current and Phase angle

angle φ The value of has a great impact on the wave current load value acting on the structure. Therefore, the wave phase angle is searched for the wave height and period, and the curve is fitted by Origin software, as shown in [Fig. 10-14].

It can be seen from Figure 3 that the wave current coupling force is the maximum when the phase angle is 53°, 2733.361 kN. This phase angle is applied as the most unfavorable working condition of wave current action in the static calculation. Input parameters in the "Water table" of ANSYS software, and the software will calculate the wave load by itself.

10.2.2.5 Analysis of static calculation results of the jacket foundation

Apply the wind, wave, current load, and gravity load on the structure through ANSYS software, calculate the Von mises equivalent stress distribution diagram, displacement, and rotation angle of the jacket foundation, and check the strength and control the displacement of the foundation. (GB/T 51308-2019海上风力发电场设计标准, 2019).

It can be seen from [Fig. 10-15] that the maximum equivalent stress of the jacket foundation is 228 MPa under the limited condition of bearing capacity. According to the Rules for Construction and Classification of Mobile Offshore Platforms issued by the China Classification Society: $[\sigma] = \sigma_s / S$, where $[\sigma]$ Is the allowable stress; σ_s is the yield strength, DH36 steel is used for the jacket foundation, and the yield

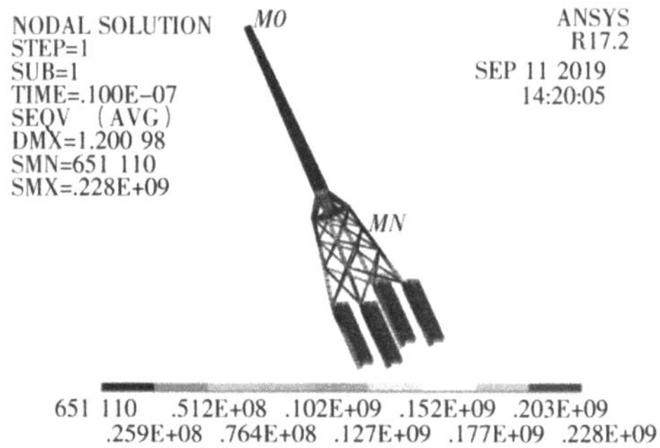

[Fig. 10-15] Equivalent stress nephogram of wind power foundation/Pa

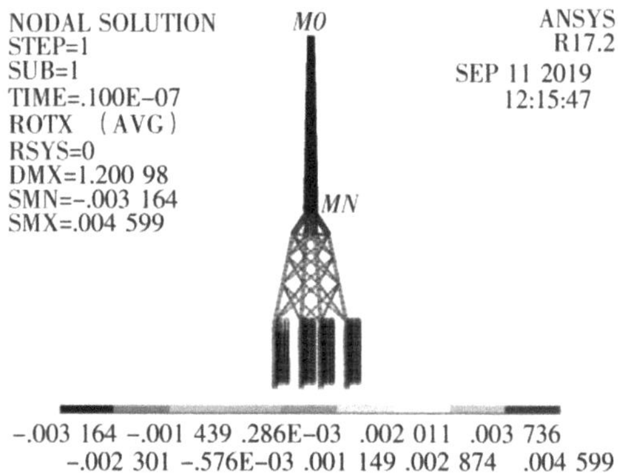

[Fig. 10-16] Corner nephogram of wind power foundation/rad

strength is 355 MPa; S is the safety factor, taken as 1.25, so the allowable stress of the fan foundation is 284 MPa. Therefore, it meets the structural strength requirements.

It can be seen from [Fig. 10-16] that the angle at the flange is 0.003 164 rad, that is, 0.18°. According to DNVGL-ST-0126 Support structures for wind turbines (Support Structures for Wind Turbines, 2021), the maximum allowable total angle at the mud surface of a single pile foundation is 0.5°. For angle control, 4/1000 (tower bottom flange surface) is taken for multi-pile foundation, and 4.36/1000 (mud surface) is taken for single pile foundation, that is, 0.25°. Therefore, it meets the requirements of corner control.

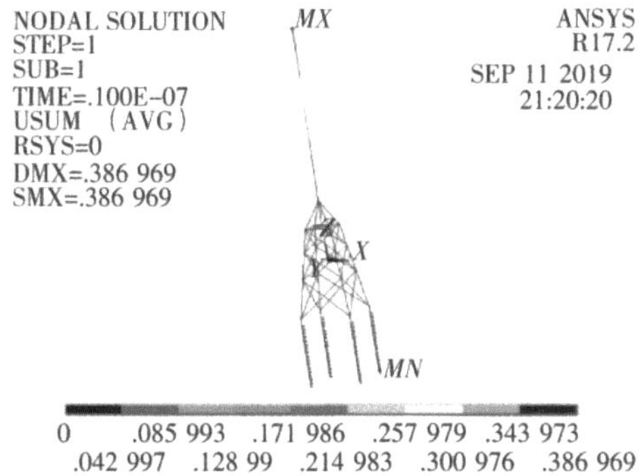

[Fig. 10-17] Displacement nephogram of wind power foundation/m

It can be seen from [Fig. 10-17] that the maximum displacement occurs at the top of the tower, which is 0.386 m. According to the reference (Nicholson, 2011), to avoid excessive displacement at the top of the tower, the maximum displacement should be between 0.25% and 1% of the tower height. The tower height is 78.03 m, which is 0.49% of the tower height. Therefore, the offset is reasonable.

10.2.3 Response of suction tube jacket considering tube soil contact under wind wave current

Pan Hongguan took an offshore wind power suction tube jacket foundation in the South China Sea as the analysis object, took the extreme working conditions with a 50-year return period as the load conditions, and considered the tube soil contact to establish a full three-dimensional finite element model to analyze the installation angle, structural strength, foundation settlement and bearing capacity of the suction tube jacket. The results show that when the relative action angle between the installation direction of the foundation and the wave current action direction is 120°, the load response of the foundation is the minimum. The foundation bearing capacity and foundation strength meet the requirements. During the installation of the foundation and fan, the maximum soil settlement of 0.431 m will occur, and sufficient settlement shall be reserved for the foundation (PAN Hongguan et al., 2022).

See the literature for the calculation and analysis theoretical formula of wind load, wave load, and current load (PAN Hongguan et al., 2022).

10.2.3.1 Establish numerical calculation model

(1) model parameter

Based on an offshore wind power project, the suction tube jacket foundation and soil are modeled. The water depth in this sea area is 27.69 m. The relevant parameters of the suction cylinder jacket and soil mass are shown in Table 1. The specific parameters are shown in the conference (PAN Hongguan et al., 2022) and Table 2. The specific parameters are shown in the conference (PAN Hongguan et al., 2022). The jacket and the tower are connected by transition sections. There are four sections of the tower, and the parameters are shown in Table 3. See the conference (PAN Hongguan et al., 2022) for specific parameters. As the lower tower needs to bear more load, the tower adopts a gradual diameter design. In Table 3 (See the conference(PAN Hongguan et al., 2022)), Item 1 of the dimension parameters refers to the diameter of the lower end of the tower, Item 2 refers to the diameter of the upper end, and Item 3 refers to the length of the section. In order to reduce the influence of soil boundary on the analysis results and balance the calculation amount, the soil calculation area is 200 m×200 m ×38 m cuboid model.

(2) Mesh generation and boundary conditions

The sweeping method is used to mesh the structure and soil, and the finite element model shown in [Fig. 10-18] is established in ANSYS. Among them, the soil mass is based on the Drucker Prager constitutive model, using Solid45 solid element; Shell 181 surface element is adopted for suction cylinder and tower cylinder; Beam189 beam element is adopted as stiffener; Pipe59 pipe unit capable of calculating hydrodynamic force is adopted for jacket structure; The mass of the main engine and blades is uniformly distributed to the nodes at the top of the tower in the form of Mass21 structural mass units.

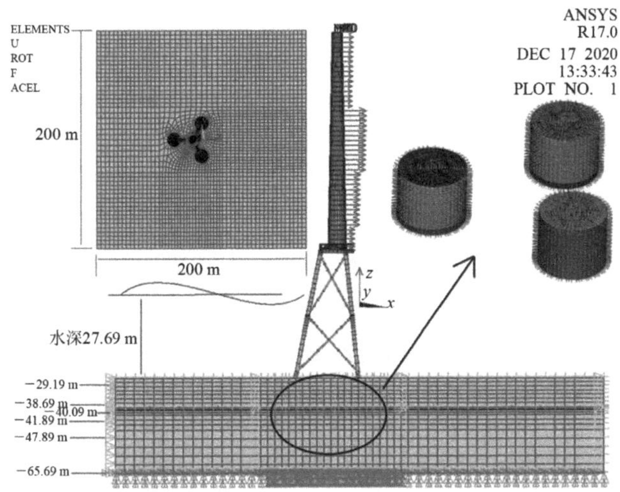

[Fig. 10-18] Finite element model of the Drucker Prager constitutive model, source:(PAN Hongguan et al., 2022)

When the stiffness of one object is much greater than that of another object, it can be regarded as a rigid flexible body contact problem. Contact analysis requires the establishment of contact pairs on contact surfaces. The interface of a suction cylinder with larger stiffness is defined as the target surface (TARGET), and the interface of soil with smaller stiffness is defined as the contact surface (CONTACT). The target element and contact element are generated in the contact area. The contact area includes the inner wall of the cylinder and the soil inside the cylinder, the outer wall of the cylinder and the soil outside the cylinder, and the inner wall of the cover plate and the soil inside the cylinder. The contact state is shown in [Fig. 10-18]. By defining the normal stiffness factor FKN, the program will estimate the contact stiffness based on the material properties of the deformable body element. After calculation, it is found that FKN of 0.2 in this model can achieve good convergence. The initial allowable penetration tolerance of the contact surface specifies the initial penetration range of the contact surface. Through iterative adjustment, when the maximum initial allowable penetration tolerance is 0.15, the entire target surface (together with the deformable body) can be moved to the specified initial penetration range, making it the initial state of closed contact. The friction resistance between the structure and the soil plays an important role in the settlement and anti-overturning stability of the foundation. Based on the engineering geological survey data and in combination with the conferences (YE Chaoliang et al., 2015), (XU Hongfa et al., 2002) the friction coefficient between the cylinder and muddy soil and soft plastic clay is 0.20, and the friction coefficient between the cylinder and saturated coarse sand is 0.28. After the grid division is completed, normal displacement constraints are imposed on the four sides of the soil, and full displacement and rotation constraints are imposed on the bottom.

(3) Calculation condition

According to the statistics of the hydrological observation station, the wave height with a 50-year return period cumulative probability of 1% is 13.9 m, and the wave period is 14.1s. The statistical values of the current velocity in the surface layer, 0.6 H layer (H is the water depth, 0.6 H layer is 0.6 times the water depth below the water surface), and bottom layer are 209.1 cm/s, 140.2 cm/s, and 99.4 cm/s respectively. The average wind speed in 10 minutes at the height of 10~120 m from the sea level is shown in Table 4, and the specific parameters are shown in the literature (PAN Hongguan et al., 2022). Based on the wind wave current theory introduced in Section 1, the load is applied to the suction cylinder jacket foundation in ANSYS APDL.

10.2.3.2 Result analysis

(1) Influence of wave phase and wave current direction

When Stokes fifth order wave theory is adopted, the value of wave phase angle φ has a great influence on the wave load acting on the structure. When the wave acts on the structure, the magnitude of its force is closely related to the wave phase angle. In the static analysis of the structure, it is assumed that the wave acts on the structure with the maximum force, to consider the most unfavorable working condition of the structure subjected to wave load.

Therefore, in the static analysis, it is necessary to conduct a 0°~360° omnidirectional search on the wave phase angle φ to determine the phase angle value when the wave current coupling force is maximum. Through the command flow function of ANSYSAPDL, prepare the corresponding wave phase angle search program, extract the wave current coupling node reaction at the bottom of the suction cylinder under the action of different phase angles, and draw the wave current coupling node reaction phase angle relationship curve as shown in [Fig. 10-19]. It can be seen from [Fig. 10-19] that when the phase angle is 34°, the reaction force at the wave current coupling node is maximum. The phase angle values of subsequent analysis are based on this calculation result.

The annual observation data of the site sea wave station shows that during the observation period, the normal wave direction is ESE (southeast by east), with a probability of 52.04%, and the probability of wave direction appearing in ESE~SSE (southeast by south) is more than 90%. The discussion of the influence of wave and current direction on the jacket foundation of the suction cylinder is helpful to find the best installation direction for the suction cylinder. The research of Xia Ke et al. (YE Chaoliang et al., 2015) shows that the load on the jacket foundation will slightly increase and then decrease with the increase of wave current angle. To simplify the calculation, the wave current angle of 0° is considered the most unfavorable condition. Due to the axial symmetry of the suction cylinder jacket

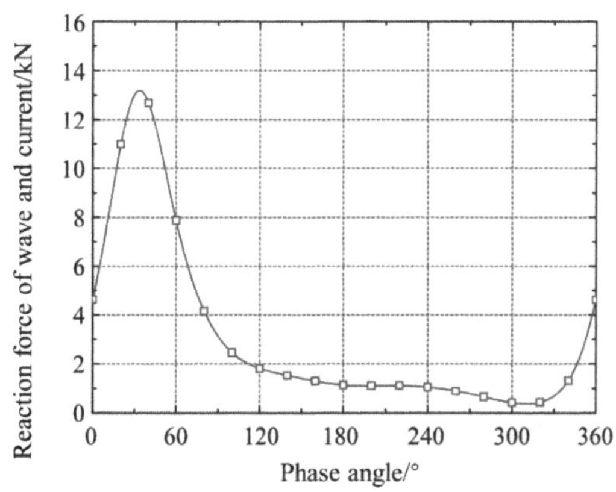

[Fig. 10-19] Wave current coupling node reaction phase angle relationship curve, source: (PAN Hongguan et al., 2022)

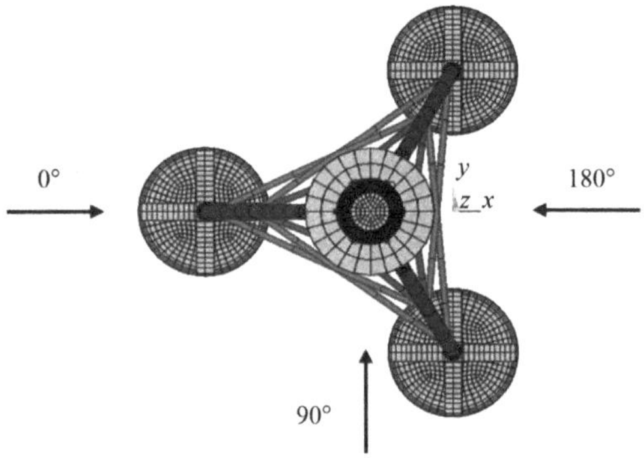

[Fig. 10-20] Action direction of the wave and current, source: (PAN Hongguan et al., 2022)

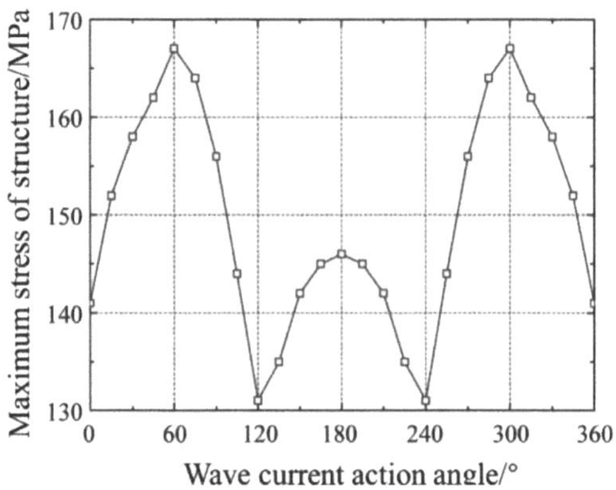

[Fig. 10-21] Maximum stress of structure under different wave current action angles, source: (PAN Hongguan et al., 2022)

foundation, only the structural stress with a relative angle of 0°~180° between the wave current and the structure is calculated. The relative action direction of wave current and suction cylinder jacket structure is shown in [Fig. 10-20].

Change the relative action direction of wave current and structure, extract the maximum stress of structure every 15°, and make statistics of the relationship between the action direction of wave current and the maximum stress of structure as shown in [Fig. 10-21]. Due to the axial symmetry of the suction cylinder jacket foundation, the maximum stress of the structure is symmetrical between about 0° and 180°. It can be seen from [Fig. 10-21] that when the wave current action angle is 120°~240°, the maximum stress of the structure is small, and when the action-angle is 120° and 240°, the maximum stress of the structure is minimum.

(2) Strength check of the suction cylinder jacket

The maximum structural equivalent stress under the extreme wind wave current load with a 50-year return period is shown in [Fig. 10-22]. It can be seen from [Fig. 10-22] that the maximum equivalent stress of the suction cylinder jacket foundation under extreme working conditions is 220 MPa,

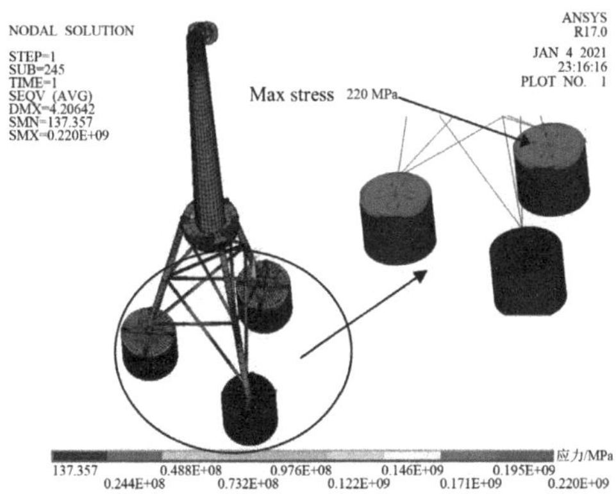

[Fig. 10-22] Maximum equivalent stress of structure under 50-year return period limit working condition, source: (PAN Hongguan et al., 2022)

which is less than the yield strength of DH36 high-strength structural steel 335 MPa, and the strength meets the requirements. The maximum stress occurs on the cylinder cover plate at the back wave side, and the stress of the cylinder cover plate at the back wave side is significantly greater than that at the upstream wave side. Under the action of wind, wave, and current load bending moment, the jacket tends to overturn towards the back wave side, resulting in the increase of pressure on the two cylinders at the back wave side and the decrease of pressure on the cylinders at the upstream wave side. Therefore, at the design stage of the suction cylinder, the wall thickness of the suction cylinder at the upstream and downstream sides can be adjusted differently according to the installation direction of the suction cylinder foundation.

(3) Foundation settlement

In practical engineering, before the construction of the fan foundation, the soil mass has completed self-weight consolidation settlement. In the finite element calculation, the self-weight of soil will cause the compression of soil. Therefore, to determine the foundation settlement caused by foundation installation, it is necessary to calculate the settlement difference between the foundation and the foundation before and after fan installation. settlement.

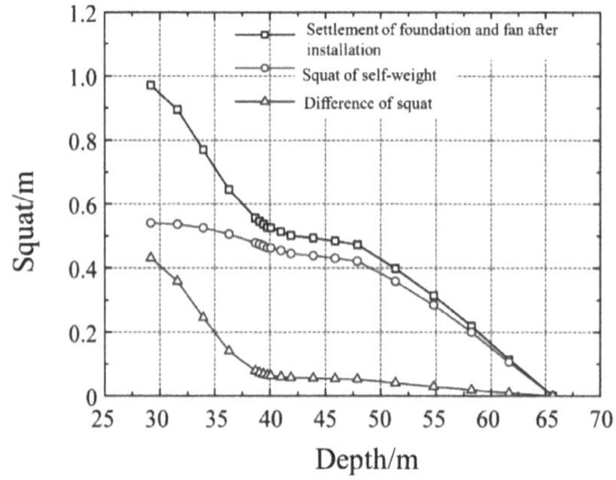

[Fig. 10-23] Soil squat changes with depth before and after foundation and fan installation, source: (PAN Hongguan et al., 2022)

The amount and settlement difference of soil mass before and after the installation of the foundation and fan with depth is shown in [Fig. 10-23].

It can be seen from [Fig. 10-23] that the maximum settlement of soil mass in the cylinder will be 0.431 m after the installation of the foundation and fan. Since the effective overburden pressure of natural silt and muddy soil layer is small, the self-weight consolidation settlement of this soil layer is relatively small. Since 90% of the soil layer in the cylinder is muddy soil, and the soil in the cylinder can only be compressed downward, the settlement difference of the soil in the cylinder is the largest. With the increase in depth, the settlement difference decreases gradually. The depth of 41.89~47.89 m is a coarse sand layer. Because the compressibility of coarse sand is small, the settlement difference of this layer is small. The depth of 47.89~65.69 m is a clay layer, the compressibility increases, and the range of settlement difference becomes larger again. It can be seen from the settlement analysis results that the soft soil layer of the project site is thick, and the sludge compressibility is large. During the foundation design and installation, the sufficient settlement must be reserved for this condition.

(4) Verification of bearing capacity of cylinder foundation

Using finite element software to calculate the relationship between load and displacement is an iterative process by finding the balance between external force and internal force. The iterative program needs to meet the convergence criteria before it can continue to run. When the calculation of the adopted load step size is not convergent, the program will automatically use the dichotomy method to reduce the step size until this step converges. If the load step size is reduced to the minimum step size, the program still cannot find a solution from the finite element equations that satisfy both the static equilibrium and the stress-strain relationship and strength criteria, then the calculation does not converge. There is a one-to-one correspondence between displacement mutation and calculation non-convergence. When the load increases slowly but the displacement suddenly changes or the calculation does not converge, the soil mass within the sliding boundary is in a plastic failure state, and the foundation loses stability. Currently, the bearing capacity is the ultimate bearing capacity.

In the process of structural load loading, the relationship between the contact pressure of the suction cylinder cover and soil mass and the compression displacement of soil mass is shown in [Fig. 10-24]. It can be seen from [Fig. 10-23] and [Fig. 10-24] that when the soil compression displacement reaches 0.971 m, the self-weight load of the structure has been loaded, and the loading continues at

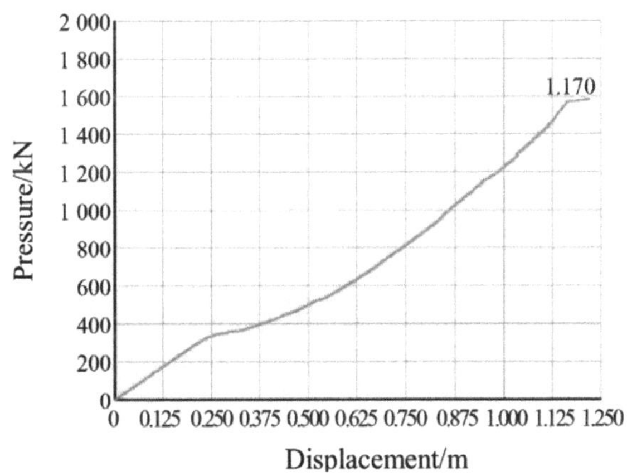

[Fig. 10-24] Contact pressure soil compression displacement, source: (PAN Hongguan et al., 2022)

this time. When the soil compression displacement reaches 1.170 m, there will be a sudden displacement change, and the applied load is the ultimate bearing capacity. The actual load of the structure is 32183 kN, which is less than 48930 kN applied when the soil yield and 56043 kN compressive bearing capacity are calculated. The bearing capacity of the barrel foundation meets the requirements. It can be seen from the calculation results that the load-displacement curve method based on finite element calculation is more conservative than the code method.

10.3 Wave load, wind load, and gravity load analysis theory on an environmental load of wind farm sea area

Li Yaya used ABAQUS numerical analysis software to establish a solid numerical finite element numerical calculation model for wind power single pile foundation in the marine environment, calculated the environmental load of the wind farm sea area using reasonable wave load, wind load, and vertical load analysis theory, and applied the environmental load equivalent to horizontal cyclic load, overturning moment load and gravity load, On this basis, part of the calculation results of the numerical model are compared with the existing research results to verify the scientific rationality of the pile-soil finite element model (LI Yazhou, 2020).

10.3.1 Establishment of numerical calculation model of offshore wind power monopile foundation under wind-wave-gravity loads

10.3.1.1 Model overview

Taking a proposed wind farm in the Xiangshui area of the East China Sea as the actual engineering environment, Li established a solid numerical finite element numerical calculation model of wind power single pile foundation in the marine environment using ABAQUS numerical analysis software. For pile foundations bearing asymmetric combined load, a three-dimensional overall modeling analysis

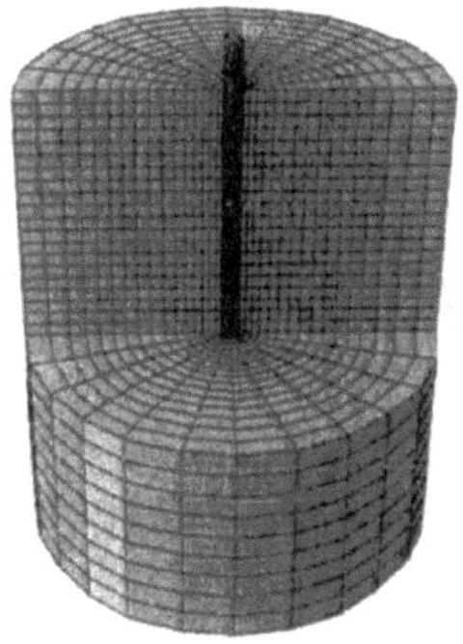

[Fig. 10-25] Finite element numerical calculation model of monopile foundation for offshore wind power, (LI Yazhou, 2020)

must be carried out. See the literature (LI Yazhou, 2020) for the geometric and physical parameters of the model, The finite element numerical calculation model of the large diameter single pile foundation of the upper wind power and the marine sandy soil layer is shown in [Fig. 10-25].

10.3.2 Analysis of numerical model calculation results of offshore wind power single pile foundation under wind-wave-gravity loads

10.3.2.1. Analysis of horizontal displacement along the pile shaft

In practical engineering, an offshore wind power single pile foundation is generally defined as a horizontally loaded pile, so the horizontal displacement of the pile shaft is an important control standard for the bearing capacity of the pile foundation. After 1000 wave load cycles in the horizontal direction of the model, according to the horizontal displacement time history curve of the pile top and mud surface and the horizontal displacement curve of the pile shaft at special moments (t=2.5s and t=7.5s during the cycle), The dynamic bearing characteristics of offshore wind power single pile foundation under the combined action of wind wave gravity load and the influence of different load changes on the dynamic bearing characteristics are analyzed.

(1) **Cumulative analysis of horizontal displacement**

When subjected to wave load in the horizontal direction, pile foundation and pile side soil will experience displacement accumulation under reciprocating cyclic loading. Through calculation and verification, the extreme value of horizontal displacement of a single pile foundation tends to be stable after 1000 cycles of horizontal loading in the horizontal direction. Therefore, the rule of horizontal displacement accumulation of offshore wind power single pile foundation under the combined action of wind wave gravity load is explored, First, take case 1.1 (LI Yazhou, 2020) and case 1.6 (see conference (LI Yazhou, 2020) as examples to analyze the time history curve of horizontal displacement at pile top and mud surface of offshore wind power single pile foundation under the combined action of wind wave gravity load and single wave cyclic load in the 20th, 200th, 500th and 1000th cycles, Then the general rule of pile displacement accumulation is summarized. Since the cumulative displacement of pile side is much smaller than the total displacement, for the convenience of observation and analysis, the time history change of displacement under the last two cycles in the corresponding cycle interval is taken for research. The time history curve of horizontal displacement at pile top and mud surface of offshore wind power single pile foundation in the Nth cycle is shown in [Fig. 10-26] and [Fig. 10-27].

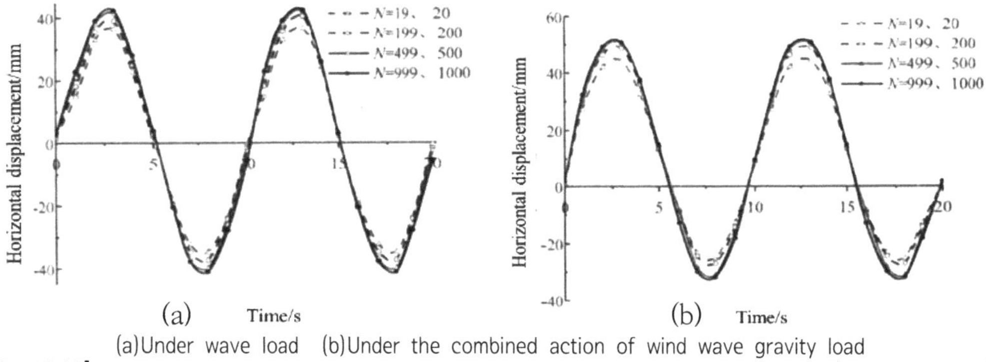

(a)Under wave load (b)Under the combined action of wind wave gravity load

[Fig. 10-26] Time-history curve of horizontal displacement of pile top in the Nth cycle, (LI Yazhou, 2020)

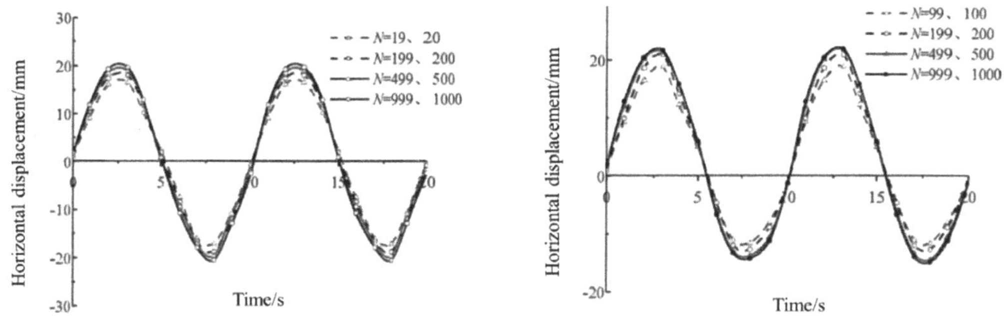

(a)The horizontal displacement in wave load

(b) The horizontal displacement in couple of wind, wave, and weight

[Fig. 10-27] Time-history curve of horizontal displacement at the mud surface of the Nth cycle (LI Yazhou, 2020)

It can be seen from [Fig. 10-26] that under the combined action of single wave load and wind wave gravity load, the overall lateral deformation law of offshore wind power single pile foundation is roughly the same, the time history curve is close to the sine function curve, and the horizontal displacement at the pile top location changes correspondingly with the change of load size and direction, and when the wave load is loaded to the positive peak value and the negative peak value, The horizontal displacement at the pile top of single pile foundation reaches the maximum and minimum values respectively. The difference is that the absolute values of the positive and negative extreme values of the pile top horizontal displacement of a single pile foundation under wave load in [Fig. 10-26] (a) are roughly equal, while the time history curve in [Fig. 10-26] (b) can be approximately regarded as the corresponding curve in [Fig. 10-26] (a) has moved up vertically, that is, there is a large gap between the absolute values of the positive and negative extreme values of the pile top horizontal displacement under the combined action of wind wave gravity load. This is because when a single pile foundation is subjected to wave load, the horizontal force is symmetrical, while under the combined action of wind wave gravity load, the single pile foundation will be subject to the action of wind-induced overturning bending moment load. Based on horizontal reciprocating deformation, the pile shaft will also have angular displacement, which plays a positive role in increasing and a negative role in reducing the horizontal displacement of the pile top.

Based on the analysis of the change rule of the pile top horizontal displacement under different cycles in [Fig. 10-26] and [Fig. 10-27], the change rule of the horizontal displacement at the mud level is consistent with the change rule of the pile top horizontal displacement under the two loading conditions of the offshore wind power single pile foundation. When the offshore wind power single pile foundation only bears wave load or the combined action of wind wave gravity load after 1000 cycles, that is, when the horizontal displacement time history changes tend to be stable, compared with the 20th cycle, the changes of the extreme value of the positive horizontal displacement of the pile top are 6.18 mm and 6.48 mm respectively, and the changes of the extreme value of the negative horizontal displacement are - 6.68 mm and - 6.72 mm respectively; The change of the extreme value of the positive horizontal displacement at the mud level is 3.10 mm and 2.61 mm respectively, and the change of the extreme value of the negative horizontal displacement is - 2.84 mm and - 2.63 mm respectively. The difference in horizontal cumulative displacement change of a single pile foundation under two different loading conditions is within 5%, and the cumulative amount of horizontal displacement is the same.

Since the horizontal displacement accumulation of wind power single pile foundation in thousands of seas under two loading conditions is most significant when the horizontal wave load reaches the positive and negative peak values, the horizontal displacement of pile shaft at t=2.5s and t=7.5s during the 20th, 20th, 500th, and 100th cycles is taken for analysis, as shown in [Fig. 10-28] and [Fig. 10-29].

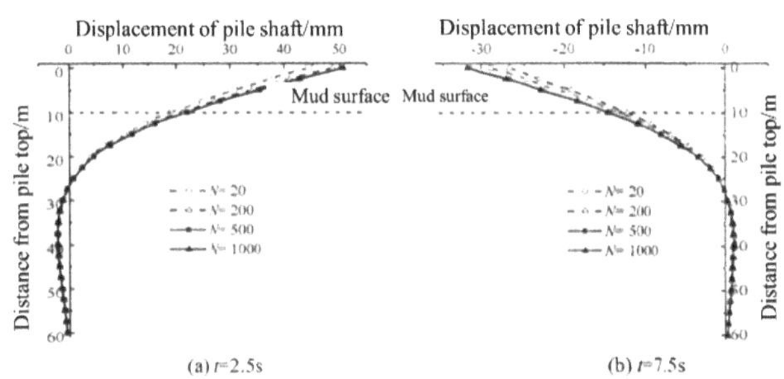

[Fig. 10-28] Horizontal displacement curve of pile body after Nth cycle under wave load, (LI Yazhou, 2020)

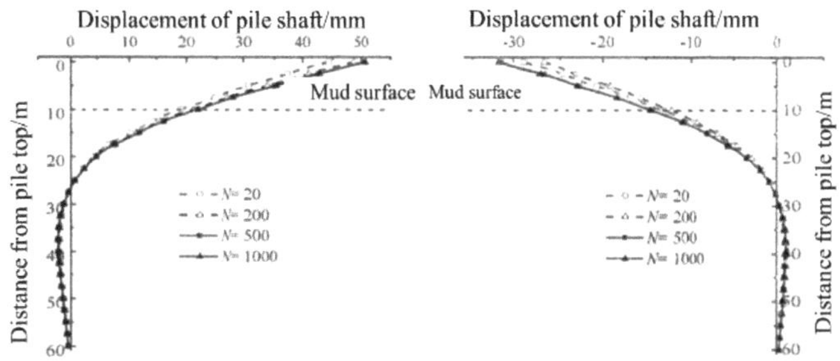

(a) t=2.5s (b) t=7.5s

[Fig. 10-29] Horizontal displacement curve of pile body after Nth cycle under the combined action of wind-wave-gravity load, (LI Yazhou, 2020)

It can be seen from [Fig. 10-28] and [Fig. 10-29] that the accumulation of horizontal displacement mainly occurs in the pile shaft section within 5 times the pile diameter from the pile top. The accumulation of horizontal displacement at the pile top is the largest. With the increasing distance from the pile top, the accumulation of horizontal displacement gradually decreases. The horizontal displacement of the pile body at the mud surface is significantly reduced compared with that at the top of the pile under two different loading conditions, showing an obvious foundation embedment effect. The ratio of the cumulative value of the horizontal displacement of a single pile foundation at the mud surface to the total water displacement is close to the ratio of the cumulative value of the horizontal displacement at the pile top under the same loading state to the total horizontal displacement, indicating that under the same loading state, the cumulative horizontal displacement caused by wave cyclic loading has roughly the same impact on the horizontal displacement of the pile shaft at different heights.

According to the analysis in [Fig. 10-26] to [Fig. 10-29], the horizontal cumulative displacement of a single pile foundation under two different loading conditions shows a numerical increase with the increase of the number of cycles, and finally tends to be stable between the 500th and 1000th cycles; The difference is that when a single pile foundation only bears wave load, the cumulative rule of horizontal displacement in positive and negative directions is the same. The cumulative displacement values after the 200th cycle and the 500th cycle respectively account for about 45% and 95% of the total cumulative displacement values, that is, the cumulative phenomenon of horizontal displacement mainly occurs in the first 500 cycles. However, when a single pile foundation is subjected to the combined action of wind, wave and gravity load, the cumulative rule of its positive and negative displacement is not consistent. After 200 cycles, the cumulative value of positive horizontal displacement accounts for more than 65% of the total cumulative value, while the cumulative value of negative horizontal displacement accounts for only about 30% of the total cumulative value; After 500 cycles, the cumulative amount of positive and negative horizontal displacement has reached more than 90% of the total cumulative amount of displacement, that is, the cumulative speed of positive horizontal displacement has been accelerated, while the cumulative amount of negative horizontal displacement mainly occurs between the 200th and 500th cycles. The reason is that the wind-induced overturning bending moment load leads to the positive horizontal displacement of the pile shaft, which makes the pile foundation and the soil around the pile subject to greater lateral pressure during the positive loading, accelerating the accumulation of positive horizontal displacement, while the accumulation of negative horizontal displacement is delayed, and the overall accumulation of positive and negative horizontal displacement does not change significantly.

From the above analysis, the change of the horizontal displacement of the pile top position after different cycles can most significantly reflect the cumulative rule of the horizontal displacement of a single pile foundation under the combined action of wind wave gravity loads. To analyze the specific effects of changes in wind load, wave load and vertical gravity load on the accumulation of horizontal displacement of a single pile foundation, the changes of the accumulated value of horizontal displacement of pile top of a single pile foundation with the number of cycles under the working condition of wind wave gravity load combined action and the reference working condition are analyzed. The relationship curve between the horizontal cumulative displacement of the pile top and the number of cycles of offshore wind power single pile foundation under working conditions 1.1~1.12 (LI Yazhou, 2020) is shown in [Fig. 10-30] ~ [Fig. 10-31].

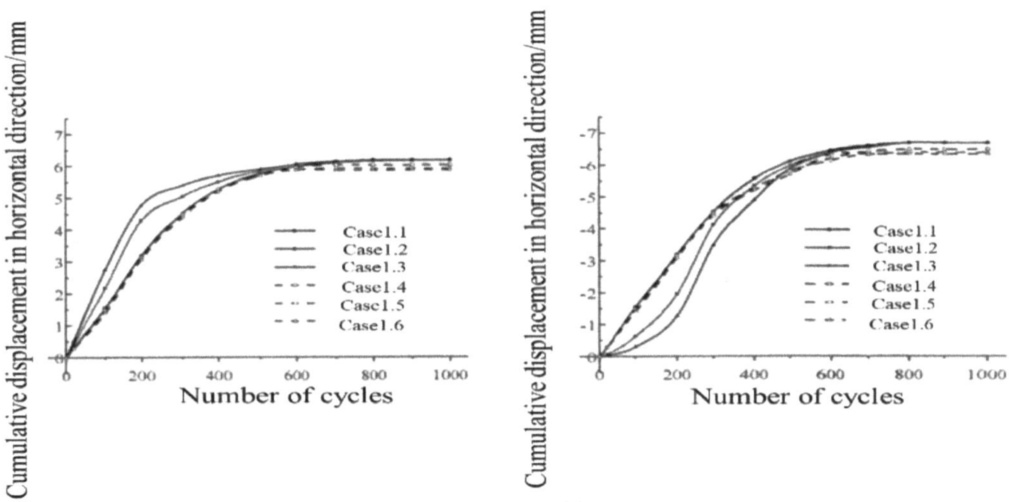

(a) Positive horizontal cumulative displacement (b) Negative horizontal cumulative displacement

[Fig. 10-30] Curve diagram of the relationship between horizontal cumulative is placemen1 of pile top and cycle times for single pile foundation under working conditions 1.1 ~ 1.6,(LI Yazhou, 2020)

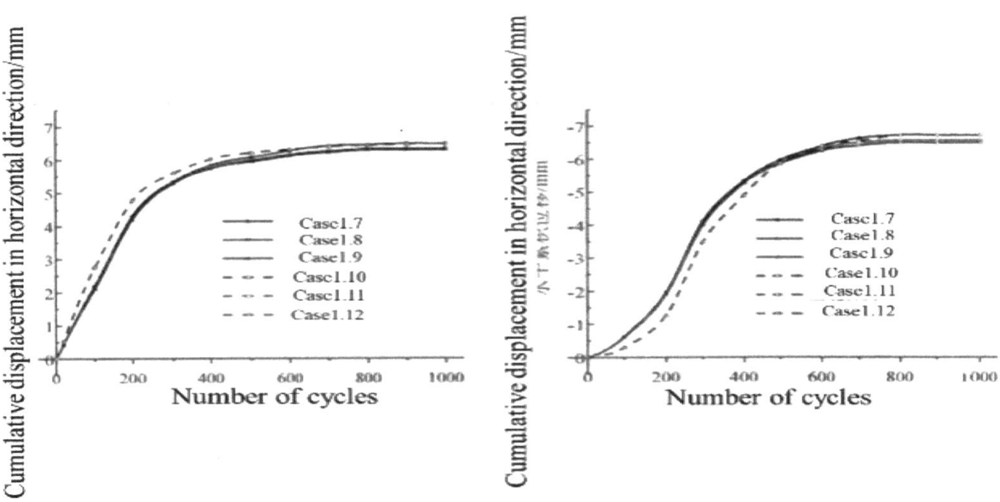

(a) Positive horizontal cumulative displacement (b) Negative horizontal cumulative displacement

[Fig. 10-31] Curve diagram of the relationship between the horizontal cumulative displacement of pile top and cycle times for single pile foundation under working conditions 1.7 ~ 1.12, (LI Yazhou, 2020)

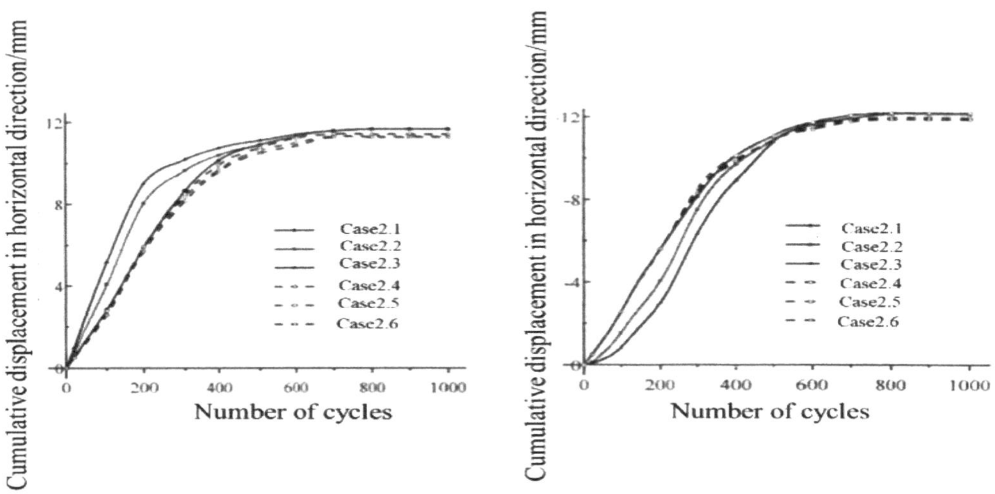

(a) Positive horizontal cumulative displacement (b) Negative horizontal cumulative displacement

[Fig. 10-32] Curve diagram of the relationship between the horizontal cumulative displacement of pile top and cycle times for single pile foundation under working conditions 2.1 ~2.6, (LI Yazhou, 2020)

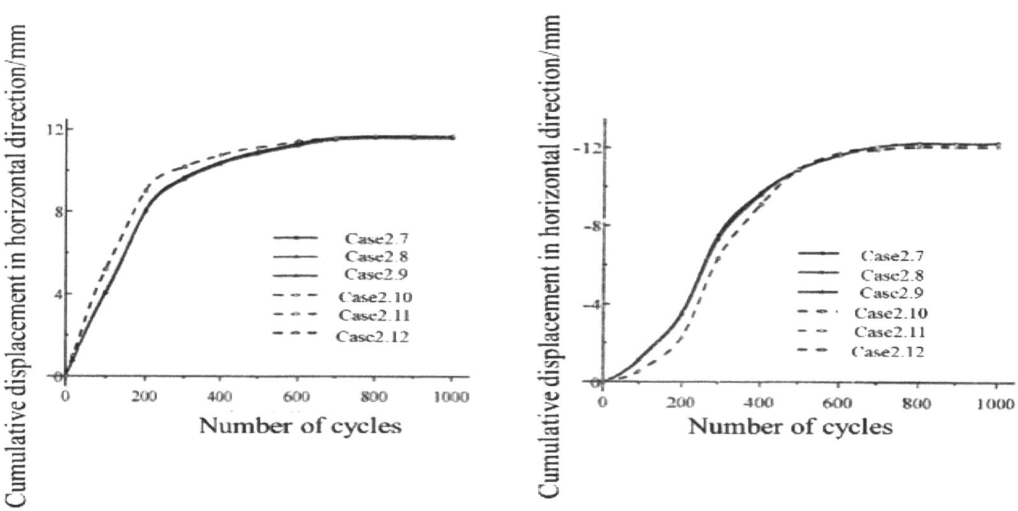

(a) Positive horizontal cumulative displacement (b) Negative horizontal cumulative displacement

[Fig. 10-33] Curve diagram of the relationship between horizontal I cumulative displacement of pile top and cycle times for sii1gle pile foundation under working conditions 2.7-2.12, (LI Yazhou, 2020)

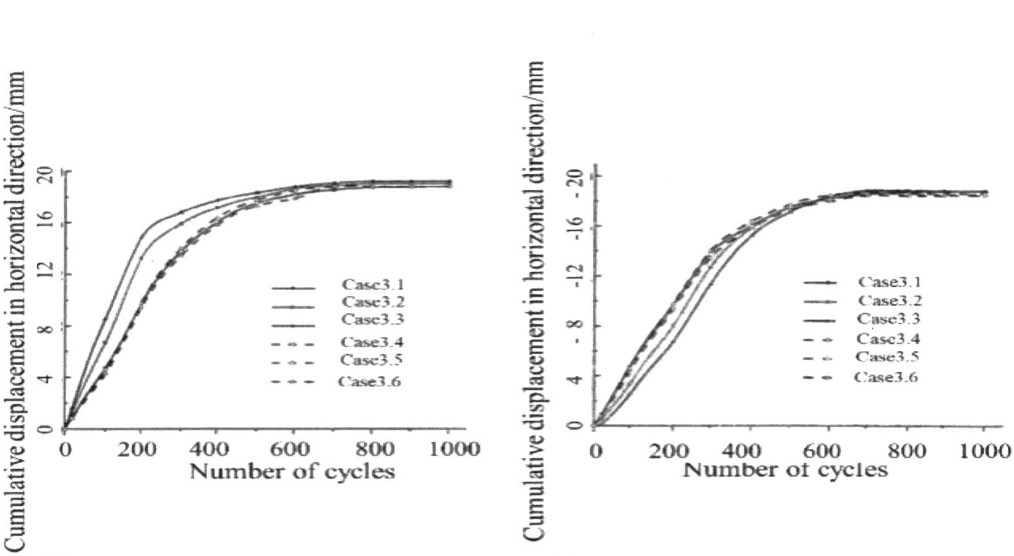

(a) Positive horizontal cumulative displacement (b) Negative horizontal cumulative displacement

[Fig. 10-34] Curve diagration of the relationship between the horizontal cumulative displacement of pile top and cycle times for si.lngle pile foundation under working conditions 3. l- 3.6,(LI Yazhou, 2020)

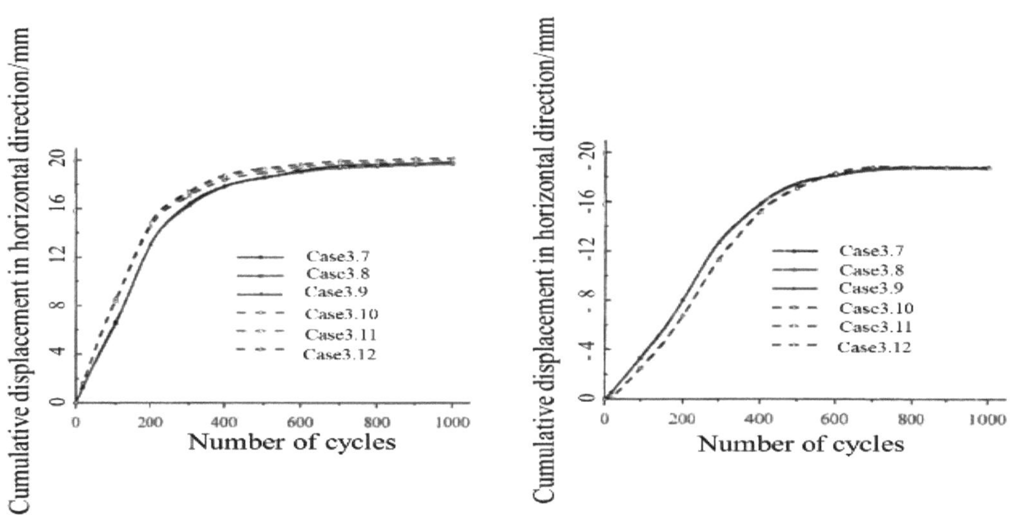

(a) Positive horizontal cumulative displacement (b) Negative horizontal cumulative displacement

[Fig. 10-35] Curve diagram of the relationship between the horizontal cumulative displacement of pile top and cycle times for single pile foundation under working conditions 3.7- 3.12, (LI Yazhou, 2020)

It can be seen from [Fig. 10-30] and [Fig. 10-31] that when the wave load wave height is 2 m, the vertical gravity load has a significant inhibition effect on the horizontal displacement accumulation. With the increase of the vertical gravity load, the final cumulative displacement in the horizontal direction gradually decreases, but the decreasing amplitude is smaller and smaller. The reason is that the vertical gravity load can improve the exertion of the side friction of the sand foundation around the pile, and have an impact on the accumulation of horizontal displacement. With the full exertion of the lateral component of the side friction, the cumulative displacement decreases. With the increase of the wind-induced overturning moment load, the cumulative value of the final horizontal displacement of the pile top of the single pile foundation has almost no change, but the development speed of the positive cumulative displacement has changed greatly: the development speed of the positive horizontal displacement of the pile top has accelerated, while the development speed of the negative horizontal displacement has decreased. However, with the further increase of the wind-induced bending moment load, the change range of the cumulative development speed of displacement decreases, indicating that the influence of overturning bending moment load on the cumulative speed of horizontal displacement has a limit. This may be related to the squeezing effect of bending moment load on the lateral soil mass of a single pile foundation. The pile side soil subjected to squeezing will enter the plastic stage ahead of time, and the cumulative speed of the horizontal displacement on the other side will be reduced accordingly.

Working condition 2.1~2.12, see conference (LI Yazhou, 2020). The relationship curve between the horizontal cumulative displacement of the pile top and the number of cycles of offshore wind power single pile foundation is shown in [Fig. 10-32] and [Fig. 10-33]. The curve of the relationship between the horizontal cumulative displacement of the pile top and the number of cycles of offshore wind power single pile foundation under working conditions 3.1~3.12 is shown in [Fig. 10-34] and [Fig. 10-35].

According to the overall analysis of [Fig. 10-30] to [Fig. 10-35] under all working conditions, with the increase of the number of cycles, the positive horizontal cumulative displacement of the pile top first shows an approximate linear growth trend, then the growth rate gradually slows down, and tends to be stable after 500 cycles; while the negative horizontal displacement accumulation of the pile top under most working conditions has the same development law as the positive horizontal displacement accumulation law, However, under some working conditions, due to the influence of changes in wind-induced bending moment load, it shows a trend of gradual growth first, then linear increase, and finally tends to be stable.

It can be seen from [Fig. 10-32] to [Fig. 10-35] that when the wave height of wave load increases, the cumulative displacement of pile top water of a single pile foundation is positively correlated with the amplitude of horizontal wave load. The larger the amplitude of the wave load is, the greater the cumulative value of horizontal displacement is. When the amplitude of wave load increases from 1.8422 MN to 3.4537 MN and 5.6794MN, the horizontal cumulative displacement also increases by about 1.8 times and 3.2 times; The cumulative velocity of displacement has also been improved slightly, but not obviously. As the wave height of wave load increases, the influence of vertical gravity load on the accumulation of horizontal displacement gradually weakens, and the ratio of the reduction of cumulative displacement caused by vertical gravity load to the total displacement decreases. With the increase of vertical gravity load, the amplitude of the reduction of cumulative displacement becomes smaller and smaller. When Hs = 8 m, the vertical gravity load has little influence on the change of the cumulative value of horizontal displacement, This shows that the friction of the soil around the pile can fully restrain the accumulation of displacement. As the wave height of wave load increases, the influence of wind-induced overturning moment load on the growth rate of cumulative displacement decreases gradually.

By comparing the positive displacement accumulation and negative displacement accumulation laws of the horizontal displacement of the pile top of the single pile foundation in [Fig. 10-31] to [Fig. 10-35], it can be seen that when the single pile foundation is subjected to the combined action of wind, wave and gravity loads, with the increase of the wind-induced bending moment load, the cumulative reduction of the horizontal displacement caused by the increase of the vertical gravity load decreases, that is, the existence of the wind-induced bending moment load will reduce the impact of the gravity load on the horizontal displacement accumulation.

To sum up, under the combined action of wave and gravity loads, the wave load and vertical gravity load will determine the final horizontal displacement accumulation of offshore wind power single pile foundation. The greater the wave load amplitude, the greater the horizontal displacement accumulation; The vertical gravity load can significantly reduce the horizontal displacement accumulation when the amplitude of the wave load is small and has little effect on the horizontal displacement accumulation when the amplitude of the wave load is large. The magnitude of wave load and wind-induced bending moment load has an impact on the cumulative speed of displacement. The increase of wave load amplitude will accelerate the cumulative speed of displacement, but it is not obvious; The wind-induced bending moment load will accelerate the cumulative speed of induced horizontal displacement, but with the increase of wave load amplitude and overturning bending moment, the increase of speed will decrease. The increase of wind-induced bending moment load will also reduce the restraining ability of gravity load on cumulative displacement, which is equal to the increase of cumulative horizontal displacement in a certain sense.

(2) Analysis of horizontal displacement of pile shaft

In the marine environment, wind power single pile foundation is subjected to thousands of wave cycles, and its pile horizontal displacement accumulation tends to be stable. Currently, the positive and negative horizontal displacement of the pile body directly reflects the dynamic bearing characteristics of the pile foundation. To study the pile horizontal displacement characteristics of offshore wind power single pile foundation under the combined action of wind wave gravity load, the pile horizontal displacement is calculated after 1000 cycles of offshore wind power single pile foundation under various working conditions, taking into account the pile horizontal cumulative displacement, Take the horizontal displacement of pile shaft when t=2.5s (wave load reaches positive amplitude and pile displacement reaches positive maximum) and t=7.5s (wave load reaches negative amplitude and pile displacement reaches negative maximum) in the 1000th cycle for analysis.

To facilitate the analysis of the pile displacement change rule of a single pile foundation under various load conditions, the working conditions are further divided in detail based on the wave height and wind speed. The horizontal displacement curve of the pile shaft of offshore wind power single pile foundation under Hs = 2 m working condition is shown in [Fig. 10-36] ~ [Fig. 10-38].

It can be seen from [Fig. 10-36] to [Fig. 10-38] that the horizontal deformation of the pile shaft of offshore wind power single pile foundation has significant nonlinear characteristics when the wave load is small. The pile shaft within 7D from the pile top has a horizontal displacement consistent with the direction of load action, while the pile shaft from the

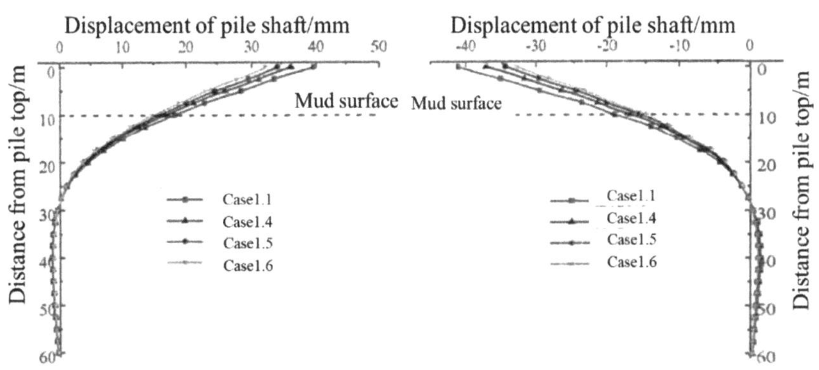

(a) Positive horizontal displacement of pile shaft (b) Negative horizontal displacement of pile shaft

[Fig. 10-36] Horizontal displacement curve of pile body of offshore wind power single pile foundation with H_s=2m, v_0=0, (LI Yazhou, 2020).000

pile top more than 7D to the bottom has a horizontal displacement opposite to the direction of load action, which is generally shown as the maximum horizontal displacement at the pile top.

The nonlinearity along the pile body decreases to zero, then the reverse displacement occurs and gradually increases, and the reverse displacement reaches the extreme value at the pile IOD from the pile top, and finally decreases to zero along the pile body. The pile shaft characteristics of a single

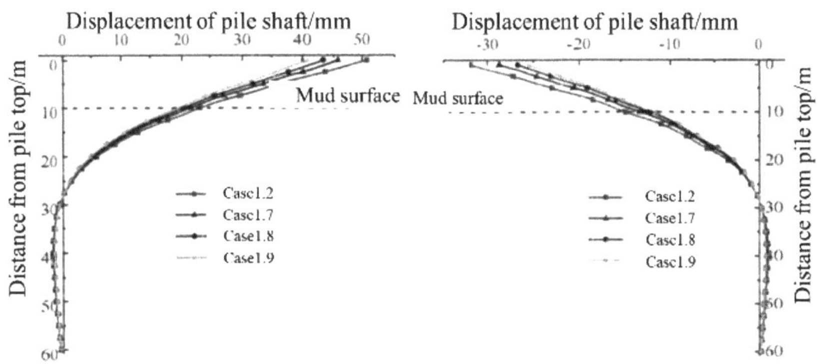

(a) Positive horizontal displacement of pile shaft (b) Negative horizontal displacement of pile shaft

[Fig. 10-37] Horizontal displacement curve of pile body of offshore wind power single pile foundation with H_s=2m, v_0=5m/s, (LI Yazhou, 2020)

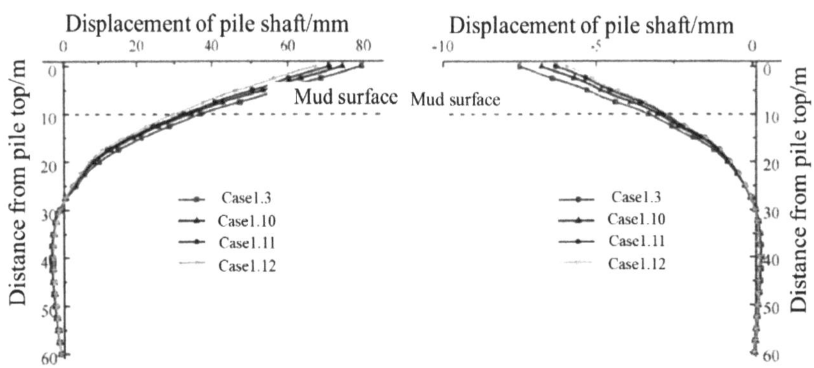

(a) Positive horizontal displacement of pile shaft (b) Negative horizontal displacement of pile shaft

[Fig. 10-38] Horizontal displacement curve of pile body of offshore wind power single pile foundation with H_s=2m, v_0=12m/s, (LI Yazhou, 2020)

pile foundation are like those of a flexible pile, but there is also horizontal displacement caused by the rotation of a point on the pile shaft axis and some characteristics of a rigid pile. When the wind speed increases, that is, the wind-induced bending moment load increases, the positive and negative horizontal displacements of the pile body will increase at the same time. The negative reduction of the horizontal displacement of the pile body is less than the positive increase. The bending moment load has a greater impact on the displacement in the direction of action. Especially when the wind speed is 12m/s, the positive horizontal displacement of the pile body increases nearly twice as compared with that under no wind load. When the wind-induced bending moment load increases, the horizontal displacement of the upper part of the pile shaft at the reverse bending point is greatly affected, while the pile shaft deflection deformation of the lower part of the reverse bending point is almost unchanged, indicating that the overturning bending moment load mainly affects the free section of the pile shaft and the location with shallow burial depth. The vertical gravity load can effectively reduce the horizontal displacement of the pile shaft. When the wind speed is 0, the vertical gravity load of 0.3 Vu, 0.6 Vu, and 0.9 Van can reduce the horizontal displacement of the positive pile shaft by 9.7%, 16.7%, and 19.7%, and reduce the horizontal displacement of the negative pile shaft by 10%, 17.2%, and 19.2%; When the wind speed is 5m/s, the vertical gravity load of 0.3 Vu, 0.6 Vu, and 0.9 Vu can reduce the horizontal displacement of positive and negative piles by 8.9%, 15.9% and 19.4% on average, and reduce the horizontal displacement of negative piles by 10.8%, 14.6%, and 20.8%; When the wind speed is 12 m/s, the vertical gravity load of 0.3 fans, 0.6 fans, and 0.9 fans can reduce the horizontal displacement of positive pile shaft by 7.9%, 12.5%, and 16.3%, and reduce the horizontal displacement of negative pile shaft by 8.3%, 13.2%, and 17.7%.

The inhibition effect of vertical gravity load on the positive and negative horizontal displacement of the pile shaft under the same working condition is the same, and the increase of gravity load will increase the inhibition

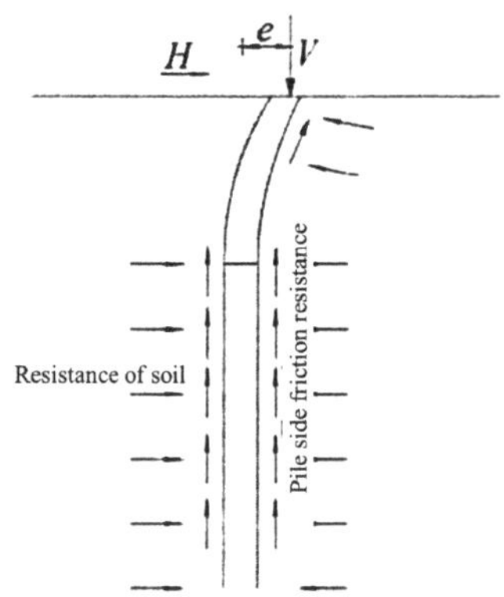

[Fig. 10-39] Schematic Diagram of Lateral Friction and Soil Resistance of Single Pile Foundation, (LI Yazhou, 2020)

effect on the horizontal displacement of the pile shaft, but the increased amplitude will continue to decrease. The reason is that the application of gravity load increases the horizontal stress in the pile foundation soil mass on the side of the load, increasing the pile side friction, which makes the pile subject to the resistance caused by the pile side friction and soil resistance, as shown in [Fig. 10-39].

However, when the vertical gravity load continues to increase and the side friction is fully exerted, the restraining effect on the horizontal displacement is no longer enhanced. With the increase of the overturning moment load, the gravity load increases the reduction of the positive horizontal displacement of the pile shaft and the reduction of the negative horizontal displacement. However, due to the corresponding change in the total displacement, the inhibition of the vertical gravity load on the positive and negative horizontal displacement of the pile shaft decreases with the increase of the wind speed.

The horizontal displacement curve of the pile shaft of offshore wind power single pile foundation under is shown in [Fig. 10-40 ~ [Fig. 10-42]. The horizontal displacement curve of the pile shaft of offshore wind power single pile foundation under Hs=8m is shown in [Fig. 10-43] ~ [Fig. 10-45].

From [Fig. 10-40] to [Fig. 10-45], under the working condition of large wave load wave height, the horizontal deformation characteristics of the pile shaft of offshore wind power single pile foundation are like those of the pile shaft when the wave height is 2 m. The pile shaft is nonlinearly deformed and rotates around a point on the pile shaft axis, reflecting the characteristics of rigid piles and flexible piles. With the increase of wave height of wave load, the amplitude of horizontal cyclic load increases correspondingly, and the deflection deformation amplitude of the pile shaft in positive and negative directions increases.

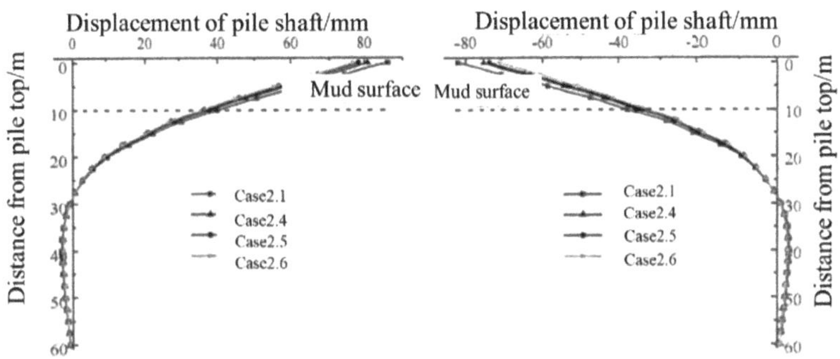

(a) Positive horizontal displacement of pile shaft (b) Negative horizontal displacement of pile body

[Fig. 10-40] Horizontal displacement curves of pile body of offshore wind power single pile foundation with $H_s = 4m, v_0 = 0$

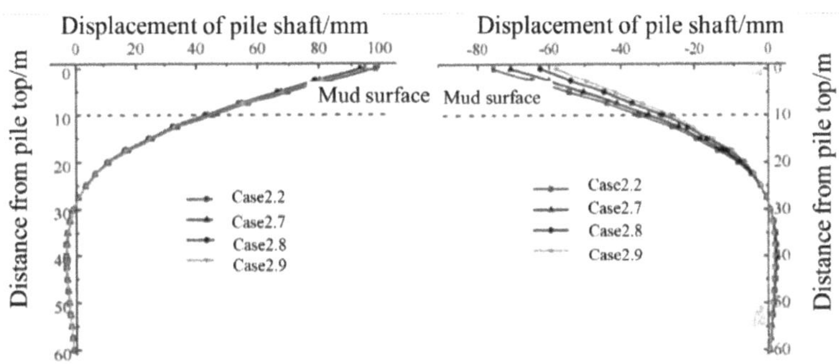

(a) Positive horizontal displacement of pile shaft (b) Negative horizontal displacement of pile body

[Fig. 10-41] Horizontal displacement curves of pile body of offshore wind power single pile foundation with $H_s = 4m, v_0 = 5m/s$

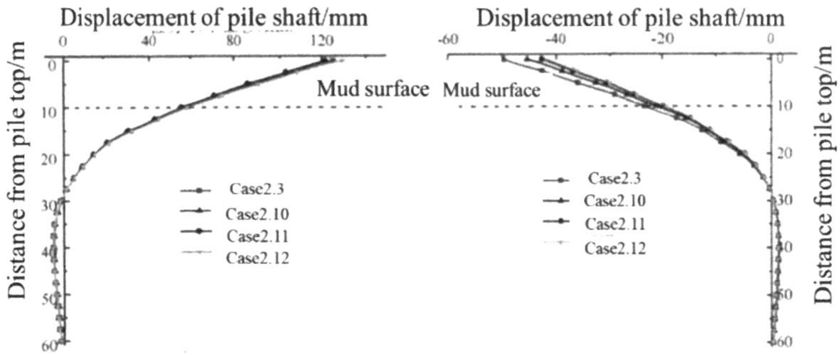

(a) Positive horizontal displacement of pile shaft (b) Negative horizontal displacement of pile body

[Fig. 10-42] Horizontal displacement curves of pile body of offshore wind power single pile foundation with $H_s = 4m, v_0 = 12m/s$

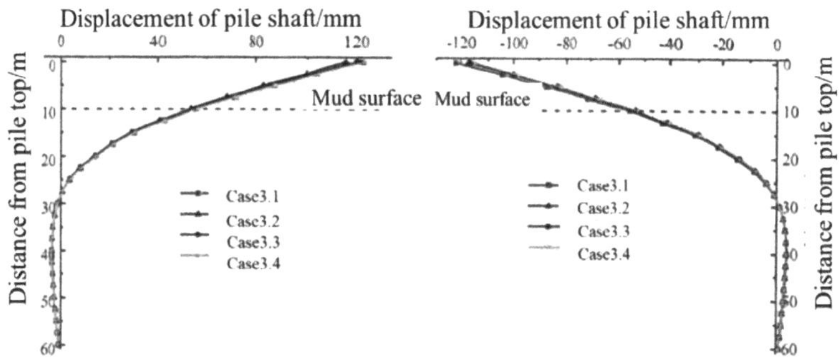

(a) Positive horizontal displacement of pile shaft (b) Negative horizontal displacement of pile body

[Fig. 10-43] Horizontal displacement curves of pile body of offshore wind power single pile foundation with $H_s = 8m, v_0 = 0$

The pile shaft presents a downward bending trend, and the absolute value of up and down displacement at the reverse bending point increases, but the overall impact on the horizontal displacement of the pile shaft at the shallow buried position above the reverse bending point is more significant. When the wind speed increases, the positive horizontal displacement of the pile body increases and the negative horizontal displacement decreases, while the positive horizontal displacement increases more than the negative horizontal displacement decreases, and the higher the wave height and wind speed, the more obvious this

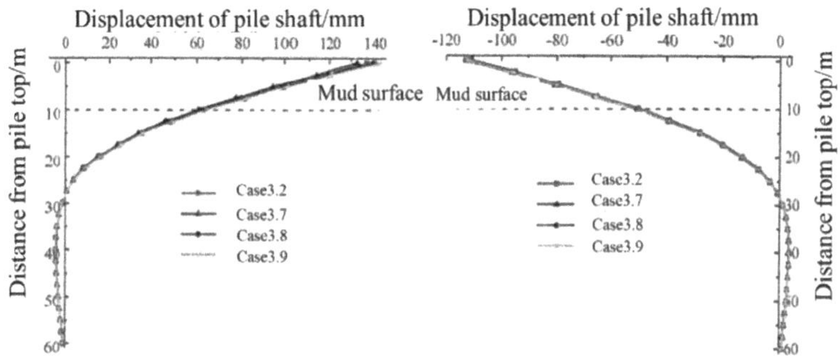

(a) Positive horizontal displacement of pile shaft (b) Negative horizontal displacement of pile body

[Fig. 10-44] Horizontal displacement curves of pile body of offshore wind power single pile foundation with $H_s = 8m, v_0 = 5m/s$

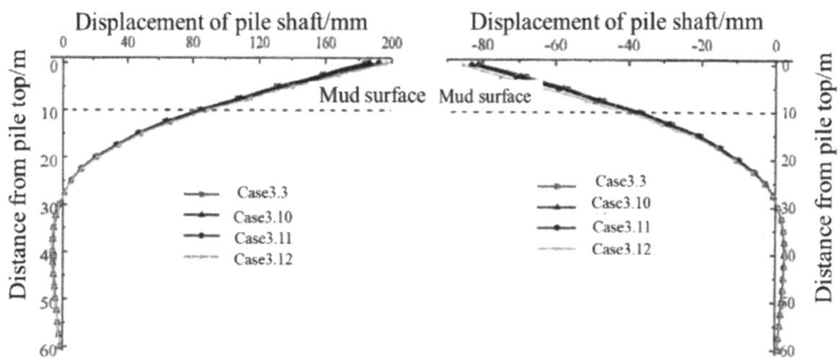

(a) Positive horizontal displacement of pile shaft (b) Negative horizontal displacement of pile body

[Fig. 10-45] Horizontal displacement curves of pile body of offshore wind power single pile foundation with $H_s = 8m, v_0 = 12m/s$

phenomenon is. The reason is that the soil around the pile on both sides begins to enter a plastic state and the bearing capacity decreases with the increase of wave height, At this time, the pile side soil extrusion caused by the wind-induced overturning bending moment causes greater deformation. The overturning bending moment load has a more obvious effect on the free section of the pile shaft and the position with shallow burial depth because the lower foundation soil is still in an elastic state and is less affected by external loads, which can effectively provide the role of embedded pile foundation and restrain the pile deformation. When the vertical gravity load changes, the change in the horizontal displacement of the pile shaft is more complex. When the wind and wave load values are small, the vertical gravity load can restrain the development of the horizontal displacement of the pile shaft and improve the bearing capacity of the pile foundation in the horizontal direction. The inhibition effect is significantly reduced compared with the working condition 1.1~1.12, as shown in the literature (LI Yazhou, 2020), and when the vertical gravity load exceeds 0.60.6 V_u, the inhibition effect no longer changed; In the case of large wind load or wave load, the application of gravity load has lost its restraining effect on the horizontal displacement of the pile shaft. With the increase of gravity load, the horizontal displacement of the pile shaft increases, and additional displacement occurs. The reason is that with the increase of the amplitude of the horizontal cyclic load or the overturning bending moment load, the pile deflection develops. Since the gravity load and the deflection deformation will form additional bending moments on the pile, which further increases the pile deformation, this phenomenon is called $p-\Delta$ effect, as shown in [Fig. 10-46]. At this time, the friction of the soil around the pile can not play a sufficient role in restraining the deformation of the pile body. In this state, the greater the gravity load, the greater the additional displacement, and the greater the damage to the bearing capacity of the pile foundation.

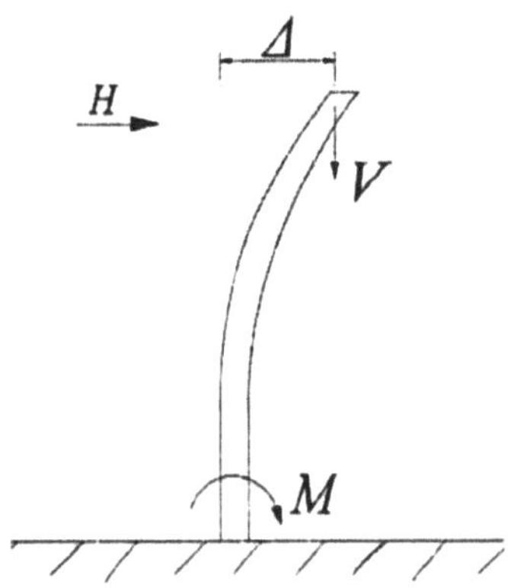

[Fig. 10-46] Diagram of "$p-\Delta$" effect of single pile, (LI Yazhou, 2020)

10.3.2.2 Analysis of bending moment along the pile shaft

The time-varying curve of the pile bending moment of offshore wind power single pile foundation under conditions 1.1 and 1.7 is shown in [Fig. 10-47] and [Fig. 10-48]. It can be seen from [Fig. 10-47] to [Fig. 10-48] that under the action of wave load, the pile bending moment of offshore wind power single pile foundation varies greatly at different times in a cycle due to the change of the horizontal load size, but the overall deformation law is similar. The bending moment of the pile shaft of a single pile foundation under positive and negative loading shows significant nonlinear characteristics. The absolute value of the bending moment of the pile shaft from the mud surface to the pile end first increases to the extreme value, and then gradually decreases to close to zero.

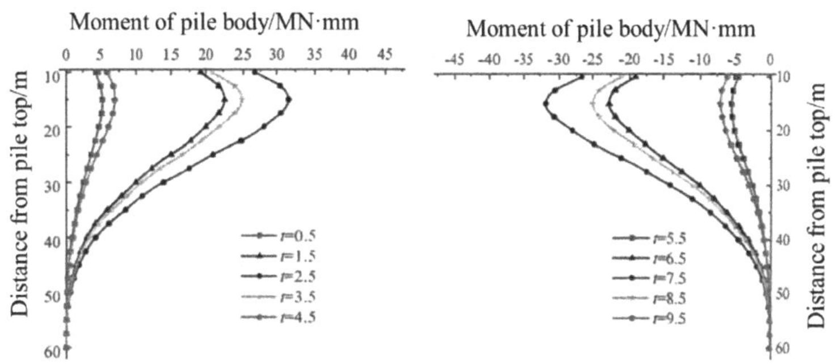

(a) Positivc load (b) Negative load

[Fig. 10-47] Bending moment curve of 1000th cycle pile und e r wave load, (LI Yazhou, 2020)

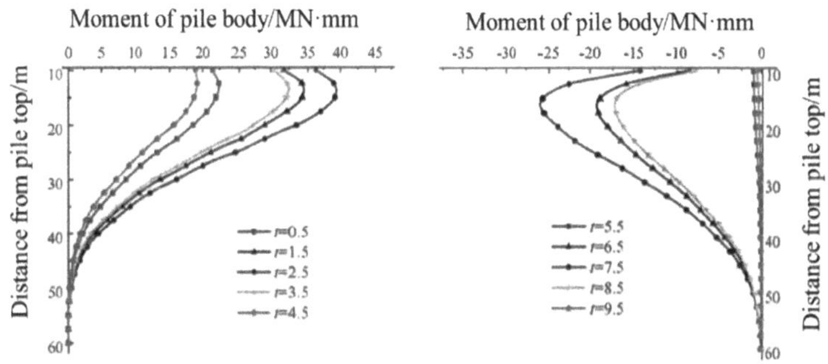

(a) Positive load (b) Negative load

[Fig. 10-48] Bending moment curve of 1000th cycle pile under wind-wave-gravity load, (LI Yazhou, 2020)

The closer to the pile end, the smaller the change amplitude. The extreme bending moment of the pile shaft under positive and negative loading occurs at. A quarter of the pile is about 4D away from the pile top. From the relationship between the bending moment and shear force, the shear force of the pile is zero. In the process of the 1000th cycle wave load from zero to extreme value in positive and negative directions, the bending moment of the pile increases with the increase of time, and the amplitude of the bending moment increase decreases in the same time interval; however, when the wave load of the 1000th cycle decreases from the extreme value to zero in the positive and negative directions, the bending moment of the pile shaft decreases with the increase of the time, and the reduction range of the bending moment in the same time interval increases. This is because the soil around the pile inhibits the development of the turning bending moment of the 1000 single pile foundation during the loading process. With the increase of the distance along the pile shaft, the change range of the bending moment of the pile shaft with time is smaller, indicating that with the increase of the burial depth of the single pile, the inhibition of the pile foundation soil on the development of the bending moment of the pile shaft is enhanced.

By comparing [Fig. 10-47] to [Fig. 10-48], it can be seen that when a single pile foundation only bears wave load, the positive and negative extreme values of the pile bending moment are close in value, and the pile bending moment at the corresponding time is roughly symmetrical. When a single pile foundation is subjected to the combined action of wind wave gravity load, the absolute value of the pile shaft bending moment increases in the first 5s of the cycle and decreases in the next 5s. The changes in the positive and negative extreme values of the bending moment of the pile shaft are approximately equal, and the magnitude of the wind-induced overturning bending moment load imposed on the value is large. The pile bending moment of a single pile foundation due to multiple loads changes significantly within 30 m of the buried depth. The closer the pile is to the pile end, the less obvious the change of the pile bending moment.

To further study the pile bending moment characteristics of offshore wind power single pile foundations under the combined action of wind wave gravity load, the pile bending moment distribution of offshore wind power single pile foundation after 1000 cycles under various working conditions is calculated, Take the pile bending moment at the 1000th cycle r-2.5s (wave load reaches the positive amplitude, and the pile bending moment reaches the maximum positive value) and t=7.5s (wave load reaches the negative amplitude, and the pile bending moment reaches the maximum negative value) for analysis. To facilitate the analysis of the pile bending moment change law of a single pile foundations under various load conditions, adopt the same mode of working condition division as that in the previous section. The horizontal displacement curve of pile shaft of offshore wind power single pile foundation under Hs=2m is shown in [Fig. 10-49] ~ [Fig. 10-51].

It can be seen from [Fig. 10-49] to [Fig. 10-51] that the horizontal bending moment of the pile shaft of offshore wind power single pile foundation under working conditions 1.1 to 1.12 has similar morphological change characteristics. The absolute value of the bending moment of the pile shaft first develops to the maximum along the buried depth of the pile, and then gradually decreases. The overall change has obvious nonlinear characteristics. The bending moment of the pile shaft of a single pile foundation under different working conditions reaches the positive and negative extreme values at the buried depth of 15 m and is close

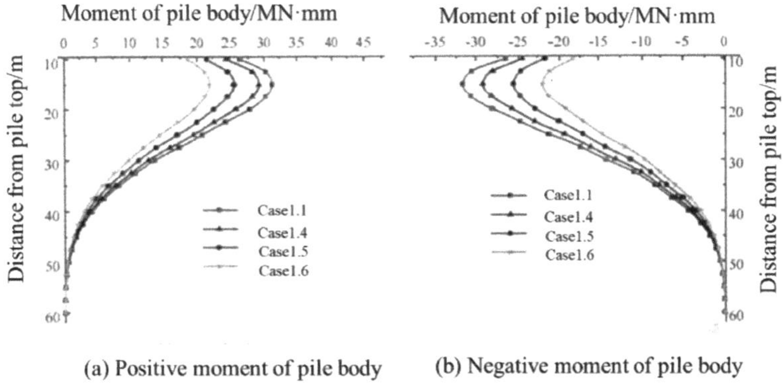

(a) Positive moment of pile body (b) Negative moment of pile body

[Fig. 10-49] Bending moment curve of pile body of offshore wind power single pile foundation with $H_s = 2m, v_0 = 0.$, (LI Yazhou, 2020)

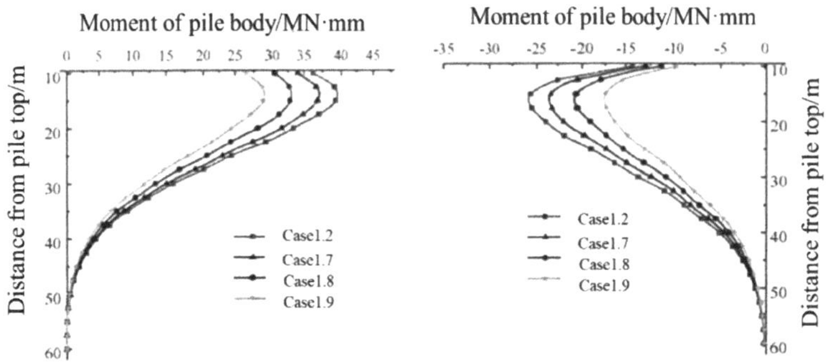

(a) Positive moment of pile body (b) Negative moment of pile body

[Fig. 10-50] Bending moment curve of pile body of offshore wind power single pile foundation with $H_s = 2m, v_0 = 5m/s.$, (LI Yazhou, 2020)

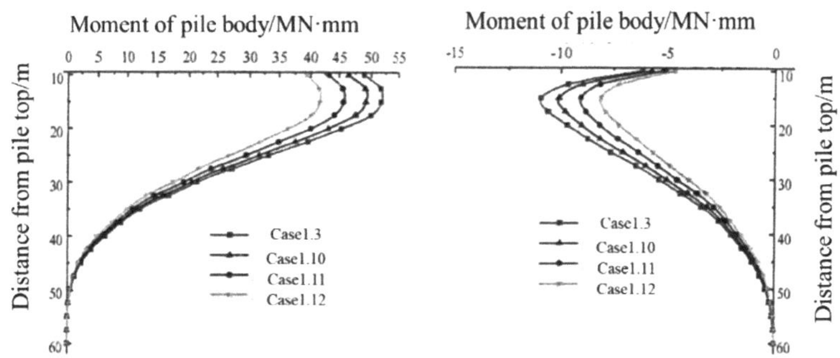

(a) Positive moment of pile body (b) Negative moment of pile body

[Fig. 10-51] Bending moment curve of pile body of offshore wind power single pile foundation with $H_s = 2m, v_0 = 12m/s.$,source: (LI Yazhou, 2020)

to zero at the pile end. With the increase of the wind-induced overturning moment load, the changes in the pile shaft bending moment values of a single pile foundation under the same wave load and vertical gravity load conditions show a positive increase and a negative decrease. The maximum change is at the extreme value of the pile shaft bending moment, where the change in the positive bending moment is slightly greater than the change in the negative bending moment, indicating that the wind-induced overturning moment has a greater impact on the pile shaft bending moment in the action direction, This is the same as the change rule of horizontal displacement of pile shaft. The change of positive and negative bending moment of the pile shaft of a single pile foundation due to bearing multiple loads is the largest at the location with a buried depth of 15 m, and the value is greater than the applied wind-induced overturning bending moment load, indicating that the influence of wind-induced overturning bending moment load on the extreme value of pile shaft bending moment of single pile foundation is not only a simple numerical addition.

Under the combined action of wind wave gravity load, the pile bending moment of offshore wind power single pile foundation decreases with the burial depth, showing a similar change law to the pile horizontal displacement. With the increase of vertical gravity load, the pile bending moment of a single pile foundation under the same wave load and wind load conditions shows a numerical reduction. At this time, the gravity load increases the limiting effect of soil on the development of pile bending moment, which gradually increases with the increase of gravity load, but the extent of the increase decreases, The vertical gravity load has a greater limit on the bending moment of the pile shaft in the process of negative loading than in the process of positive loading. The reason is that the longitudinal force effectively promotes the exertion of the lateral friction of the foundation soil and improves the horizontal bearing capacity of the soil around the pile. When the gravity load of the same size acts at the same time with the wind-induced overturning bending moment load of different sizes, with the increase of the wind-induced overturning bending moment load, the numerical reduction of the vertical gravity load on the pile shaft bending moment of offshore wind power single pile foundation continues to decrease, that is, the inhibition effect on the pile shaft bending moment development continues to weaken, which is more significant in the case of positive loading, At this time, the wind-induced overturning bending moment makes the plastic deformation of the soil around the pile on the side of the load develop, and the lateral friction has been fully exerted at this time, and the bearing capacity of the soil around the pile is affected, which cannot provide sufficient restraint.

The horizontal displacement curve of the pile shaft of offshore wind power single pile foundation under Hs = 4 m is shown in [Fig. 10-49] ~ [Fig. 10-51]. The horizontal displacement curve of the pile shaft of offshore wind power single pile foundation under Hs = 8 m is shown in [Fig. 10-52] ~ [Fig. 10-53].

By [Fig. 10-49] ~ [Fig. 10-54], It can be seen that under the working condition of large wave load wave height, the overall development characteristics of the pile bending moment of offshore wind power single pile foundation are similar to those when the wave height is 2 m. The absolute value of the bending

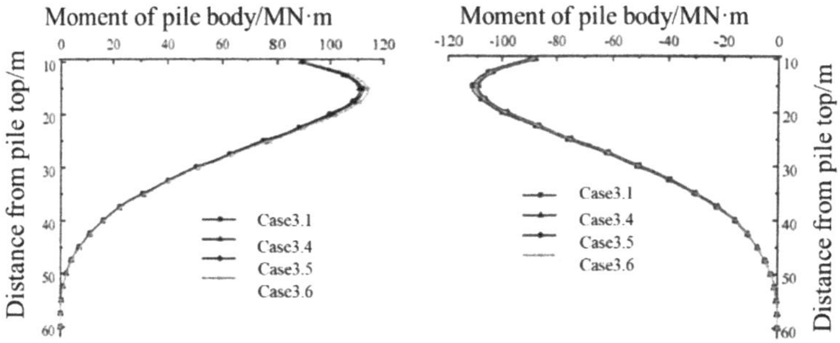

(a) Positive bending moment of pile body (b) Negative bending moment of pile body

[Fig. 10-52] Bending moment curves of pile body of offshore wind power single pile foundation with $H_s = 8m, v_0 = 0$.

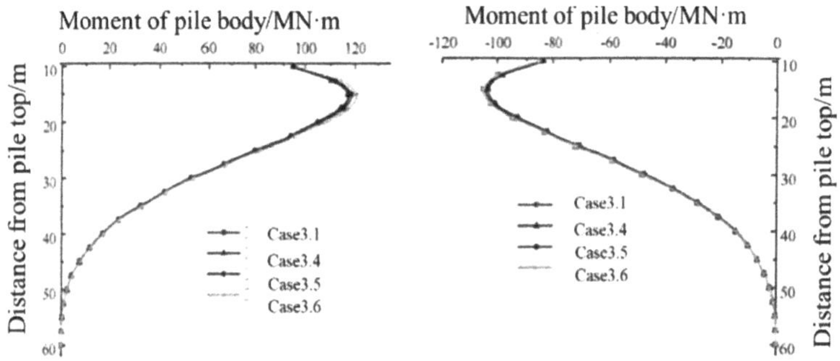

(a) Positive bending moment of pile body (b) Negative bending moment of pile body

[Fig. 10-53] Bending moment curves of pile body of offshore wind power single pile foundation with $H_s = 8m, v_0 = 5m/s$.

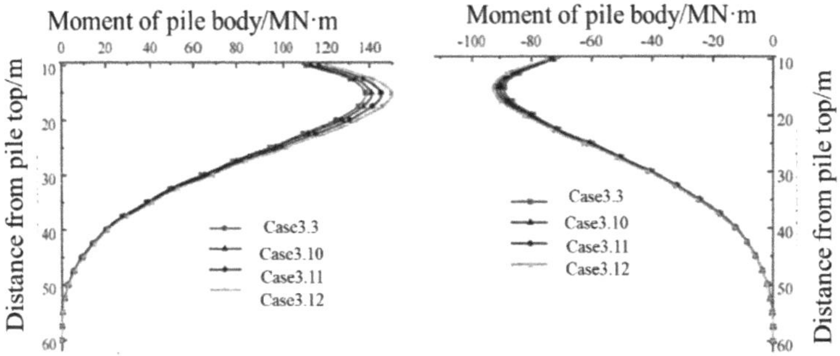

(a) Positive bending moment of pile body (b) Negative bending moment of pile body

[Fig. 10-54] Bending moment curves of pile body of offshore wind power single pile foundation with $H_s = 8\ m,\ v_0 = 12\ m/s$.

moment first increases to the extreme value along the burial depth, and then gradually decreases. The absolute value of the bending moment at the pile end is close to zero. The pile bending moment under different working conditions develops nonlinearly along the burial depth and reaches the extreme value at the burial depth of 15 m. With the increase of wave height of wave load, the amplitude of horizontal cyclic load increases correspondingly, and the positive bending moment of the pile increases significantly, while the negative bending moment decreases significantly. The amplitude of increase and decrease is roughly proportional to the multiple of load amplitude increase. The change of the bending moment of the pile at the location with shallow burial depth is more significant, indicating that the soil around the pile at a deeper depth can provide a stronger restraining effect on the development of the bending moment of the pile. When the wind speed increases, the positive bending moment of the pile body increases while the negative bending moment decreases. The development law is the same as that when the horizontal load changes, but the impact is weak. This shows that the bending moment of a single pile foundation is more sensitive to the change of wave height of wave load than the wind speed change of wind load. The increase of the positive bending moment of the pile shaft caused by the influence of the wind-induced overturning bending moment is greater than the decrease in the negative bending moment, and the greater the wave height and wind speed, the more obvious this phenomenon is.

The reason is that with the gradual increase of the transverse stress, the plastic zones of the soil on both sides of the pile foundation continue to develop, and the bearing capacity continues to weaken. The pile shaft bending moment is larger under the positive compression of the pile side caused by the wind-induced overturning bending moment load. The changing amplitude of the overturning bending moment load on the bending moment of the pile shaft at the deeper embedded position is less than, which indicates that the bearing and embedding action of the lower foundation soil is disturbed by the overturning bending moment load, and is less affected by external loads, which can effectively embed the pile foundation and restrain the development of the bending moment of the pile.

When the vertical gravity load changes, the development law of the pile bending moment is similar to the change law of the pile horizontal displacement. The influence of gravity load under different working conditions shows two sides. Under working conditions where the wind wave load values are small, the vertical gravity load can inhibit the development of the pile bending moment, reducing the absolute values of the positive and negative bending moments of the pile. With the increase of the gravity load, the inhibition effect gradually increases, However, the amplitude of enhancement is getting smaller and smaller, and when the gravity load changes from $0.6V_u$ increased to $0.3V_u$ When u, the increase of restraining effect is less than that of vertical load by $0.3V_u$ increased to $0.6V_u$; When the wind load or wave load value is large, the gravity load has no obvious inhibition effect on the horizontal displacement of the pile shaft. When the gravity load increases, the bending moment of the pile shaft decreases first, then increases, or even directly increases. Especially when the wind and waves are strong, the additional bending moment caused by the increase of the vertical gravity load is more serious, which is similar to the change law of the horizontal displacement of the pile shaft, The reason is also due to the $p-\Delta$ an effect caused by gravity load and deflection deformation, which forms a large additional bending moment. It can be seen from the generation principle of $p-\Delta$ effect that when the horizontal deflection and vertical gravity load exist in the pile body at the same time, $p-\Delta$ the effect will be generated. When the lateral load is small, the additional bending moment and additional displacement are small, and the role of soil friction on the pile side gradually plays. In the macro view, the inhibition of pile-side friction on pile displacement and bending moment is the main performance; However, when the lateral load increases and the deflection exceeds a certain limit, the pile-side soil friction effect can be fully exerted to the extreme, while the additional bending moment and additional displacement caused by the $p-\Delta$ effect increase significantly. The inhibition effect beyond the contribution of the pile side soil friction is shown to further develop the pile shaft displacement and bending moment with the increase of gravity load in macro. Therefore, in the actual project, when the lateral load and overturning moment load are large, the hazards caused by $p-\Delta$ should be fully considered, and the additional displacement and bending moment should be paid attention to

Li Yaya also analyzed the pile top acceleration and pile foundation settlement under the combined action of wave load, wind load, and gravity load, which we will not elaborate in this book. For the readers' interests, please refer to the conference (LI Yazhou, 2020).

10.4 Influence of combined action of wind, wave, ice, and current

Liao Yingdi studies the combined effects of environmental loads (wind, wave, ice, and current) on offshore wind power structures in complex marine environments (LIAO Yingdi et al., 2021). Taking the tripod foundation offshore wind power structure as the research object, the overall model of the tripod foundation offshore wind power structure is established using ANSYS finite element software, and the static check under the combined action of loads (wind, wave, current, and wind, ice, current) is carried out. The transient analysis of the tripod foundation offshore wind power structure under the combined action of environmental loads is carried out.

10.4.1 Establishment of finite element model of integral structure

For the tripod foundation of offshore wind power foundation selected by Liao Yingdi (LIAO Yingdi et al., 2021), refer to the article of ZHANG (J. Zhang et al., 2018) The wind turbine is the traditional windward, variable speed, horizontal axis wind turbine - NREL 5 mW wind turbine. Because this paper mainly carries out the relevant research on the dynamic response characteristics of offshore wind power structures based on tripod foundations, especially focusing on the dynamic response of tower structures, the following simplifications are made for this tripod foundation structure before the overall structure modeling: the fan blades, wheel hubs, engine rooms and other components at the top of the tower are simplified with concentrated mass points; The fictitious anchor point method (*《Building Structure Load Code》: GB 50009—2011*, 2012) was used to fix the pile foundation at 6 times the pile diameter below the seabed mud line.

The mud line elevation of the sea area where the offshore wind power tripod foundation structure is located is -30.00 m, the water surface elevation is 0.00 m, the top elevation of the foundation is +10.00 m, and the fan elevation is +90.00 m. The structure uses 3 steel pipe piles as the foundation, with a pile diameter of 2.3 m. The pile foundation is arranged as an equilateral triangle

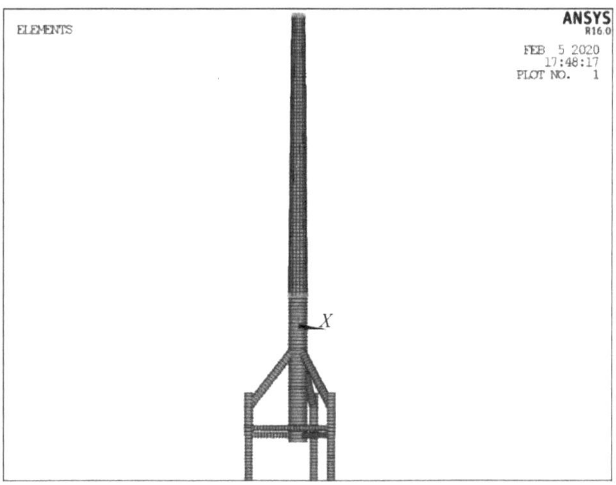

[Fig. 10-55] Integral structure finite element model, source: (LIAO Yingdi et al., 2021)

with a side length of 26 m. Composition of the upper tripod structure of the pile foundation: the casing at the pile top relates to the middle column through diagonal braces, and horizontal braces are set between the casings to further improve the structural stability. The diameter of the pile top casing is 2.6 m, and its elevation range is -30.00 ~ -16.00 m. The structure includes the upper tower and the lower tripod foundation. Use the reasonable element type in the ANSYS software to establish the overall structure model of the simplified model. The established ANSYS finite element model is shown in [Fig. 10-55].

10.4.2 Static check

10.4.2.1 Load combination and action direction

Concerning the《*Design Requirements for Offshore Wind Turbines*》(IEC 61400-3) (Nicholson, 2011) and the Det Norske Veritas(LOU Chunjuan, 2017), when considering the combined action of wind, wave, current, and ice loads, it is assumed that these environmental loads have the same direction of action and are all used for load combination. The tripod-type wind power foundation structure is a 120° symmetrical structure. Through calculation, the most unfavorable load joint action direction in static check is 0°, as shown in [Fig. 10-56].

The wind, wave, current, and ice load parameters adopted by Liao Yingdi are listed in [Tab.10-5] (LIAO Yingdi et al., 2021). The main purpose of static analysis on the overall structure of offshore wind power is to ensure that the strength and stiffness of the overall structure are within the safe range. The combined actions of the two loads in the static analysis are as follows::

(1) Wind load (fan load+tower wind load)+wave load+current load.

(2) Wind load (fan load+tower wind load)+sea ice load+current load.

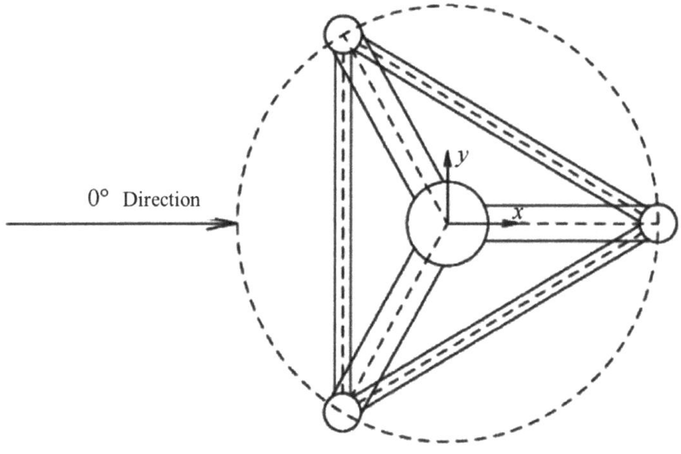

[Fig.10-56] The direction of the most unfavorable combined action of loads, source: (LIAO Yingdi et al., 2021)

[Tab. 10-5] The load conditions of wind, wave, current and ice are adopted in this paper, (LIAO Yingdi et al., 2021)

wind	10min average wind speed at 10m height: $U_{10,50-yr} = 35.4m/s$
wave	Significant wave height: $H_{s,50-yr} = 4.7m$, $T_s = 10s$
current	Bottom flow rate $v_{bottom,5-yr} = 0.85m/s$, Middle-velocity $v_{middle,5-yr} = 1.05m/s$, Surface velocity $v_{surface,5-yr}$=1.25m/s
ice	Ice thickness $H_{ice,50-yr} = 0.38m$, Uniaxial compression strength of ice $\sigma_c = 2Mpa$, bending strength $\sigma_f = 700kPa$

10.4.2.2 Verification results

The static calculation results of the structure are shown in [Fig. 10-57] and [Fig. 10-58].

It can be seen from [Fig. 10-57] that under the combined action of wind, ice, and current loads, the maximum equivalent stress of the structure tripod foundation is 143 MPa, and the maximum displacement of the tower top is 0.72 m, which is greater than the combined action of wind, wave, and current loads (see [Fig. 10-58], the maximum equivalent stress is 102 MPa, and the tower top displacement is 0.67 m). Sea ice load cannot be ignored in the design of wind power infrastructure in ice area.

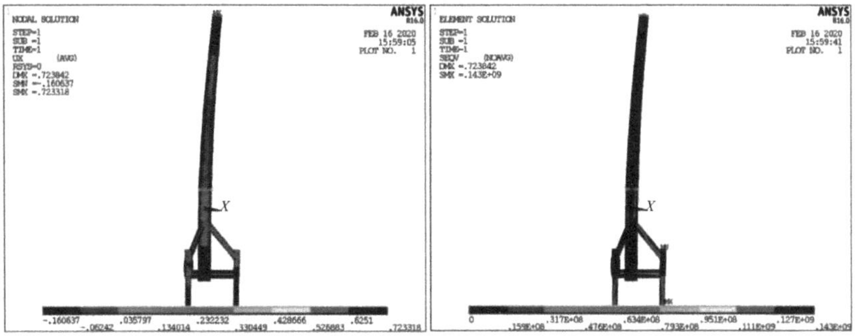

(a) Displacement nephogram of structure/m (b) Equivalent stress nephogram of structure/Pa

[Fig. 10-57] Diagram of static analysis results under the combined action of wind, ice, and flow loads, source: (LIAO Yingdi et al., 2021)

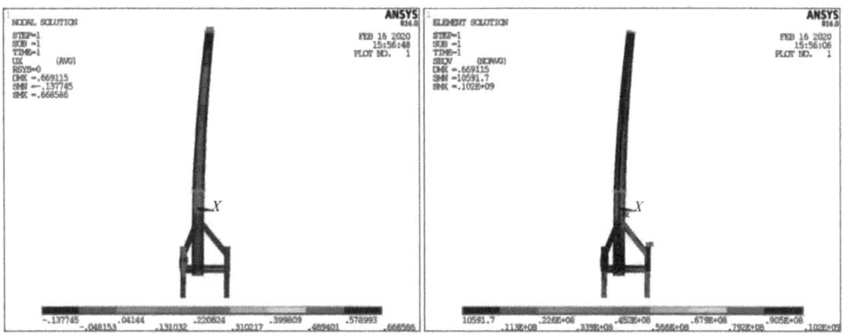

(a) Displacement nephogram of structure/m (b) Equivalent stress nephogram of structure/Pa

[Fig. 10-58] Diagram of static analysis results under the combined action of wind, wave, and flow loads, source: (LIAO Yingdi et al., 2021)

The static check of the tripod foundation adopts the allowable stress method to verify whether the strength of the structure meets the requirements. The verification process is carried out according to the following formula(《*Building Structure Load Code*》: *GB 50009—2011*, 2012)

$$\sigma_{\max} \leq [\sigma]$$

$$[\sigma] = 0.8 \times \sigma_s$$

where, $\sigma_{\max}$ is stress extreme value of structure under load (MPa); $[\sigma]$ is the allowable stress value of materials used in a structure (MPa); is the yield strength of materials (MPa). It can be seen from the calculation that the allowable stress of DH36 steel used for the tripod structure in this section is 288 MPa, which is greater than the extreme value of the equivalent stress of the structure under the combined load of 143 MPa, so the strength of the offshore wind power tripod foundation meets the requirements.

The top displacement of the tower barrel is checked according to the checking method of the structural rigidity of the suspended wall beam (Nicholson, 2011). The ratio of the top displacement of the tower barrel to the height of the tower barrel should be limited to 0.25%~1%. The extreme value of the tower top displacement is 0.72 m, and the cantilever length is 80 m of the tower in the vertical direction, so the extreme value of the tower top displacement is 0.9% of the cantilever length, which is within the required range, so the tower stiffness meets the requirements.

10.4.3 Dynamic response analysis

The tower of an offshore wind power structure needs relatively stable conditions for the normal operation of its supporting fan, so the size of its top dynamic response must be controlled within a certain range to ensure the safety of the structure and the stability of the power generation function(Feyzollahzadeh et al., 2016). The top of the tripod foundation is connected to the bottom of the tower, and the severity of its vibration under dynamic load will directly affect the stability of the tower. Therefore, this paper focuses on the displacement and acceleration response of the tower top and the tripod foundation top under the combined action of two environmental loads.

This section analyzes the transient response of the structure through the ANSYS software. Although it is a numerical simulation based on the actual situation, the simulation results are somewhat different from the actual due to the randomness of wind, wave, current, and ice loads. Especially for the early loaded structure, the environmental load action is a continuous and relatively stable process, and there is no sudden increase in the structural response caused by sudden load application, However, in general, after about 40~70 seconds of loading, the structural response at the junction is very large, which is inconsistent with the actual situation. The acceleration amplitude of the top of the wind power tripod foundation under the combined action of wind, wave, and current loads is less than the superposition of the wind load and wave and current loads at this stage.

10.4.3.1 Combined action of wind, wave, and current loads

The difference between the structural response under the combined action of wind, wave, and current load and the coupling action of wind load, wave, and current load is shown in [Fig. 10-59] and [Fig. 10-60].

It can be seen from the time history curve in [Fig.10-59] and [Fig.10-60] that the response at the top of the tower starts to fluctuate steadily in about 60s, which better reflects the actual situation. Under the combined action of wind, wave, and current loads, the dynamic response of the tower top is the largest, the

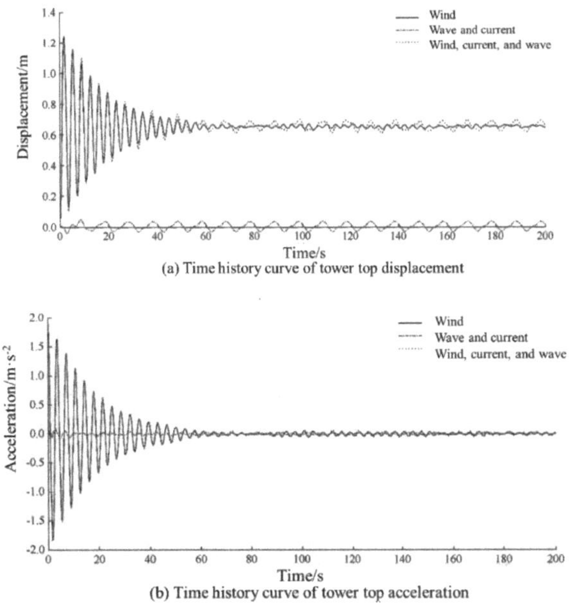

[Fig. 10-59] The time history curves of displacement and acceleration at the top of the tower, source: (LIAO Yingdi et al., 2021)

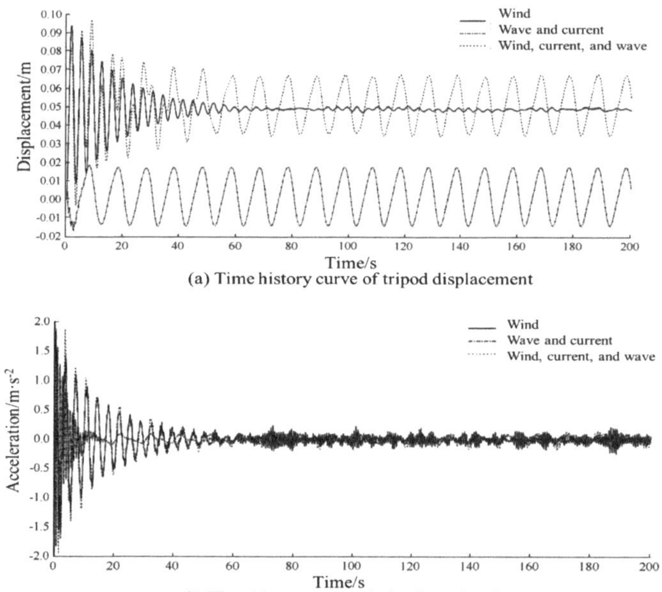

(a) Time history curve of tripod displacement

(b) Time history curve of tripod acceleration

[Fig. 10-60] The time history curves of displacement and acceleration at the top of the tripod foundation, source:(LIAO Yingdi et al., 2021)

displacement fluctuates stably within 0.61~0.69 m, and the acceleration fluctuates stably within - 0.036~0.057 m/s^2 which is consistent with the law when the wind load acts alone, but the value is slightly larger, indicating that the wind load dominates the dynamic response of the tower top. [Fig. 10-60] shows the dynamic response law of the top of the tripod foundation: at about 50s, the displacement of the top of the tripod foundation starts to fluctuate steadily within the range of 0.036~0.067 m. It can be seen that the displacement of the top of the tripod foundation under the combined action of wind, wave, and current loads is almost the superposition of the displacement under the separate action of wind load and wave and current load; At about 43s, the acceleration at the top of the tripod foundation began to fluctuate steadily within the range of -0.036 ~ 0.057 m/s^2. Analysis shows that the acceleration amplitude at the top of the tripod foundation under the combined action of wind, wave, and current loads in the stable fluctuation stage is less than the superposition of the acceleration under the action of wind load and wave and current loads in this stage.

10.4.3.2 Combined action of wind, ice, and current loads

The difference between the structural response under the combined action of wind, ice, and current loads and the separate action of wind and ice loads is shown in [Fig. 10-61] and [Fig. 10-62].

It can be seen from [Fig. 10-61] that the displacement and acceleration response laws of the tower top under the combined action of wind, ice, and current loads are similar. The steady vibration starts in about 70s, the displacement fluctuation range is 0.50~0.94 m, and the acceleration fluctuation range is - 0.63~0.68 m/s^2. The dynamic response of the top of the tower is mainly controlled by the ice load to reach a steady state of vibration. From the numerical point of view, the dynamic response of the top of the tower under the combined action of wind, ice, and current loads is the superposition of the loads under separate action. It can be seen from [Fig. 10-62] that the displacement

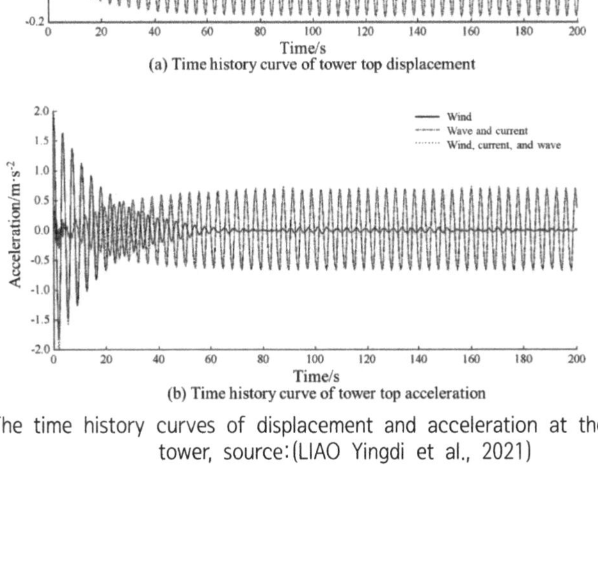

[Fig. 10-61] The time history curves of displacement and acceleration at the top of the tower, source:(LIAO Yingdi et al., 2021)

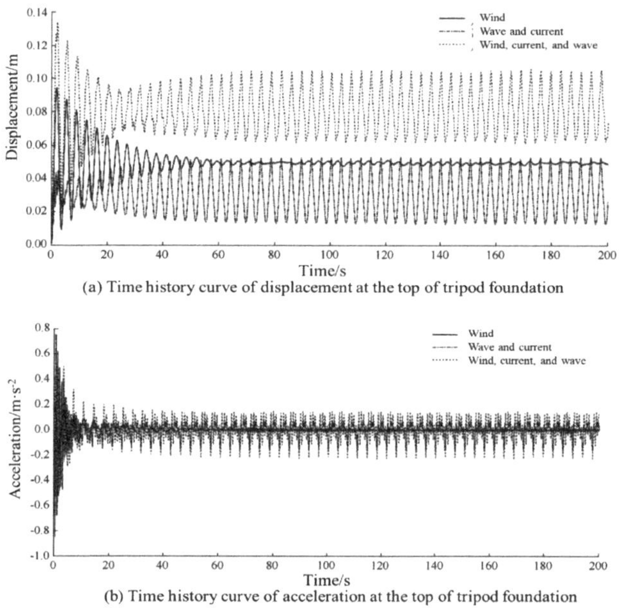

[Fig. 10-62] The time history curves of displacement and acceleration at the top of the tripod foundation, source:(LIAO Yingdi et al., 2021)

response law of the top of the tripod foundation under the combined action of wind, ice, and current loads is similar to that of the top of the tower. The steady vibration starts in about 60 seconds, and the displacement fluctuation range is 0.062 ~ 0.104 m; The acceleration response law is different from the displacement response law. The acceleration starts to respond steadily in about 40s, and the fluctuation range is - 0.199 ~ 0.154 m/s^2, which is greater than the acceleration response under the ice load alone.

Appendix: Wind Power Software

1. Design software for deep-sea floating wind turbine
2. Turbine design, Software

1. Design software for deep-sea floating wind turbine

Modern wind turbines are becoming larger and more intelligent, which means that the original engineering design methods and empirical control methods will be gradually replaced by modern wind turbine design software and control software. As a carrier embodying design ideas, design methods, and design tools, wind turbine design software has been at the core of the modern wind turbine industry. Developing offshore floating wind turbine design software with independent intellectual property rights is one of the basic conditions for research in this field (GAO Wei et al., 2014).

The existing design software for deep-sea floating wind turbines mainly includes GH. Bladed software of GL and NWTC of NREL FAST software and MSC ADAMS portfolio, SIMO/RIFLEX software of Norwegian Institute of Marine Engineering, HAWC/HAWC2 or FLEX5 portfolio of Danish University of Science and Technology, and SESAM of DNV DeepC software, 3Dfloat software of Norwegian University of Life Sciences, etc. See the literature for other similar design software or combinations (Cordle & Jonkman, 2011)(Cordle A , Jonkman J. State of the art in floating wind turbine design tools [R]. NREL/TP-500-50543. USA : Golden , 2011.).

Although these software functions are different. However, it is composed of four functional modules, namely, the pneumatic module, hydrodynamic module, structural dynamics module, and control module (as shown in [Tab. AP-2]. Most of the algorithms adopted are similar, but there is no in-depth and comprehensive research on the coupling mechanism and decoupling mechanism of the overall dynamics of the deep-sea floating wind turbines. All the above development and design software or their combinations adopt the interface coupling method, that is, set constraints on the wind turbine aerodynamic platform structure and platform structure hydrodynamic interface respectively, force coupling, and each system calculates independently, or uses the iterative calculation method to calculate one effect first, and then convert it into an additional mass or additional load method to consider calculating another effect, Therefore, the essence and influence of the coupling mechanism cannot be accurately reflected. It is necessary to develop more effective fully coupled algorithms and simulation models for the design and certification of deep-sea floating wind turbines.

[Tab. AP- 1] Table 1 Technical specifications of floating wind turbine hydrodynamic experiments

Name	FLOAT	wind	Windfloat	Japanese semi-submersible	Japanese floating column type
Pool	BMT No.2 pool	Norwegian Institute of Marine Engineering(MARINRTEK)Pool	During the ship model test at the University of California, Berkeley	Wave-making basin of Japan National Maritime Research Institute	Wave-making basin of Japan National Maritime Research Institute
Model scale ratio	1:48	1:47	1:105	1:150	1:22.5
Design depth/m	100~300	200~700	150	100	75~200
Design draft /m	27.0	120.0	22.9	20	60
Floating body size /m	D=12.0	D=8.3	span 56.4	span 184.0	$D_{up}=4.8$ $D_{middle}=12$ $D_{down}=8.4$
Mooring line/piece	8	3	6	6	2
Wind turbine power /MW	1.4	5.0	5.0	3×2.4	2
Diameter of wind turbine /m	60	123	126	92	80
Height from wheel hub to water surface /m	45	81.5	90	70	55

[Tab. AP- 2] Software architecture and module of offshore wind turbine design and simulation (Jonkman J M , Larsen T , Hansen A , et al. Offshore code comparison collaboration within IEA wind task 23 : phase IV results regarding floating wind turbine modeling [C]// EWEC 2010.Warsaw , Polang , 2010)

Software	Research institutions	Pneumatic module	Structure Dynamic Module	Hydrodynamic module	Mooring Module	Control Module
FAST	NREL	(BEM/GDW) +DS	(Modal/MBS) +FEM	Airy+ME,Airy+PF+ ME	QSCE	DLL
GH.Bladed	GH	(BEM/GDW) +DS	(Modal/MBS) +FEM	(Airy/Stream)+ME	UDFD	DLL
ADAMS	MSC+NREL+L UH	(BEM/GDW) +DS	MBS	Airy+ME,Airy+PF+ ME	QSCE, UDFD	DLL,U DF
HAWC2	Ris-DTU	(BEM/GDW) +DS	MBS+FEM	Airy+ME,Airy+PF+ ME	QSCE, UDFD	DLL,U DF
SIMO	MARINTEK	BEM	MBS	Airy+ME	QSCE,M BS	DELL
SESAM.D eepC	DNV	None	MBS	Airy+ME, Airy+PF+ME	QSCE,F EM	None
3Dfloat	IFE-UMB	BEM/GDW	FEM	Airy+ME	FEM, UDFD	UDF

In [Tab. AP-2], the momentum blade element theory and the generalized dynamic wake theory are used for the aerodynamic module. Some software considers the dynamic stall correction, but there is generally no algorithm reflecting the aeroelasticity of long flexible blades and high towers; The hydrodynamic module mostly adopts the method of combining linear wave theory with Morrison equation, and individual software (GH. Bladed) further considers the influence of ocean current. Some of them are from the ship and offshore platform calculation software, and the linear potential flow theory with wave radiation and scattering correction is combined with the method of the Morrison equation. However, due to the particularity of the superstructure of the floating wind turbine, the effectiveness of this algorithm is questionable; The wind turbine and floating platform in the structural dynamics module adopt the finite element method (FEM) or multi-body dynamics method, and the mooring system adopts the quasi-static catenary equation. The special structural dynamics software for wind turbines (NWTC. FAST and GH. Bladed) adopts the technology of combining the assumed mode method with multi-body dynamics, while the ship and ocean engineering calculation software uses the user-defined force-displacement relationship method to calculate the load of the mooring system due to the high requirements for the calculation of the mooring system, which requires a lot of work and is difficult to master; The control modules adopt GH Dynamic link library (DLL) in Bladed format is used to import simulation modules, and some software provides control files in the user-defined format (UDF).

2. Turbine design, Software, https://en.wikipedia.org/wiki/Wind_energy_software

Wikipedia, the free encyclopedia

Specialized wind energy software applications aid in the development and operation of wind farms.

2.1 Pre-feasibility and feasibility analysis

The RETScreen software wind power model is designed to evaluate energy production and savings, costs, emission reductions, financial viability and risk for central-grid, isolated-grid and off-grid wind energy projects, for multi-turbine and single-turbine hybrid systems. Developed by the Government of Canada, the software is free, multilingual, and includes links to wind energy resource maps.

The Wind Data Generator (WDG) is a Wind Energy Software tool capable of running the WRF (Weather Research and Forecasting) model to create a wind atlas and to generate wind data at resolutions of 3 km to 10 km.

2.2 Turbine design

The software helps design wind turbines. Several aero-elastic packages are used in this design process.

FOCUS6 aids in the design of wind turbines and turbine components such as rotor blades. It was developed by the Knowledge Centre Wind turbine Materials and Constructions (WMC) and the Energy Research Centre of the Netherlands (ECN).

The National Wind Technology Center (NWTC), a division of the U.S. National Renewable Energy Laboratory (NREL), has developed many packages which are used by turbine manufacturers and researchers. NWTC has developed a suite of turbine design and performance prediction codes that rely on Blade Element Momentum (BEM) theory. WT_Perf uses steady BEM theory to model turbine performance. FAST is a comprehensive aero-elastic simulator that uses unsteady BEM theory to model a turbine as a collection of rigid and flexible bodies in a spatiotemporal field of turbulent flow. Germanischer Lloyd found FAST suitable for "the calculation of onshore wind turbine loads for design and certification."

The open-source software QBlade developed by the wind energy research group of Hermann Föttinger Institute of TU Berlin (Chair of Fluid Dynamics) is a BEM code coupled with the airfoil simulation code XFOIL. It allows the user to develop/import airfoil shapes, simulate them and use them for the design and simulation of wind turbine blades/rotors with the use of steady-state BEM theory. The software is built with the Qt framework thus it includes a graphical user interface.

The open-source software Vortexje, developed by Baayen & Heinz GmbH in Berlin, is an unsteady 3D panel method implementation suitable for the dynamic simulation of vertical and horizontal axis wind turbines. Easily coupled with other simulation environments such as Simulink and Dymola, it is suitable for aerodynamic optimization, fluid-structure interaction problems, and unsteady control system simulation.

Ashes is a software package for analyzing aerodynamic and mechanical forces for onshore and offshore horizontal axis wind turbines. It is based on research done at the Norwegian University of Science and Technology in Trondheim, Norway.

Table. Turbine design engineering software

Software	Publisher	Latest stable version	License	Certified
Ashes	SIMIS as	3.15	proprietary	No
Bladed	DNV	4.12	proprietary	Yes
Fast	NREL	8.12	open source	Yes
Flex	DTU Lyngby	5	proprietary	Yes
FOCUS6	WMC	6	proprietary	No
HAWC2	DTU Wind Energy	12.2	proprietary	No
QBlade	TU Berlin	0.8	open source	No
Simpack	Simpack	2017.2	proprietary	No
Vortexje	Baayen & Heinz GmbH	-	open source	No

2.3 Flow modeling

Wind flow modeling software predicts important wind characteristics at locations where measurements are not available. Furrow is a software that offers a lineal flow model and a Computational fluid dynamic model in the same software. WAsP was created at Denmark's Risø National Laboratory. WAsP uses a potential flow model to predict how wind flows over terrain at a site. Meteodyn WT, Windie, WindSim, WindStation and the open-source code ZephyTOOLS use computational fluid dynamics instead, which are potentially more accurate, but more computationally intensive.

Table. Flow modeling software

Software	Publisher	Latest stable version	License	Certified
Continuum	Cancalia/One Energy	2.2.5	proprietary	-
Furow	Etulos Solute	2.0	proprietary	Yes
WAsP	DTU Wind Energy	12	proprietary	Yes
WindSim	Vector AS	8.0	opensource	Yes
Meteodyn WT	Meteodyn	6.1	proprietary	Yes
ZephyCFD	Zephy-Science	19.12	GPL3	Yes
WindStation	menzio GmbH	1.3.21	proprietary	-
Windie	Windie Lda.	-	proprietary	-
SimWorks	IdealSimulations	20.12	free software	-

2.4 Farm modeling

This software simulates wind farm behavior, most importantly to calculate its energy output. The user can usually input wind data, height and roughness contour lines (topography), turbine specifications, and background maps, and define environmental restrictions. Processing this information produces the design of a wind farm that maximizes energy production while accounting for restrictions and construction issues. Packages include Furow, Meteodyn WT, openWind, WindFarm, WindFarmer: Analyst, WindPRO, WindSim and WindStation. WakeBlaster is a specialised CFD service for modelling the wind farm wake losses.

2.5 Farm visualization

Wind farm visualization software graphically presents a proposed wind farm, most importantly for the purpose of obtaining building permits. The primary techniques include photomontages, zone-of-visual-impact maps and three-dimensional visualization (perspective views of the landscape often incorporating aerial photography and including turbines and other objects).

Table. Farm visualization software

Software	Publisher	Latest stable version	License	Certified
Furow	Etulos Solute	2.0	proprietary	Yes
openWind	AWS Truepower		proprietary	Yes
windPRO	EMD International A/S	3.5	proprietary	
Meteodyn WT	Meteodyn	6.1	proprietary	Yes
ZephyCFD	Zephy-Science	19.12	GPL3	Yes
Windplanner	The Imagineers	v3	proprietary	
WindFarmer: Analyst	DNV GL	1.0	proprietary	

2.6 Farm monitoring

Wind farm monitoring software is software that allows people to see if the wind turbines are running well or are going to become broken. Other functions of monitoring software are reporting, analysis of measurement data (power curve) and tools for monitoring environmental constraints (bat control, etc.).

Table. Farm monitoring software

Software	Publisher	Latest stable version	License	Certified
Clir	Clir		proprietary	
Meteodyn SPA	Meteodyn	2.2	proprietary	Yes
GPM Horizon	GreenPowerMonitor			
Greenbyte Energy Cloud	Greenbyte AB		proprietary	
ROTORsoft	DrehPunkt GmbH		proprietary	
windPRO - PERFORMANCE CHECK module	EMD International A/S	3.5	proprietary	
windPRO - TR10 module	EMD International A/S	3.5	proprietary	Yes
windOPS	EMD International A/S	2	proprietary	

2.7 Prediction Software

Main article: Wind power forecasting

For existing wind farms, several software systems exist which produce short and medium-term forecasts for the generated power (single farms or complete forecast regions) using existing numerical weather prediction data (NWP) and live (SCADA) farm data as input. Examples of numerical weather prediction models used for this purpose are the European HiRLAM (High Resolution Limited Area Model) and the GFS (Global Forecast System) from NOAA.

References

[1] Abadie, C. (2016). *Cyclic lateral loading of monopile foundations in cohesionless soils* [Http://purl.org/dc/dcmitype/Text, University of Oxford]. https://ora.ox.ac.uk/objects/uuid:b3002ca4-742d-4b08-836f-1c15eafb7156

[2] All Daily. (2017, June 30). *The pre-research phase of Portugal's new offshore wind power station has come to a perfect conclusion. Let children not know what smog is!* The Pre-Research Phase of Portugal's New Offshore Wind Power Station Has Come to a Perfect Conclusion. Let Children Not Know What Smog Is! http://www.inengjing.com/NewEnergy/Wind/201706/20231.html

[3] Arany, L., Bhattacharya, S., Macdonald, J., & Hogan, S. J. (2017). Design of monopiles for offshore wind turbines in 10 steps. *Soil Dynamics and Earthquake Engineering*, 92, 126–152. https://doi.org/10.1016/j.soildyn.2016.09.024

[4] Arapogianni, A., Genachte, A.-B., Manzanas, R., Vergara, J. P., Castell, C., Rodriguez Tsouroukdissian, A., Korbijn, J., Bolleman, N. C. F., Huera-Huarte, F., Schuon, F., Ugart, A., Sandberg, J., Laleu, V., Maciel, J., Tunbjer, A., Roth, R., Gueriviere, P., Coulombeau, P., Jedrec, S., & Grubel, H. (2013). Deep water. The next step for offshore wind energy. *Eur. Wind Energy Assoc.*

[5] Astariz, S., Vazquez, A., Sánchez, M., Carballo, R., & Iglesias, G. (2018). Co-located wave-wind farms for improved O&M efficiency. *Ocean & Coastal Management, 163,* 66–71. https://doi.org/10.1016/j.ocecoaman.2018.04.010

[6] Bakmar, C. L., Byrne, B. W., & Houlsby, G. T. (2010). Response of stiff piles to random two-way lateral loading. *Geotechnique, 60*(9), 715–721. https://doi.org/10.1680/geot.09.T.011

[7] Bhattacharya, S. (2014). Challenges in design of foundations for offshore wind turbines. *Eng. Technol. Ref, 1*(1), 922.

[8] Bhattacharya, S., & Adhikari, S. (2011). Experimental validation of soil–structure interaction of offshore wind turbines. *Soil Dynamics and Earthquake Engineering, 31*(5), 805–816. https://doi.org/10.1016/j.soildyn.2011.01.004

[9] Bhattacharya, S., Lombardi, D., & Muir Wood, D. (2011). Similitude relationships for physical modelling of monopile-supported offshore wind turbines. *International Journal of Physical Modelling in Geotechnics, 11*(2), 58–68. https://doi.org/10.1680/ijpmg.2011.11.2.58

[10] Bottasso, C. L., Croce, A., Gualdoni, F., & Montinari, P. (2016). Load mitigation for wind turbines by a passive aeroelastic device. *Journal of Wind Engineering and Industrial Aerodynamics, 148*, 57~69. https://doi.org/10.1016/j.jweia.2015.11.001

[11] 《Building Structure Load Code》: GB 50009—2011. (2012). China Architecture & Building Press. https://www.waizi.org.cn/bz/131451.html

[12] CCS. (2021). *海上浮式风机平台指南*. CCS. https://www.ccs.org.cn/ccswz/specialDetail?id=202111181033885482

[13] Chen, X., Li, C., & Xu, J. (2015). Failure investigation on a coastal wind farm damaged by super typhoon: A forensic engineering study. *Journal of Wind Engineering and Industrial Aerodynamics, 147*, 132~142. https://doi.org/10.1016/j.jweia.2015.10.007

[14] Chen, X., & Xu, J. Z. (2016). Structural failure analysis of wind turbines impacted by super typhoon Usagi. *Engineering Failure Analysis, 60*, 391~404. https://doi.org/10.1016/j.engfailanal.2015.11.028

[15] Chen Zhaohui, & Tang Haitao. (2008). Distribution models of extreme typhoon winds based on numerical simulation of wind data. *Journal of Chongqing University, 11*, 1285~1289.

[16] CHI Yongning, LIANG Wei, ZHANG Zhankui, LI Yan, JIN Shuanglong, CAI Xu, HU Junhui, & TIAN Wei. (2016). An overview on key technologies regarding power transmission and grid integration of large scale offshore wind power. *Proceedings of the CSEE, 36*(14), 3758~3771.

[17] China Classification Society. (2005). *Rules for Classification and Construction of Offshore Mobile Platforms*. China Communications Press Co., Ltd.

[18] Choo, Y. W., & Kim, D. (2016). Experimental Development of the p-y Relationship for Large-Diameter Offshore Monopiles in Sands: Centrifuge Tests. *Journal of Geotechnical and Geoenvironmental Engineering*, 142(1), 04015058. https://doi.org/10.1061/(ASCE)GT.1943-5606.0001373

[19] Choo, Y. W., Kim, D., Park, J.-H., Kwak, K., Kim, J.-H., & Kim, D.-S. (2014). Lateral Response of Large-Diameter Monopiles for Offshore Wind Turbines from Centrifuge Model Tests. *GEOTECHNICAL TESTING JOURNAL*, 37(1). https://doi.org/10.1520/GTJ20130081

[20] CHU Chenhui, & CHEN Shaolin. (2021). Wind Vibration Response and Wind Vibration Coefficient of Wind Turbine Tower Under Different Wind Speed Spectrum Excitation. *Journal of Water Resources and Architectural Engineering, 19*(01), 39~45.

[21] Cordle, A., & Jonkman, J. (2011). State-of-the-art in Floating Wind Turbine Design Tools. *Proceedings of the International Offshore and Polar Engineering Conference.*

[22] Cuéllar, P. (2011). *Pile foundations for offshore wind turbines: Numerical and experimental investigations on the behaviour under short-term and long-term cyclic loading.*

[23] Daily wind power. (2019, April 12). *break the record! Hoisting of China's largest offshore wind turbine jacket completed.* http://www.chinawindnews.com/6313.html

[24] DENG Tao, LIN Cong-yu, LIU Zhi-peng, HUANG Ming, & CHEN Wenjing. (2020). A simplified elastoplastic method for laterally loaded single pile with large displacement. *Rock and Soil Mechanics*, 41(01), 95~102. https://doi.org/10.16285/j.rsm.2018.2346

[25] Ding, H., Lian, J., Li, A., & Zhang, P. (2013). One-step-installation of offshore wind turbine on large-scale bucket-top-bearing bucket foundation. *Transactions of Tianjin University, 19*(3), 188~194. https://doi.org/10.1007/s12209-013-2099-0

[26] Ding, H., Liu, Y., Zhang, P., & Le, C. (2015). Model tests on the bearing capacity of wide-shallow composite bucket foundations for offshore wind turbines in clay. *Ocean Engineering, 103*, 114~122. https://doi.org/10.1016/j.oceaneng.2015.04.068

[27] Ding Hongyan, Guo Weibo, Zhang Puyang, & Le Conghuan. (2013). Response of Large-Scale Bucket Foundation Used in Offshore Wind Turbine in Shoal Areas Under Wind Load. *Journal of Tianjin University(Science and Technology), 46*(02), 121~126.

[28] Dinh, V. N., & Basu, B. (2013). *Floating offshore wind turbines with spar-type support.* https://doi.org/10.13140/RG.2.2.34353.99686

[29] Dismukes, D. E., & Upton, G. B. (2015). Economies of scale, learning effects and offshore wind development costs. *Renewable Energy*, 83, 61~66. https://doi.org/10.1016/j.renene.2015.04.002

[30] Design of Offshore Wind Turbine Structures, Pub. L. No. DNV-OS-J101, DNV-OS-J101 (2014). https://docplayer.net/21082564-Design-of-offshore-wind-turbine-structures.html

[31] Support structures for wind turbines, (2021). https://www.dnv.com/Default

[32] Dongfang Wind Power Network. (2022). *One step ahead, Elite provides tower flanges to European floating wind power.* http://www.eastwp.net/news/show.php?itemid=62450

[33] DU Jie, YU Hongyan, LIU Jianhui, & ZHANG Chao. (2005). Research on Boundary Selection of Soil of Bucket Foundation with Finite Element A nalysis. OCEA N T ECHN OLOGY, 24(02), 109~113.

[34] Enevoldsen, P., & Valentine, S. V. (2016). Do onshore and offshore wind farm development patterns differ? *Energy for Sustainable Development, 35,* 41~51. https://doi.org/10.1016/j.esd.2016.10.002

[35] Erbrich, C. T., & Tjelta, T. I. (1999, May 3). *Installation of Bucket Foundations and Suction Caissons in Sand—Geotechnical Performance.* Offshore Technology Conference. https://doi.org/10.4043/10990-MS

[36] Esteban, M. D., Diez, J. J., López, J. S., & Negros, V. (2009). Integral Management Applied to Offshore Wind Farms. *Journal of Coastal Research,* 1204~1208.

[37] Esteban, M. D., Espada, J. M., Ortega, J. M., López-Gutiérrez, J.-S., & Negro, V. (2019). What about Marine Renewable Energies in Spain? *Journal of Marine Science and Engineering, 7*(8), Article 8. https://doi.org/10.3390/jmse7080249

[38] Esteban, M. D., López-Gutiérrez, J. S., Diez, J. J., & Negro, V. (2011a). Methodology for the Design of Offshore Wind Farms. *Journal of Coastal Research,* 496~500.

[39] Esteban, M. D., López-Gutiérrez, J. S., Diez, J. J., & Negro, V. (2011b). Offshore Wind Farms: Foundations and Influence on the Littoral Processes. *Journal of Coastal Research,* 656~660.

[40] Fang Tao. (2014). The development model of UK offshore wind power and its reference significance. *Overseas Energy, 36*(12), 26~30.

[41] Fathabadi, H. (2016). Novel high efficient speed sensorless controller for maximum power extraction from wind energy conversion systems. *Energy Conversion and Management, 123,* 392~401. https://doi.org/10.1016/j.enconman.2016.06.046

[42] Feyzollahzadeh, M., Mahmoodi, M. J., Yadavar-Nikravesh, S. M., & Jamali, J. (2016). Wind load response of offshore wind turbine towers with fixed monopile platform. *Journal of Wind Engineering and Industrial Aerodynamics, 158,* 122~138. https://doi.org/10.1016/j.jweia.2016.09.007

[43] GAO Wei, LI Chun, & YE Zhou. (2014). The current situation and latest research of deep-sea floating wind turbine. *中国工程学, 16*(2), 79~87.

[44] GB/T 51308-2019海上风力发电场设计标准, (2019). http://www.cnmhg.com/Standard-Download/1654095323223232.html

[45] Gil-García, I. C., García-Cascales, M. S., Fernández-Guillamón, A., & Molina-García, A. (2019). Categorization and Analysis of Relevant Factors for Optimal Locations in Onshore and Offshore Wind Power Plants: A Taxonomic Review. *Journal of Marine Science and Engineering, 7*(11), Article 11. https://doi.org/10.3390/jmse7110391

[46] Global wind Energy Council. (2019). *GWEC Global Wind Report 2018*. Global wind Energy Council: Brussels, Belgium. https://periscope-network.eu/analyst/global-wind-report-2019

[47] GUO Mengyuan, & FAN Lichan. (2019). Structural characteristics analysis of jacket foundation under wind, wave and current loads. *Shanxi Architecture, 45*(22), 73~75.

[48] HAN Chaoshuai. (2014). *Research for fatigue performance of jacket offshore wind turbine support structure* [Master's Thesis]. Haerbin University of Technology.

[49] HAN Jiwei, LIU Xiaoming, YAG Yuehong, YANG Weidong, & LI Bing. (2022). Research on Horizontal Bearing Characteristics and Measures to Improve Bearing Capacity of Single Fan Pile Foundation. *Journal of Inner Mongolia Agricultural University (Natural Science Edition), 43*(02), 55~60.

[50] Hanjia-Xin. (2015). *Offshore Engineering of China Recycle Resourses of Ocean*. China Ocean Press.

[51] HE Ben, WANG Huan, HONG Yi, WANG Lizhong, ZHAO Changjun, & QIN Xiao. (2016). Effect of vertical load on lateral behavior of single pile in clay. *Journal of Zhejiang University (Engineering Science), 50*(07), 1221~1229.

[52] HE Yanping, & TAN Jiahua. (2002). The development history and typical application of bucket foundation. *China Offshore Platform, 06*, 11~15.

[53] Henderson, A. R., & Witcher, D. (2010). Floating Offshore Wind Energy—A Review of the Current Status and an Assessment of the Prospects. *Wind Engineering, 34*(1), 1~16. https://doi.org/10.1260/0309-524X.34.1.1

[54] Heronemus, W. E. (1972). *Pollution-free energy from the offshore winds*. Marine Technology Society.

[55] Hevia-Koch, P., & Klinge Jacobsen, H. (2019). Comparing offshore and onshore wind development considering acceptance costs. *Energy Policy, 125*, 9~19. https://doi.org/10.1016/j.enpol.2018.10.019

[56] Houlsby, G., Ibsen, L., & Byrne, B. (2005). Suction Caissons for Wind Turbines. *Int Symp Front Offshore Geotech, 75*. https://doi.org/10.1201/NOE0415390637.ch4

[57] Houlsby, G. T., Kelly, R. B., Huxtable, J., & Byrne, B. W. (2005). Field trials of suction caissons in clay for offshore wind turbine foundations. *Géotechnique, 55*(4), 287~296. https://doi.org/10.1680/geot.2005.55.4.287

[58] HU Anfeng, XIAO Zhirong, JIANG Jinhua, FU Peng, & NAN Bowen. (2020). Cumulative Lateral Displacement of Single Pile Foundation Under Lateral and Vertical Cyclic Loadings. *Journal of Shanghai Jiaotong University, 54*(01), 85~91. https://doi.org/10.16183/j.cnki.jsjtu.2020.01.011

[59] HU Chunhong, & ZHANG Zhihao. (2011). Relationship between adjustment of section configuration and flow-sediment of tail channels in the yellow river estuary. *Journal of Basic Science and Engineering, 19*(04), 543~553.

[60] Hu, K., & Chen, Q. (2011). Directional spectra of hurricane-generated waves in the Gulf of Mexico. *Geophysical Research Letters, 38*(19). https://doi.org/10.1029/2011GL049145

[61] HUANG Shen, ZHAI Endi, XU Chengshun, HU Zhongbo, & SUN Yanguo. (2018). Research on the Side Resistance in the Offshore Wind Farm Monopile Foundation. *Journal of Ocean Technology, 37*(06), 101~108.

[62] Petroleum and natural gas in-dustries-arctic offshore structures: ISO 19906 [S] . Geneva: Interna-tional Standard Organization , 2010., (2010). https://www.iso.org/cms/render/live/en/sites/isoorg/contents/data/standard/06/54/65477.html

[63] 风力发电机组 设计要求, (2001). https://www.bz39.com/archives/183557

[64] Jeng, D. S. (1997). Wave-induced seabed instability in front of a breakwater. *Ocean Engineering, 24*(10), 887~917. https://doi.org/10.1016/S0029-8018(96)00046-7

[65] JIA Xiaohui & GUAN Hui. (2017). Discussion on the basic types and development trends of offshore wind power at home and abroad. *China Plant Engineering, 12*, 131~132.

[66] Jiang Zhenqiang, HE Ben, Shan Zhigang, Lai Yongqing, Wang Huan, & Hong Yi. (2021). CUMULATIVE DEFORMATION AND FATIGUE BEHAVIOUR OF

OFFSHORE WIND TURBINE STRUCTURE SUBJECTED UNDER EXTREME LOADING IN YELLOW SEA—A COMPARATIVE STUDY BETWEEN THREE TYPICAL FOUNDATIONS. *ACTA ENERGIAE SOLARIS SINICA, 42*(04), 386~395. https://doi.org/10.19912/j.0254-0096.tynxb.2018-1313

[67] JIN Weiliang. (2006). *Engineering Load Combination Theory and Application.* China Machine Press.

[68] Kaldellis, J. K., Apostolou, D., Kapsali, M., & Kondili, E. (2016). Environmental and social footprint of offshore wind energy. Comparison with onshore counterpart. *Renewable Energy*, 92, 543~556. https://doi.org/10.1016/j.renene.2016.02.018

[69] Karimirad, M., & Moan, T. (2012). Feasibility of the Application of a Spar-type Wind Turbine at a Moderate Water Depth. *Energy Procedia, 24*, 340~350. https://doi.org/10.1016/j.egypro.2012.06.117

[70] Klinkvort, R. T. (2013). *Centrifuge modelling of drained lateral pile -soil response: Application for offshore wind turbine support structures.* Technical University of Denmark.

[71] Knut O. Ronold, Einar Landet, Erik R. Jørgensen, & Johan Sandberg. (2011). *Design standards for floating wind turbine structures.* Conference: European Wind Energy ConferenceAt: Brussels, Belgium, Brussels, Belgium. https://www.researchgate.net/publication/279197071_Design_standards_for_floating_wind_turbine_structures

[72] Kumar, Y., Ringenberg, J., Depuru, S. S., Devabhaktuni, V. K., Lee, J. W., Nikolaidis, E., Andersen, B., & Afjeh, A. (2016). Wind energy: Trends and enabling technologies. *Renewable and Sustainable Energy Reviews, 53*, 209~224. https://doi.org/10.1016/j.rser.2015.07.200

[73] Lacal-Arántegui, R., Yusta, J. M., & Domínguez-Navarro, J. A. (2018). Offshore wind installation: Analysing the evidence behind improvements in installation time. *Renewable and Sustainable Energy Reviews, 92*, 133~145. https://doi.org/10.1016/j.rser.2018.04.044

[74] LI Feng, & Yue Qianjin. (2000). Failure of ice sheet under simultaneously longitudinal and transverse load on slope structures. *Journal of Hydraulic Engineering, 31*(9), 44~48.

[75] LI Feng, & Yue Qianjin. (2001). Analysis of ice force for narrow conical structures. *Journal of Dalian University of Technology, 41*(2), 223~227.

[76] Li, J., & He, D. (2020). *Strategies of Sustainable Development in China's Wind Power Industry*. Springer Singapore Pte. Limited. http://ebookcentral.proquest.com/lib/kunsankr/detail.action?docID=6012485

[77] Li, J., & Yu, X. (Bill). (2018). Onshore and offshore wind energy potential assessment near Lake Erie shoreline: A spatial and temporal analysis. *Energy, 147*, 1092~1107. https://doi.org/10.1016/j.energy.2018.01.118

[78] LI Wei, ZHANG Min, LIU Zhenya, & ZHAO Shengxiao. (2015). FUNDAMENTAL STRUCTURAL FREQUENCYANALYSIS FOR TRIPOD-TYPE OFFSHORE WIND TURBINE. *ACTA ENERGIAE SOLARIS SINICA, 36*(01), 90~95.

[79] LI Yazhou. (2020). *Study on dynamic bearing characteristics of offshore wind power monopile foundation under wind-wave-gravity loads* [Master's Thesis]. Shandong University of Science and Technology.

[80] LI Zhenglin & HU Jinpeng. (2020). Numerical Simulation of Hydrodynamic Load Calculation of High-cap Composite Pile Foundation of Offshore Wind Turbine. GUANGDONG SHIPBUILDING, 39(02), 21~24.

[81] Lian, J., Ding, H., Zhang, P., & Yu, R. (2012). Design of large-scale prestressing bucket foundation for offshore wind turbines. *Transactions of Tianjin University, 18*(2), 79~84. https://doi.org/10.1007/s12209-012-1661-5

[82] Lian, J., Sun, L., Zhang, J., & Wang, H. (2011). Bearing capacity and technical advantages of composite bucket foundation of offshore wind turbines. *Transactions of Tianjin University, 17*(2), 132~137.

[83] LIAO Yingdi, WANG Lu, ZHANG Pengcheng, & CHEN Da. (2021). Dynamic response analysis of offshore wind turbine structures under combined environmental loads. *Water Resources and Hydropower Engineering*, 52(03), 206~216. https://doi.org/10.13928/j.cnki.wrahe.2021.03.024

[84] LIN Heyun, GUO Yujing, SUN Beibei, JIANG Yanlong, HUANG Yunkai, ZHANG Jianrun, & LU Xi. (2011). Overview of offshore wind power key technologies. *Journal of Southeast University(Natural Science Edition), 41*(4), 882~888.

[85] LIN Jiongzeng, & HU Jinpeng. (2019). Wave Forces Acting on Foundation of Offshore Wind Turbine Subjected to Freak Waves. *GUANGDONG SHIPBUILDING, 38*(01), 21~24.

[86] LIU Baiqiong, XU Min, & LIU Qing. (2015). Main problems and countermeasures for the development of offshore wind power in China. *Developent and Management, 32*(03), 7~12.

[87] LIU Hongjun & LI Hongjiang. (2014). A new sunction anchor foundation of the yellow river delta offshore wind power. *Periodical of Ocean University of China, 44*(07), 71~76. https://doi.org/10.16441/j.cnki.hdxb.2014.07.011

[88] LIU Hongjun, SUN Pengpeng, HU Rungeng, & RUAN Hao. (2019). Horizontal bearing capacity of offshore wind power pile foundation under different sea conditions and geological conditions. *Periodical of Ocean University of China, 49*(S1), 51~58. https://doi.org/10.16441/j.cnki.hdxb.20190010

[89] LIU Jianxiu. (2018). Dynamic response analysis and bearing performanceon of offshore wind power foundation under the combined wind wave and current loadings [Master's Thesis]. Beijing University of Technology.

[90] LIU Jianxiu, LIU Qi, QIAO Chaonan, & LIU Guangxue. (2022). Lateral bearing capacity of large-diameter offshore wind turbine pile foundation. *Energy and Energy Conservation, 07*, 18-20+135. https://doi.org/10.16643/j.cnki.14-1360/td.2022.07.038

[91] LIU Junwei, LV Wei, & YU Xiuxia. (2022). Review on Dynamic Response Characteristics of Offshore Wind Turbine Pipe Pile Foundation in Typhoon Environment. *Journal of Water Resources and Architectural Engineering, 20*(01), 1~7.

[92] LIU Xizhu & LIU Zhifeng. (2013). Analysis of flow-induced vibration response of offshore wind power tower-tube foundation structure. *Water Conservancy Science and Technology and Economy, 19*(07), 8~10.

[93] LIU Yonggang, YU Hongyan, & ZHANG Puyang. (2016). Model tests on the bearing capacity of composite bucket foundation in clay. *Chinese Journal of Geotechnical Engineering, 38*(12), 2315~2321.

[94] LIU Zan-qiang, ZHANG Ning-chuan, YU Yu-xiu, & PEI Yu-guo. (2010). The generation of freak waves based on a modified phase modulation method: I-Theory and validation. *Chinese Journal of Hydrodynamics A*, 25(03), 383~390.

[95] LIU Zhenya. (2015). *Dynamic response and patigue damage analysis for floating offshore wind turbine structure* [Master's Thesis]. Ocean University of China.

[96] LOU Chunjuan. (2017). S*tudy on the Design of Simple Offshore Wind Turbine Foundation* [Master's Thesis]. Dalian University of Technology.

[97] LU Jinliang, ZHANG Yi, & REN Min. (2012). Grouting technology of gravity base for offshore wind power. *Electric Power Constructure, 33*(07), 95~98.

[98] LU Xiaobing, ZHENG Zhemin, & ZHANG Jinlai. (2003). Progress in the study on the bucket foundations of offshore platform. *Advances in Mechanics, 01*, 27~40.

[99] LU Zhiyue, MA Hongwang, LI Yutao, CHEN Shen, & YANG Jun. (2019). Experimental Study on the Effect of Typhoons on Accumulated Deformation of the Monopile Foundation for Offshore Wind Turbines. *JOURNAL OF OCEAN TECHNOLOGY, 38*(06), 75~82.

[100] LUO Chao, & CAO Wensheng. (2013). The impact of typhoon on my country's offshore wind power development. *Energy and Environment, 03*, 2~3.

[101] Ma, K., Wu, Y., Stolen, S. F., Bello, L., ver der Horst, M., & Luo, Y. (2021, October 11). *Mooring Designs for Floating Offshore Wind Turbines Leveraging Experience From the Oil & Gas Industry*. ASME 2021 40th International Conference on Ocean, Offshore and Arctic Engineering. https://doi.org/10.1115/OMAE2021-60739

[102] Mahdy, M., & Bahaj, A. S. (2018). Multi criteria decision analysis for offshore wind energy potential in Egypt. *Renewable Energy, 118*, 278~289. https://doi.org/10.1016/j.renene.2017.11.021

[103] Martínez-Chaluisant, V., Schneider, J. A., & Fratta, D. (2012). *Physical Modeling of the Dynamic Response of Offshore Wind Turbines Founded on Monopiles*. 2079~2086. https://doi.org/10.1061/41096(366)190

[104] Matlock, H. (1970, April 21). *Correlation for Design of Laterally Loaded Piles in Soft Clay*. Offshore Technology Conference. https://doi.org/10.4043/1204-MS

[105] MHI Vestas. (2019). *MHI Vestas Achieves Final Turbine Installation at Horns Reef 3*. https://www.vestas.com/en/media/mwow-press-releases/2019/Final-Turbine-Installation-Horns-Reef-3

[106] Musial, W. (2007). *Engineering Challenges for Floating Offshore Wind Turbines*.

[107] Nicholson, J. C. (2011). *Design of wind turbine tower and foundation*

systems: Optimization approach [M.S.]. https://www.proquest.com/docview/880288246/abstract/9A1CC95FBD98403FPQ/1

[108] Novaes Menezes, E. J., Araújo, A. M., & Bouchonneau da Silva, N. S. (2018). A review on wind turbine control and its associated methods. *Journal of Cleaner Production, 174*, 945~953. https://doi.org/10.1016/j.jclepro.2017.10.297

[109] Ochi, M. (2003). *Hurricane Generated Seas*. Elsevier.

[110] *Offshore Wind in Europe Key Trends and Statistics 2020*. (2021, November 12). The European Maritime Spatial Planning Platform. https://maritime-spatial-planning.ec.europa.eu/practices/offshore-wind-europe-key-trends-and-statistics-2020

[111] *OVERVIEW - World's first floating wind farm moves towards completion in Scotland*. (2017). https://renewablesnow.com/news/overview-worlds-first-floating-wind-farm-moves-towards-completion-in-scotland-580153/

[112] PAN Chenchen, WANG Quandi, ZHANG Peng, & LIU Hongjun. (2019). The horizontal bearing capacity and improvement of monopile foundation for offshore wind power. *Periodical of Ocean University of China, 49*(S2), 99~106. https://doi.org/10.16441/j.cnki.hdxb.20170319

[113] PAN Hongguan, CHEN Chaohe, NI Daojun, & FAN Tianhui. (2022). Response of sunction bucket jacket considering contact between bucket and soil under wind, wave and current. *China Offshore Platform, 37*(02), 1~27.

[114] People's Daily Online. (2013, November 12). *The world's first floating wind turbine is put into use*. 全球首座浮动风力发电机投入使用. http://scitech.people.com.cn/n/2013/1112/c1057-23518180.html

[115] Pérez-Collazo, C., Greaves, D., & Iglesias, G. (2015). A review of combined wave and offshore wind energy. *Renewable and Sustainable Energy Reviews, 42*, 141~153. https://doi.org/10.1016/j.rser.2014.09.032

[116] Pradhan, D. L. (2012). Development of P-Y Curves for Monopiles in Clay using Finite Element Model Plaxis 3D Foundation. 74. https://ntnuopen.ntnu.no/ntnu-xmlui/handle/11250/232178

[117] Qu Yan, Yue Qianjin, Bi Xiangjun, & Kärnä Tuomo. (2006). A random ice force model for narrow conical structures. *Cold Regions Science and*

Technology, 45(3), 148~157.
https://doi.org/10.1016/j.coldregions.2006.05.008

[118] IEC 61400-3: Wind turbines - Part 3: Design requirements for offshore wind turbines, International Electrotechnical Commission, (2005). https://global.ihs.com/doc_detail.cfm?document_name=IEC%2061400%2D3&item_s_key=00517157

[119] RUAN Shengfu. (2010). *Study on the Dynamic Response for Floating Foundation of Offshore Wind Turbine* [Master's Thesis]. TianJin University.

[120] Rudolph, C., Grabe, J., & Albrecht, I. (2014). Simple shear tests with a varying shearing direction during cyclic shearing. *Géotechnique Letters, 4*(2), 102~107. https://doi.org/10.1680/geolett.13.00088

[121] Sánchez, S., López-Gutiérrez, J.-S., Negro, V., & Esteban, M. D. (2019). Foundations in offshore wind farms: Evolution, characteristics and range of use. Analysis of main dimensional parameters in monopile foundations. *Journal of Marine Science and Engineering, 7*(12), 441.

[122] Schwanitz, V. J., & Wierling, A. (2016). Offshore wind investments - Realism about cost developments is necessary. *Energy, 106*, 170~181. https://doi.org/10.1016/j.energy.2016.03.046

[123] Shen Xiaolei, Wang Kuangshan, Shen Jiayi, Chen Qian, Wu Ronghui, & Shang Jin. (2021). Numerical Simulation on the Vertical Bearing Capacity of the High-pile Cap Foundation of the Substation. *BULLETIN OF SCIENCE AND TECHNOLOGY, 37*(07), 91~95.
https://doi.org/10.13774/j.cnki.kjtb.2021.07.017

[124] Shi, W., Park, H. C., Chung, C. W., Shin, H. K., Kim, S. H., Lee, S. S., & Kim, C. W. (2015). Soil-structure interaction on the response of jacket-type offshore wind turbine. *International Journal of Precision Engineering and Manufacturing-Green Technology, 2*(2), 139~148.
https://doi.org/10.1007/s40684-015-0018-7

[125] SHI Xiaochun, 晓春, XU Riqing, GONG Xiaonan, CHEN Guoxiang, & YUAN Zhongli. (2000). Introduction of bucket foundation. *China Civil Engineering Journal, 04*, 68-73+92. https://doi.org/10.15951/j.tmgcxb.2000.04.014

[126] Sirnivas, S., Musial, W., Bailey, B., & Filippelli, M. (2014). *Assessment of Offshore Wind System Design, Safety, and Operation Standards* (NREL/TP-5000-60573). National Renewable Energy Lab. (NREL), Golden, CO (United States). https://doi.org/10.2172/1122306

[127] SONG Yang, & WANG Ying. (2018). Installation and construction technology of 171m diameter impeller offshore wind turbine. *Development and Management, S1 vo 35*, 40~45.

[128] 《*Standards for Geotechnical Test Methods*》*(GB/T50123-2019)*. (1999). China Planning Press. https://www.waizi.org.cn/bz/66863.html

[129] Su, D., Wu, W. L., Du, Z. Y., & Yan, W. M. (2014). Cyclic Degradation of a Multidirectionally Laterally Loaded Rigid Single Pile Model in Compacted Clay. *Journal of Geotechnical and Geoenvironmental Engineering, 140*(5), 06014002. https://doi.org/10.1061/(ASCE)GT.1943-5606.0001084

[130] Su, D., & Yan, W. M. (2013). A multidirectional p-y model for lateral sand-pile interactions. *Soils and Foundations, 53*(2), 199~214. https://doi.org/10.1016/j.sandf.2013.02.002

[131] SUN Xuting, YANG Danliang, & MA Chunjie. (2019). The research status of offshorewind power foundation and the sustainable development analysis. *SHANXI ARCHITECTURE, 45*(18), 64~65. https://doi.org/10.13719/j.cnki.cn14-1279/tu.2019.18.032

[132] TAO Ye, & SHI Jingli. (2013). Research on my country's Offshore Wind Power Development Situation and Electricity Price Policy. *Research and Approach, 35*(06), 25~29.

[133] Thiagarajan, K. P., & Dagher, H. J. (2014). A Review of Floating Platform Concepts for Offshore Wind Energy Generation. *Journal of Offshore Mechanics and Arctic Engineering, 136*(2). https://doi.org/10.1115/1.4026607

[134] TOU Jinzhou, CHEN Qian, ZHOU Maoqiang, & ZHANG Qiang. (2022). Numerical Simulations on Horizontal Bearing Capacity of High-Pile Cap Foundation of the Offshore Wind Power in Radial Sandbank Area. *Bulletin of Science and Technology, 38*(04), 39~44.

[135] Veigas, M., & Iglesias, G. (2015). A Hybrid Wave-Wind Offshore Farm for an Island. *International Journal of Green Energy, 12*(6), 570~576. https://doi.org/10.1080/15435075.2013.871724

[136] Velarde, J., & Bachynski, E. E. (2017). Design and fatigue analysis of monopile foundations to support the DTU 10 MW offshore wind turbine. *Energy Procedia, 137*, 3~13. https://doi.org/10.1016/j.egypro.2017.10.330

[137] Wang Guojun, Zhang Dayong, Wang Shuaifei, YU Gaosong, & Yue Qianjin. (2020). Dynamic ice load of conical offshore wind turbine foundation in cold region. *Science Technology and Engineering, 20*(31), 12820~12826.

[138] WANG Jianhua, LIU Zhenwen, & QIN Chongren. (1999). Pseudo-Static Elasto-Plastic Cyclic Creep Model and Cyclic Stability of Offshore Soft Foundation. C*hian Ocean Engineering, 13*(3), 247~256.

[139] WANG Jun, CAI Yuanqiang, & PAN Linyou. (2009). Degradation of stiffness of soft clay under bidirectional cyclic loading. *Chinese Journal of Geotechnical Engineering, 31*(02), 178~185.

[140] WANG Tao, ZHANG Qi, & YE Guanlin. (2022). Influence of cyclic wave load and soil characteristics on horizontal deformation of wind power monopile foundation. *The Ocean Engineering, 40*(1), 93~104.

[141] WANG Xiuqiong, & CUI Jianfeng. (2002). Formula of Coeddicient kin expression of davenport spectrum and its engineering application. *JOURNAL OF TONGJI UNIVERSITY, 30*(07), 849~852.

[142] WANG Xuyue. (2017). *Research on Dynamic Characteristics of Composite Bucket Foundation for Offshore Wind Turbine* [Doctor]. TianJin University.

[143] WEN Feng. (2016a). Developments and Characteristics of Offshore Wind Farms in China. *ADVANCES IN NEW AND RENEWABLE ENERGY, 4*(02), 152~158.

[144] WEN Feng. (2016b). Developments and characteristics of offshore wind farms in China. *Advances in New and Renewable Energy, 4*(2), 152~158.

[145] Wilson, R. E., Lissaman, P. B. S., & Walker, S. N. (1976). *Aerodynamic performance of wind turbines.* Oregon State Univ., Corvallis (USA). https://www.osti.gov/biblio/7117144

[146] WU Ke, SUN Yan, & LI Shucai. (2008). *Numerical analysis of horizontally loading of monopile foundation of large scale offshore wind turbine.* 351~354.

[147] WU Shaobo, YIN Haiqing, & ZHANG Kaihua. (2017). Status and future development direction of Japan floating wind turbine theology. *China Harbour Engineering, 37*(6), 108~114.

[148] Wüstemeyer, C., Madlener, R., & Bunn, D. W. (2015). A stakeholder analysis of divergent supply-chain trends for the European onshore and offshore wind installations. *Energy Policy,* 80, 36~44. https://doi.org/10.1016/j.enpol.2015.01.017

[149] XU Hongfa, WU Huajie, GUO Shaoping, & LIAO Tieping. (2002). Element parameter analysis of pile-soil contact surface. *ExplorationEngineering(Rock&SoilDrillingandTunneling), 05,* 10~12.

[150] XU Honglu & PENG Ya. (2021). Wave dynamic response of jacket offshore wind turbine under random waves. *China Water Transport, 21*(04), 57~59.

[151] XU Liangliang, & HU Xiaozhen. (2018). Analysis and suggestion of offshore wind power industry patent. *China Invention &Patent, 15*(2), 58~66.

[152] Xu Ning. (2001). *Research on ice force of conical offshore structures* [硕士]. Dalian University of Technology.

[153] XU Ning. (2011). *Research on ice force of conical offshore structures* [Doctor's Thesis]. Dalian University of Technology.

[154] XU Rongbin. (2011). On Fan Infrastructure Form of Wind Power Station at Sea. *Building Materials Technology and Application, 07*, 7~9. https://doi.org/10.13923/j.cnki.cn14-1291/tu.2011.07.021

[155] YANG Fang, YIN Ming, & LIU Lin. (n.d.). European offshore wind power integration: Technique analysis and policy interpretation. *Energy Technology and Economics, 23*(10), 51~55.

[156] Yang, J., Chang, Y., Zhang, L., Hao, Y., Yan, Q., & Wang, C. (2018). The life-cycle energy and environmental emissions of a typical offshore wind farm in China. *Journal of Cleaner Production, 180*, 316~324. https://doi.org/10.1016/j.jclepro.2018.01.082

[157] YANG Wei, LIN Yifeng, & ZHANG Quan. (2018). Review of Gravity Foundation Development for Offshore Wind Turbines. *Wind Energy*, 09, 38~42.

[158] YE Chaoliang, GAO Xinqiang, & SONG He. (2015). Research on pipe-soil friction coefficient of FRPM pipe by laboratory model test. *Journal of Shijiazhuang Tiedao University (Natural Science)*, 28(3), 30-34+39.

[159] Young, A. G., Kraft, L. M., Jr., & Focht, J. A., Jr. (1976). Geotechnical Considerations in Foundation Design of Offshore Gravity Structures. *Journal of Petroleum Technology*, 28(08), 925~937. https://doi.org/10.2118/5683-PA

[160] YU Hongyan, ZHAI Shaohua, & ZHANG Puyang. (2013). FINITE ELEMENT ANALYSIS OF BEARING CAPACITY BEHAVIOR OF COVER-LOAD-BEARING LARGE-SCALE BUCKET FOUNDATION FOR OFFSHORE WIND TURBINES. *ENGINEERING MECHANICS, 30*(06), 124~132.

[161] Yu, L.-Q., Wang, L.-Z., Guo, Z., Bhattacharya, S., Nikitas, G., Li, L.-L., & Xing, Y.-L. (2015). Long-term dynamic behavior of monopile supported offshore wind turbines in sand. *Theoretical and Applied Mechanics Letters, 5*(2), 80~84. https://doi.org/10.1016/j.taml.2015.02.003

[162] YU Tongshun. (2016). *Research on Dynamic Response and Local Scour of Composite Bucket Foundation* [Doctor's Thesis]. TianJin University.

[163] Yu Xiao. (2001). *Research on ice force of conical structures in Bohai* [硕士]. Dalian University of Technology.

[164] Yu Yonghai, & Yue Qianjin. (2000). Prediction methods of sheet ice load on conical structures. *Acta Oceanologica Sinica, 22*(3), 74~85.

[165] YUAN Ruhua, HUANG Hailong, SUN Daoqing, DAI Wanbao, YIN Shiwei, TIAN Runnan, & PU Yingjie. (2018). Research on Basic Structure and Installation Technology of Offshore Wind Turbine. *Energy and Energy Conservation, 12*, 59~61. https://doi.org/10.16643/j.cnki.14-1360/td.2018.12.023

[166] ZE Yankang. (1981). ExPerimental Study on Stiffenss and Crack of Reinforeed Cocnerre Ring Section Membersin Bending. *Journal of Building Structures, 04*, 79-80+33. https://doi.org/10.14006/j.jzjgxb.1981.04.008

[167] Zhang, C., White, D., & Randolph, M. (2011). Centrifuge Modeling of the Cyclic Lateral Response of a Rigid Pile in Soft Clay. *Journal of Geotechnical and Geoenvironmental Engineering, 137*, 717~729. https://doi.org/10.1061/(ASCE)GT.1943-5606.0000482

[168] Zhang, J., Kang, W.-H., Zhang, C., & Sun, K. (2018). Reliability based analysis and design of a tripod offshore wind turbine structure assuring serviceability performance. *Polish Maritime Research*, 139~148. https://doi.org/10.2478/pomr-2018-0140

[169] ZHANG Puyang & HUANG Xuanxu. (2018). Application Research on Suction Bucket Foundation for Offshore Wind Power. *SOUTHERN ENERGY CONSTRUCTION, 5*(04), 1~11. https://doi.org/10.16516/j.gedi.issn2095-8676.2018.04.001

[170] ZHAO Jing. (2009). *Study of hydrodynamic loads on offshore wind turbines* [Master's Thesis]. Haerbin University of Technology.

[171] ZHOU Xuhong, WANG Yuhang, & DENG Ran. (2020). Review on Floating Foundation Structures for Offshore Wind Turbines. *ELECTRIC POWER, 53*(07), 100-105+112.

[172] Zhu, B., Kong, D., Chen, R., Kong, L., & Chen, Y. (2011). Installation and lateral loading tests of suction caissons in silt. *Canadian Geotechnical Journal, 48*(7), 1070~1084. https://doi.org/10.1139/t11-021

[173] ZHU Bin, XIONG Gen, LIU Jinchao, SUN Yongxin, & CHEN Renpeng. (2013). Centrifuge modelling of a large-diameter single pile under lateral loads in sand. *Chinese Journal of Geotechnical Engineering, 35*(10), 1807~1815.

[174] ZHU Yanrong. (1991). Ocean Engineering *Wave Mechanics.* Tianjin University Press.

Editor

Chief Editor

Shengdong Zhang
Jining University, China, 273199
sdzluck@163.com

Associate Editor

Xiuying Yang / Kunsan National University, 54150 / yangxiuying@kunsan.ac.kr
Kaicheng Yan / Kunsan National University, 54150 / kaichengyan@126.com
Jungkeun Oh / Kunsan National University, 54150 / jkoh@kunsan.ac.kr
Daewon Seo / Kunsan National University, 54150 / dwseo@kunsan.ac.kr
Jackyou Noh / Kunsan National Uiversity, Korea, 54150/ snucurl@kunsan.ac.kr
Kiweon Kang / Kunsan National Uiversity, Korea, 54150/ kwkang68@kunsan.ac.kr

Foundation of offshore wind turbine

인쇄발행 2024년 1월 4일 인쇄발행

저 자 Shengdong Zhang, Xiuying Yang, Kaicheng Yan, Jungkeun Oh, Daewon Seo, Jackyou Noh, Kiweon Kang

발 행 인 송기수

발 행 처 도서출판 GS인터비전

편 집 인 김명신

편 집 처 도서출판 GS인터비전

인 쇄 처 트윈벨 미디어

등록번호 제 25100-2016-000050 호

전 화 02-976-7898

팩 스 02-6468-7898

이 메 일 gsinter7@gmail.com

홈페이지 gsintervision.kr

주 소 서울 은평구 증산로 15길 69, 2층

I S B N 979-11-5576-431-2(93500)

정가 30,000원

도서출판 GS인터비전